Lecture Notes in Mathematics

Edited by A. Dold, Heidelberg and B. Eckmann, Zürich

347

J. M. Boardman
Johns Hopkins University, Baltimore, MD/USA

R. M. Vogt
Mathematisches Institut der Universität des Saarlandes, Saarbrücken/BRD

Homotopy Invariant Algebraic Structures on Topological Spaces

Springer-Verlag
Berlin · Heidelberg · New York 1973

AMS Subject Classifications (1970): 55-02, 55D10, 55D15, 55D35, 55D45, 55F35, 18C10, 18C15

ISBN 3-540-06479-6 Springer-Verlag Berlin · Heidelberg · New York
ISBN 0-387-06479-6 Springer-Verlag New York · Heidelberg · Berlin

Offsetdruck: Julius Beltz, Hemsbach/Bergstr.

TABLE OF CONTENTS

INTRODUCTION

Originally, we developed the theory of homotopy invariant structures to obtain a machine for proving that the stable groups O,U,SO,SU,F, Top,PL, their various coset spaces, and their classifying spaces are infinite loop spaces. But soon we realized that the homotopy invariant structures in themselves were the main subject of our research. The idea of using categories of operators (called PROPs in these notes) and to identify topological spaces with algebraic structures with functors from a suitable PROP to the category of topological spaces was implicitly contained in a talk of Stasheff given in a seminar of MacLane at the University of Chicago in 1967. He suggested to look for a topological analogue of the notion of a PACT as developed by Adams and MacLane [29] and to use it in the theory of infinite loop spaces. The topological version of the conjecture following Theorem 25.1 of [29] gave some hope for an application: If a topological PACT, whose homotopies satisfies all higher coherence conditions, acts on a space X, then it also acts on its classifying space.

The coherence conditions for higher homotopies naturally lead us to consider homotopy invariance, which later turned out to be useful for the application to infinite loop spaces.

After we had announced our results in [8], Beck pointed out to us that our PROPs are just subcategories of topological theories as known from categorical universal algebra. This motivated us to consider general topological-algebraic theories, too, although in most of our investigations we had to restrict our attention to the previously treated PROPs, which now cropped up as "spines" of theories closely related to the theory of commutative monoids.

Shortly after the appearance of [8] several other authors could show by different methods that the stable groups and their classifying spaces listed above are infinite loop spaces. Their approaches avoid the theory of homotopy invariant structures so that they reach the required result more easily and more directly. Therefore we want to stress the point that infinite loop spaces are just one field of application of our theory and, as we will show, not the only one.

We briefly want to compare our method with the most interesting other approaches to infinite loop spaces. Using his construction of the classifying space of a category [44], Segal was able to show that a topological category $\mathfrak{C}$ with an appropriate bifunctor $\mathfrak{C} \times \mathfrak{C} \longrightarrow \mathfrak{C}$ gives rise to a spectrum of simplicial spaces, the realizations of which form an infinite loop space [45]. He then showed that the groups under consideration determine such categories and that the associated spectrum makes the group into an infinite loop space. In contrary to this, our method is to investigate the internal algebraic-topological structure of the groups and to show that certain structures, we call them E-structures, characterize infinite loop spaces.

A second direct proof is due to Beck [3]. He also starts with E-structures on topological spaces. He then extends the suspension and loop space functor to adjoint endofunctors of the category of spaces with E-structures and shows that the front adjunction map $X \longrightarrow \Omega SX$ is a weak homotopy equivalence if the E-structure on X makes $\pi_0(X)$ into a group.

A third approach, due to May [34], is very closely related to our method. His theory is geared towards applications to Dyer-Lashof homology operations. He first develops a theory of operads with free action of the symmetric groups. They are slight specializations of our PROPs. Then using the operads obtained from our little cube categories $\underline{Q}_n$ and $\underline{Q}_\infty$ of [8] (see also (2.49) of these notes) and generalizing an n-stage classifying space construction of Beck [2], he was able to prove a recognition principle for iterated and infinite loop spaces, which resembles much of our recognition principles of chapter VI, §3. His approach has two advantages over ours. Firstly, the category of operads has products, which essentially substitute our tensor products of PROPs. Since the topological and algebraic structures of these products are far more transparent than the structure of the tensor products, one need not be reluctant to work with them. Secondly, his n-stage classifying space construction, which is quite interesting in itself, makes an inductive proof of the recognition principle for n-fold loop spaces redundant, although the proof of consistency requirements for infinite loop spaces boils down to an argument similar to an induction.

On the other hand, our approach has some advantages over May's. First of all, we admit all PROPs and not only PROPs with free actions of the cyclic group, which correspond to Σ-free operads. Thus important PROPs, such as the PROP $\mathfrak{S}$ associated with the theory of commutative monoids, are allowed in our theory but not in May's. (It has been known for some time [37] that a connected abelian monoid is of the weak homotopy type of an infinite loop space). Secondly, taking PROPs and tensor products instead of operads and products keeps us in closer connection with general algebraic-topological theories, so that generalizations to more complicated algebraic structures suggest themselves. Thirdly, once the theory of homotopy invariant structures has been made available, a few more or less elementary facts of the standard classifying space construction imply a unified proof of the recognition principle for both n-fold and infinite loop spaces. Moreover, for any E-space X, we have maps preserving the structure up to coherent homotopies

$$\Omega^n B^n X \longleftarrow \Omega^{n-1} B^{n-1} X \longleftarrow \dots \qquad \dots \longleftarrow \Omega BX \longleftarrow X ,$$

which are weak homotopy equivalences if the E-structure on X makes $\pi_0(X)$ a group, while in May's approach there are structure preserving maps

$$\Omega^n B^n X \xleftarrow{\ f\ } \bar{B}X \xrightarrow{\ g\ } X$$

such that g is a deformation retraction and f is a weak homotopy equivalence if X is connected. So the maps do not go one way.

Besides the points which allow a direct comparison, we also treat the theory of maps which preserve the algebraic-topological structures up to coherent homotopies in great detail, thus obtaining a delooping result for maps between E-spaces which are not quite homomorphisms. Moreover, an additional analysis of Milgrams classifying space construction allows us to show that the weak homotopy equivalences above are strict homotopy equivalences for a wider class of topological spaces than CW-complexes. This side of the theory is, of course, unnecessary for the purpose of May's notes, which are thought of as a basis for the development of a theory of homology operations; but they are of great interest from the homotopy point of view.

A short idea of what we are going to do and a recollection of existing results on homotopy invariant structures on topological spaces is given in chapter I. The second chapter is a self-contained treatment of multi-coloured theories generalizing constructions of Bénabou [4]. Many ideas of the section on multi-coloured triples (II,§4) are taken from papers of Beck [2] and others. In (II,§5) we define the topological analogue of Adams' and MacLane's notion of a PROP and PACT and put them into relation with general theories. We complete the chapter with a list of PROPs, which will be used in the characterization and recognition of n-fold and infinite loop spaces, or which define algebraic-topological structures occurring in the literature.

In the third chapter we define the bar construction for theories and PROPs and prove its important properties. It is the main tool for the development of the theory of homotopy invariant structures, which is given in chapter IV. In the first parts of chapter IV we construct categories of spaces with homotopy invariant structures and homotopy classes of maps which preserve such structures up to coherent homotopies. We offer three definitions for such maps, which turn out to be more or less equivalent, and continue to work with two of them. This side of the theory indicates that the last word has not been spoken yet. Relationships of such maps to homomorphisms are studied in sections 4,5, and 6. The main results on homotopy invariance are given in section 3. In section 7 we prove a homotopy invariance result for general theories, thus indicating possible generalizations of our results. Chapter V lists the modifications which are necessary to prove the more important results of chapter III and IV in the category of based topological spaces.

As a first application of our theory we in chapter VI study n-fold and infinite loop spaces. We start with a detailed treatment of Milgram's classifying space construction [37] in the frame-work of our theory before we in section 3 prove the recognition principle for n-fold and infinite loop spaces mentioned above. In section 4 we extend this recognition principle to arbitrary E-structures on a space X and show that X cannot be of the homotopy type of an abelian monoid if it has non-trivial Dyer-Lashof operations. We list a number of infinite loop spaces in section 5. Chapter IV and VI include the proofs of the announcements in [8].

In the final chapter VII we briefly indicate how our theory can be used in other branches of homotopy theory. We define homotopy colimits, prove some elementary properties, and illustrate some applications.

A summarized version of parts of our results has already appeared in [6], and we try to stick to its terminology. The first chapters constitute a vastly improved version of [54]. The second author takes the full responsibility of the present exposition and he is to be blamed for cryptic formulations, frequent violations of standard rules of the English language, and mathematical inaccuracies.

It is our pleasure to thank our friends who in various discussions helped us to clarify one or the other point of our theory. In particular, we are indebted to Jon Beck for explaining to us the notion of an algebraic theory and who pointed out that the homotopy-everything H-spaces of [8]are not really homotopy-everything H-spaces (compare VI,§4), and to Tammo tom Dieck for suggesting to us the use of numerable principal G-spaces and for many invaluable conversations. We also want to thank Mrs. Victoria Löffler for her quick and neat typing of the major part of these notes.

During our research the first author was partly supported by the National Science Foundation under the grant number GP-19481, while the second author received partial financial support from the Studienstiftung des deutschen Volkes and the Deutsche Forschungsgemeinschaft.

I. CHAPTER

MOTIVATION AND HISTORICAL SURVEY

In these notes we study homotopy-associative and homotopy-commutative H-spaces, where the homotopies satisfy "higher coherence conditions", and maps that preserve such structures up to homotopy, and again we require that the homotopies satisfy "higher coherence conditions". For the time being call those H-spaces "structured spaces" and these maps "structured maps". Our aim is to define structures which approximate the structure of a monoid or a commutative monoid and which live in homotopy theory. To make the last remark precise, our structured spaces and maps should satisfy the following statements:

If X is a structured space and f : X —> Y a homotopy equivalence, then Y can be structured such that f becomes a structured map.

If f : X —> Y is a structured map of structured spaces and g homotopic to f, then g can be structured.

If f : X —> Y is a structured map of structured spaces and a homotopy equivalence, then any homotopy inverse can be structured.

We are interested in structures that approximate a monoid structure or the structure of a commutative monoid because a reasonable monoid is of the homotopy type of a loop space and a reasonable commutative monoid of the homotopy type of an infinite loop space. Precise definitions and statements of this remark will follow in this chapter.

We now want to give a short survey of the existing literature on this subject and introduce the main ideas of our approach. Some of the statements of this cnapter will be rather vague and made precise later on in these notes. In most cases we refrain from giving proofs.

Throughout these notes we work in the category of k-spaces, i.e. the category of compactly generated spaces (see Appendix I).

1. "DELOOPING" VIA THE CLASSIFYING SPACE CONSTRUCTION

Definition 1.1: An H-space is a topological space X with base point e and a multiplication map $m : X \times X \longrightarrow X$ such that e is a homotopy unit, i.e the maps $x \to m(x,e)$ and $x \to m(e,x)$ are homotopic to the identity rel(e,e) (Where reasonable we write xy for m(x,y)).

If $(X \times X, X \vee X)$ has the homotopy extension property, the multiplication map can be deformed to one for which e is a strict unit. Since we are concerned with "homotopy invariant structures" it is more natural to work with homotopy units.

Obviously, the concept of an H-space is a natural generalization of that of a topological group, and since there are many spaces which admit an H-space structure which is not a group structure, e.g. loop spaces, it is worth considering.

Because of the lack of structure, many interesting and important constructions which apply to topological groups cannot be applied to H-spaces. For example, there is no H-space analogue to Milnor's classifying space of a group unless the H-space in question has some additional structure. Such a construction is rather important for algebraic topology, because it implies that any reasonable topological group is of the homotopy type of a loop space, which has further consequences. From the homotopy theoretical point of view, the distinguishing feature is the lack of associativity (and commutativity) rather than the lack of a continuous inverse (see Prop. 1.5 below). So associativity and commutativity, both strict and up to homotopy, will play a significant role in our development.

A close investigation of Milnor's construction of the classifying space of a topological group (Milnor [39]) shows that one can do without the existence of a continuous inverse.

Definition 1.2: An H-space with strictly associative multiplication and strict unit is called a monoid.

Proposition 1.3 (Dold-Lashof [16]): If X is a monoid such that right translation is a weak homotopy equivalence, then there is a space BX and a weak homotopy equivalence $X \longrightarrow \Omega BX$ which respects the multiplication up to homotopy.

The condition that right translation is a weak homotopy equivalence is necessary. Since $\pi_0(\Omega Y)$ is a group for any space Y, the statement can only be true if $\pi_0(X)$ is a group. It is easy to see that this implies that right translation is even a homotopy equivalence.

Fuchs has modified the Dold-Lashof construction. He obtains a homotopy equivalence and not just a weak one.

Proposition 1.4 (Fuchs [22]): If X is a monoid with a homotopy inverse, then there exists a space BX and a homotopy equivalence $X \longrightarrow \Omega BX$.

Proposition 1.3 yields a homotopy equivalence if X is a CW-complex. But in this case, X admits a homotopy inverse, because it is numerably contractible [13; Prop.6.7].

Proposition 1.5 (tom Dieck-Kamps-Puppe [12]): Let Y be a homotopy-associative H-space such that right translation is a homotopy equivalence. If Y is numerably contractible then it admits a homotopy inverse.

We have seen that monoids are closely related to loop spaces. The following result shows that loop spaces are related to topological groups.

Proposition 1.6 (Milnor [39]): If X has the homotopy type of a connected countable CW-complex, then there is a topological group G(X) of the homotopy type of ΩX.

2. A_∞-SPACES

The last section showed that monoids may replace the topological groups in homotopy theory. Since the loop space on X is a SDR (= strong deformation retract) of a monoid, namely the Moore-loops on X, we have added a large class of spaces to the class of topological groups. The disadvantage of topological groups and monoids is that they do not live in homotopy theory, i.e. if M is a monoid and $f : M \longrightarrow X$ a homotopy equivalence then we cannot expect that X has a monoid structure such that f becomes a homomorphism or only a homomorphism up to homotopy.

The "weakest" approxiamtion of a monoid structure which is in homotopy theory is a homotopy-associative H-space structure. Such an H-space is in general far away from being of the homotopy type of a monoid. This motivates to look for richer structures which are better approximations. J. Stasheff solved this problem with his A_n-spaces.

Let us investigate the best homotopy-invariant approximation, namely a space X of the homotopy type of a monoid. Then X inherits a homotopy-associative H-space structure from the monoid via the homotopy equivalences. Let

$$M_3 : I \times X^3 \longrightarrow X$$

be the canonical associating homotopy (I is the unit interval), and $M_3(x,y,z)$ the corresponding path from (xy)z to x(yz). Considering the various ways of associating four factors we obtain five maps from X^4 to X, each of which is homotopic to two others by a single application of homotopy associativity. For each quadrupel (x,y,z,w) of elements in X we can construct a loop S(x,y,z,w) in X,

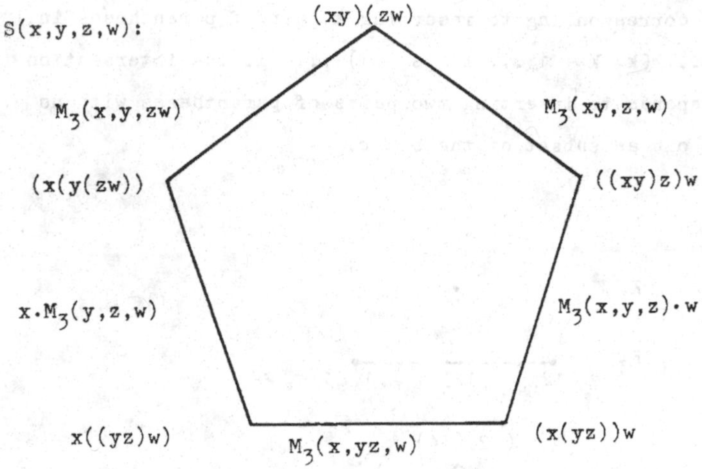

continuous in each variable, and hence a map $S^1 \times X^4 \longrightarrow X$. This map can be extended to a map

$$M_4 : D^2 \times X^4 \longrightarrow X$$

where D^2 is a 2-cell with boundary S^1. If we take five factors, we can construct a map $S^2 \times X^5 \longrightarrow X$ using the multiplication M_2 and the maps M_3 and M_4. If M_4 is chosen properly, this maps can be extended to a map

$$M_5 : D^3 \times X^5 \longrightarrow X$$

and so on. We end up with a sequence of maps $M_n : D^{n-2} \times X^n \longrightarrow X$, $n \geq 2$, such that M_n extends the map $S^{n-3} \times X^n \longrightarrow X$ which is induced by $M_2, \ldots, M_{n-1}$. We can obtain such a sequence of maps for any space of the homotopy type of a monoid. Later on we shall see that this structure , which we call an A_∞-structure, leads to a homotopy-invariant characterization of spaces of the homotopy type of a monoid. To be able to give precise statements, we must know how to obtain the maps $S^{n-3} \times X^n \longrightarrow X$ from $M_2, \ldots, M_{n-1}$.

<u>Definition 1.7</u> (Stasheff [46]): Let K_i denote the complex constructed inductivity as follows: $K_2 = D^0$, the 0-cell. Let K_i be the cone CL_i on L_i which is the union of various copies $(K_r \times K_s)_k$ of $K_r \times K_s$,

$r + s = i + 1$, corresponding to inserting a pair of parentheses in i symbols $1\ 2 \ldots . (k\ \ k+1 \ldots k+s-1) \ldots .\ i$. The intersection of copies corresponds to inserting two pairs of parentheses with no overlap or with one as subset of the other.

Examples

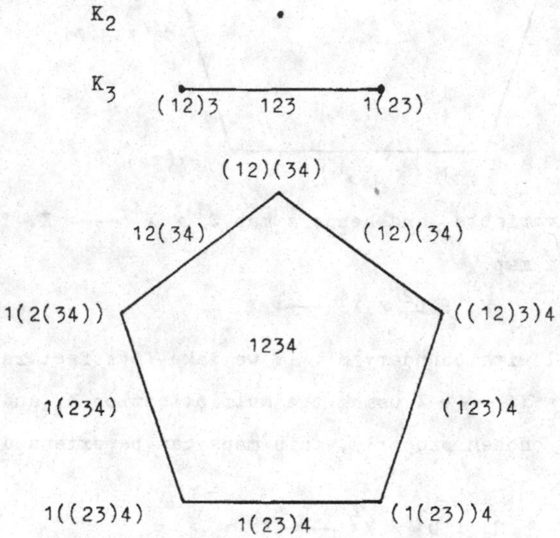

One can show [46; Prop. 3] that K_i is an $(i-2)$-cell. It will take the place of the $(i-2)$-cell in the definition of an A_∞-structure.

Let $\partial_l(r,s) : K_r \times K_s \longrightarrow K_i$ denote the inclusion of the copy indexed by $1\ 2 \ldots (l\ \ l+1 \ldots l+s-1) \ldots i$.

<u>Definition 1.8</u> (Stasheff [46]): An A_n-space $(X; \{M_i\})$ consists of a space X and a collection of maps

$$M_i : K_i \times X^i \longrightarrow X \qquad i = 2,3,\ldots, n$$

such that

(i) M_2 is a multiplication with unit

ii) $M_i(\partial_l(r,s)(k_1,k_2),x_1,\ldots, x_i) =$
$= M_r(k_1,x_1,\ldots, x_{l-1},M_s(k_2,x_l,\ldots, x_{l+s-1}),\ldots, x_i)$ where
$(k_1,k_2) \in K_r \times K_s$.

If M_i exists for all $i \geq 2$ and satisfies the conditions (i),(ii), then $(X;\{M_i\})$ is called an A_∞-space.

So an A_2-space is an ordinary H-space with strict unit, an A_3-space a homotopy-associative H-space with strict unit etc. We are especially interested in A_∞-spaces because they turn out to be of the homotopy type of a monoid. Now it is a natural question to ask whether we can do with less than an A_∞-structure. Since we know that any space of the homotopy type of a monoid admits an A_∞-structure, this amounts to asking whether or not an A_n-structure extends to an A_∞-structure. Counterexamples were given by Adams [1] and Stasheff [46].

Proposition 1.9 (Adams and Stasheff): If Y is a Moore space of type $(G,2p + 1)$ where G is an abelian group in which division is possible for all primes q less than the prime p, then Y admits an A_{p-1}-structure but not an A_p-structure.

Stasheff succeeded to generalize the classifying space construction of Dold and Lashof. Using this tool he could show that any A_∞-space is of the homotopy type of a loop space, and hence of a monoid. On the other hand, it is easy to show that a loop space admits an A_∞-structure. So we obtain

Proposition 1.10 (Stasheff [48]): A connected CW-complex X admits the structure of an A_∞-space iff X has the homotopy type of a loop space.

Since Stasheff uses the long exact homotopy sequence of the quasi-fibration associated with his classifying space construction, he needs the requirement tnat X is a CW-complex. The second disadvantage of his construction is that he has to use strict units. Building up a monoid with help of the models K_i, Adams gave an alternative proof of a stronger result, which makes no use of units.

<u>Proposition 1.11</u> (Adams, unpublished): If X admits maps $M_i : K_i \times X^i \longrightarrow X$ for $i \geq 2$ satisfying 1.8(ii), then X is a SDR of a space Y with an associative multiplication m such that $m|X \times X$ is homotopic in Y to M_2.

Adams original proof is a little tedious. We shall give a simple proof of a stronger version. It turns out that the inclusion map $X \subset Y$ preserves not only the multiplication up to homotopy but up to homotopy and higher coherence conditions. This leads us to the investigation of maps which are homomorphisms up to homotopy and higher coherence conditions.

3. A_∞-MAPS

<u>Definition 1.12</u>: A map $f : X \longrightarrow Y$ between A_n-spaces $(X;\{M_i\})$ and $(Y;\{N_i\})$ is called a homomorphism, if the following diagram commutes for all i, $2 \leq i \leq n$.

$$\begin{array}{ccc} K_i \times X^i & \xrightarrow{\;M_i\;} & X \\ \downarrow{\scriptstyle 1 \times f^i} & & \downarrow{\scriptstyle f} \\ K_i \times Y^i & \xrightarrow{\;N_i\;} & Y \end{array}$$

Analogously for homomorphisms of A_∞-spaces.

If we try to define maps between A_n-spaces which respect the structure up to homotopy and higher coherence conditions in the same way as we defined A_n-structures, i.e. homotopy-associative multiplications with higher coherence conditions, the details became more and more complicated with increasing n. For example, respecting an A_2-structure

involves a 1-cell

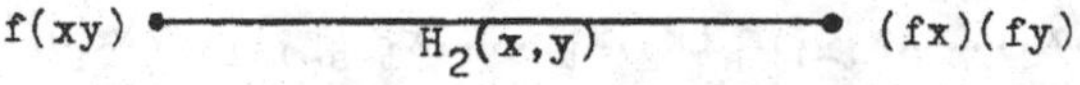

respecting an A_3-structure involves a 2-cell subdivided as a hexagon

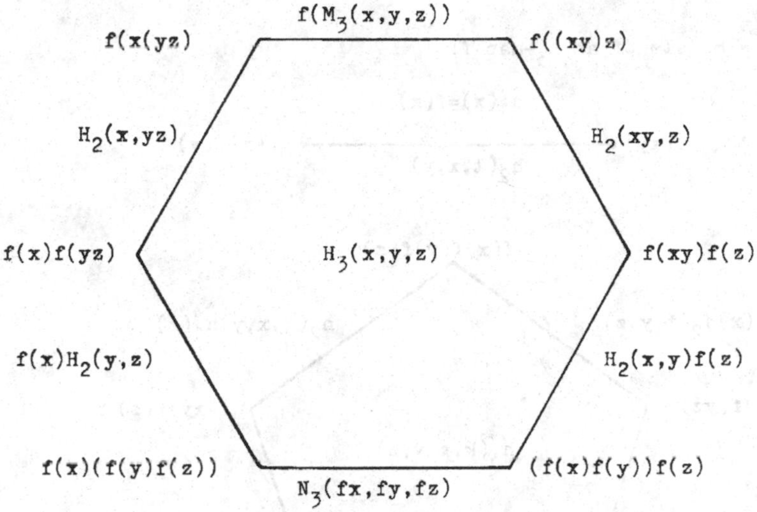

respecting an A_4-structure involves a 3-cell whose boundary is subdivided in a complicated way; it contains products of 1-cells H_2, 2-cells H_3 and of copies of the models K_2, K_3 and K_4. For a picture see [48;p.53]. So it does not surprise that such maps have not been studied up to now. In fact, maps which respect the structure up to homotopy and higher coherence conditions have been investigated to a larger extend for monoids only, although maps of an A_n-space into a monoid are manageable too.

<u>Definition 1.13</u> (Stasheff [48;p.54]): Let $(X;\{M_i\})$ be an A_n-space and Y a monoid. A map $f : X \longrightarrow Y$ is an A_n-<u>map</u> if there exists a family of maps

$$h_i : K_{i+1} \times X^i \longrightarrow Y \qquad i = 1,2,\ldots, n$$

such that $h_1 = f$ and

$$h_i(\partial_l(r,s)(k_1,k_2),x_1,\ldots,x_i) =$$
$$= \begin{cases} h_r(k_1,x_1,\ldots,x_{l-1},M_s(k_2,x_l,\ldots,x_{l+s-1}),\ldots,x_i) & l < r \\ h_{r-1}(k_1,x_1,\ldots,x_{r-1})h_{s-1}(k_2,x_r,\ldots,x_i) & l = r \end{cases}$$

for $r + s = i + 2$.

Example: The models of an A_3-map f:

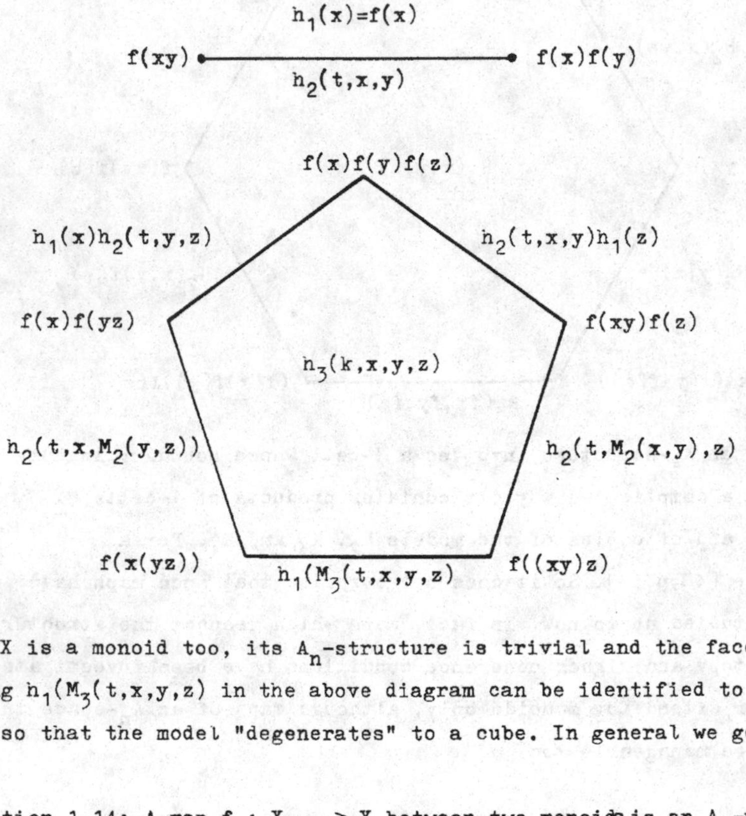

If X is a monoid too, its A_n-structure is trivial and the face involving $h_1(M_3(t,x,y,z)$ in the above diagram can be identified to a point so that the model "degenerates" to a cube. In general we get

Definition 1.14: A map $f : X \longrightarrow Y$ between two monoids is an A_n-map if there exists a family of maps

$$h_i : I^{i-1} \times X^i \longrightarrow Y \qquad i = 1,2,\ldots, n$$

such that $h_1 = f$ and

$$h_i(t_1,\ldots,t_{i-1},x_1,\ldots,x_i) = \begin{cases} h_{i-1}(t_1,\ldots,\hat{t}_j,\ldots,t_{i-1},x_1,\ldots,x_j\, x_{j+1},\ldots x_i) & \text{if } t_j=0 \\ h_j(t_1,\ldots,t_{j-1},x_1,\ldots,x_j)h_{i-j}(t_{j+1},\ldots,t_i,x_{j+1},\ldots,x_i) & \text{if } t_j=1 \end{cases}$$

<u>Example</u>: **The models of an A_3-map f between monoids:**

$$h_1(x)=f(x)$$

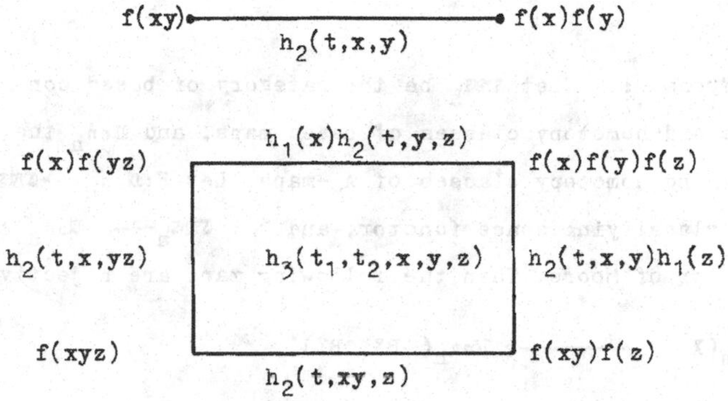

Sugawara [53] was to our knowledge the first to define A_∞-maps between monoids while Stasheff [47] was the first to investigate A_n-maps between monoids and later [48] between A_n-spaces and monoids. They were looking for maps which are no homomorphisms but nevertheless induce maps between the Dold-Lashof classifying spaces respectively their n-th approximation. We know that any reasonable A_∞-space X has a classifying space BX such that X is of the homotopy-type of ΩBX, i.e. an A_∞-space can be delooped. It seems natural to try the same for A_∞-maps.

<u>Proposition 1.15</u> (Sugawara [53]): Let $f : X \longrightarrow Y$ be an A_∞-map between monoids. Then f induces maps $f' : E(X) \longrightarrow E(Y)$ and $\bar{f} : B(X) \longrightarrow B(Y)$ such that

$$\begin{array}{ccc} E(X) & \xrightarrow{f'} & E(Y) \\ {\scriptstyle p(X)}\downarrow & & \downarrow{\scriptstyle p(Y)} \\ B(X) & \longrightarrow & B(Y) \end{array}$$

commutes, where $p(X)$ and $p(Y)$ are the universal quasifibrations of the Dold-Lashof construction.

Stasheff succeeded to proof a similar result for A_n-maps and the n-the stage of the Dold-Lashof construction.

The most detailed study of A_∞-maps can be found in an article of Fuchs [21] where the question mentioned above found its complete solution.

<u>Proposition 1.16</u> (Fuchs) : Let $\mathfrak{CW}_h$ be the category of based connected CW-complexes and homotopy classes of based maps, and $\mathfrak{Mon}_h$ the category of monoids and homotopy classes of A_∞-maps. Let $B:\mathfrak{Mon}_h \longrightarrow \mathfrak{CW}_h$ be the Dold-Lashof classifying space functor, and $\Omega : \mathfrak{CW}_h \longrightarrow \mathfrak{Mon}_h$ the loop space functor of Moore. Then the following maps are bijective

$$\Omega B : \mathfrak{Mon}_h(X,Y) \longrightarrow \mathfrak{Mon}_h(\Omega BX, \Omega BY)$$

$$B : \mathfrak{Mon}_h(X,Y) \longrightarrow \mathfrak{CW}_h(BX,BY)$$

The first statement in particular shows that A_∞-maps can be "delooped". To be precise, let $f : X \longrightarrow Y$ be an A_∞-map between monoids and $i(X) : X \longrightarrow \Omega BX$ the homotopy equivalence of Proposition 1.3. Fuchs showed that this map is an A_∞-map and that there exists a map $Bf : BX \longrightarrow BY$ such that the diagram

$$\begin{array}{ccc} X & \xrightarrow{\quad f \quad} & Y \\ {\scriptstyle i(X)}\downarrow & & \downarrow{\scriptstyle i(Y)} \\ \Omega BX & \xrightarrow{\quad \Omega Bf \quad} & \Omega BY \end{array}$$

commutes up to homotopy.

Although the models of an A_∞-map are simple they are still difficult to work with. This became apparent in the following important result of Fuchs.

Proposition 1.17 (Fuchs [21]) : A homotopy equivalence between monoids is an A_∞-map iff any inverse is.

In fact, a complete proof of this proposition has never been published, because the details are to messy. So a new approach was necessary, especially if one tried to study A_∞-maps between A_∞-spaces.

4. A DIFFERENT APPROACH

In this section we want to describe the main points of our constructions which will be developed in full generality in the coming chapters. We show how to apply them by giving a simple proof of a stronger version of Proposition 1.11.

There are essentially two ingredients, one coming from categorical algebra and one from physics. In categorical algebra a monoid structure is not given by an multiplication $m : X \times X \longrightarrow X$ satisfying certain identities, but as follows:

Definition 1.18: A monoid structure on a space X consists of a family of maps, called operations,

$$\lambda_i : X^i \longrightarrow X, \qquad i = 0,1,2,...$$

such that

(i) $\lambda_1 = 1_X$

(ii) $\lambda_n \circ (\lambda_{r_1} \times ... \times \lambda_{r_n}) = \lambda_m$, where $m = r_1 + ... + r_n$

The map λ_i corresponds to $(x_1,...,x_i) \longrightarrow x_1x_2 ... x_i$, so that λ_2 is an associative multiplication with unit $e = \lambda_o(X^o)$.

To get a better grip on composites of operations, we consider each λ_i as an electrical box with i inputs and one output. (In the general case, we have more than one operation $X^i \longrightarrow X$. In this case we label the box by the operation it represents.)

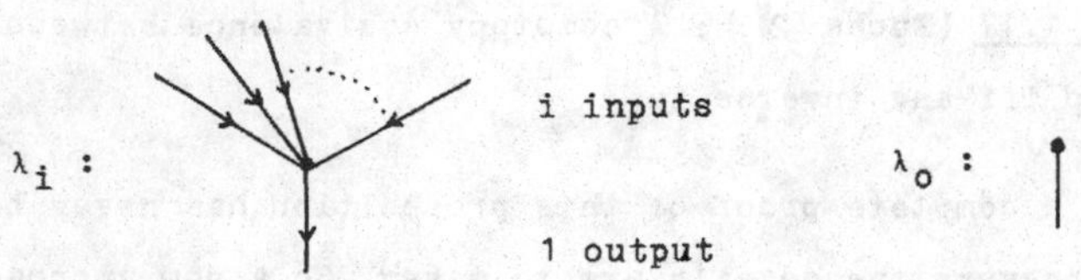

The box representing λ_o has no input but an output. A composite operation is obtained by wiring together, e.g. wiring together the operation $\lambda_2 : X^2 \longrightarrow X$ and $\lambda_2 \times \lambda_3 : X^5 \longrightarrow X^2$ we obtain

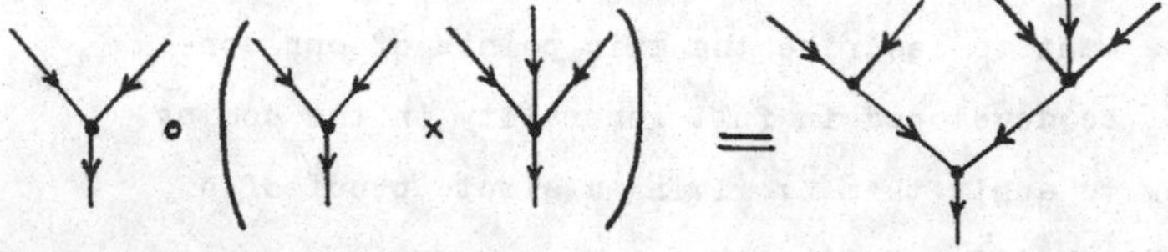

So a composite operation is represented by a certain kind of directed planar tree. The edges need not have vertices on both ends; the inputs have no beginning vertex, we call them <u>twigs</u>, the output, called <u>root</u>, has no end vertex. Edges with vertices on both ends are called <u>internal</u>. We call the box representing λ_o a <u>stump</u>, and to be able to cope with occuring relations, we introduce the <u>trivial tree</u> which has no vertex

and represents the identity operation $X \longrightarrow X$.

To compute the value of a composite operation represented by a tree with n twigs on an n-tuple $(x_1,\ldots,x_n) \in X^n$ we proceed inductively by labelling each edge with a point in X starting with $x_1,\ldots,x_n$ for the n twigs. At each vertex with k inputs, we apply λ_k to the values of the inputs to obtain the label of the output. The value of the composite operation is given by the label of the root.

<u>Examples</u>:

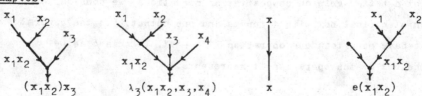

The relations 1.18 (i) and (ii) expressed in tree form read

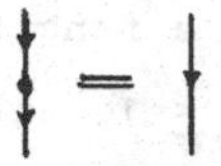

and each tree with n twigs represents the same operation as

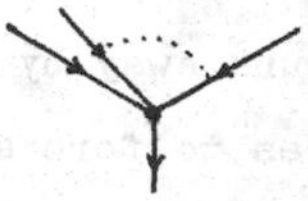

Now let X be of the homotopy type of a monoid. We want to find the collection of operations which are induced by the monoid structure under the homotopy equivalence, i.e. we look for a sort of A_∞-structure. Combining the operations λ_i of the monoid with the homotopy equivalence and its inverse, we obtain operations which we denote for simplicity reasons by λ_i too. The only difference is that the relations 1.18 (ii) hold up to homotopy only (here we assume that λ_1 is chosen to be the identity). The trees enable us to give a reasonable description of these homotopies. Before we consider the general situation, let us illustrate what we want to do by an example. We know that the induced multiplication on X is homotopy associative. Disregarding the stumps, there are three trees with three twigs, namely

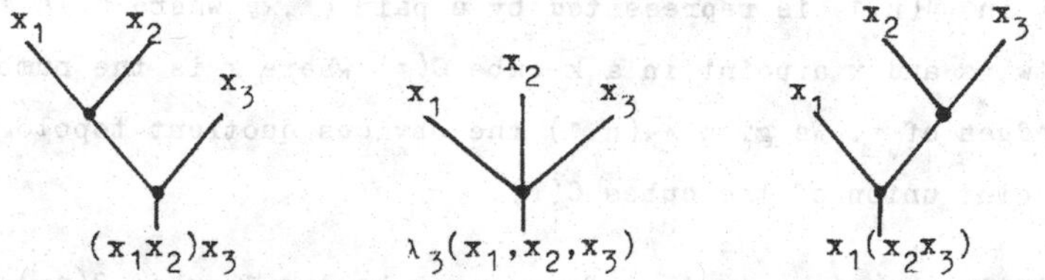

(the direction of the edges is given by gravity). We see the outer trees represent the two different ways of multiplying three elements. Instead of joining the left composite operation directly to the right one and thus obtaining the associating homotopy we join each of them to the middle tree. We do that by giving the internal edges a "length" between 0 and 1 and allow edges of length 0 to be shrunk away, i.e. the two vertices of the outer trees are identified to give the middle tree if the internal edge has length 0.

<u>Definition 1.19</u>: We define the space $W\mathfrak{A}(n,1)$. A point of this space is a tree with n twigs together with a function assigning to each internal edge a real number t, $0 \leq t \leq 1$, called its <u>length</u>, subject to the relations

(a) Any edge of length 0 may be shrunk away by removing it and amalgamating its two end vertices to form a new vertex;

(b) Any vertex with only one input may be removed. We give the resulting edge the length $t_1 * t_2 = t_1 + t_2 - t_1t_2$, where t_1 and t_2 are the lengths of the edges above and below this vertex. (Here we assume that the root and the twigs have the fictive length 1.)

Pictorially, the relations are

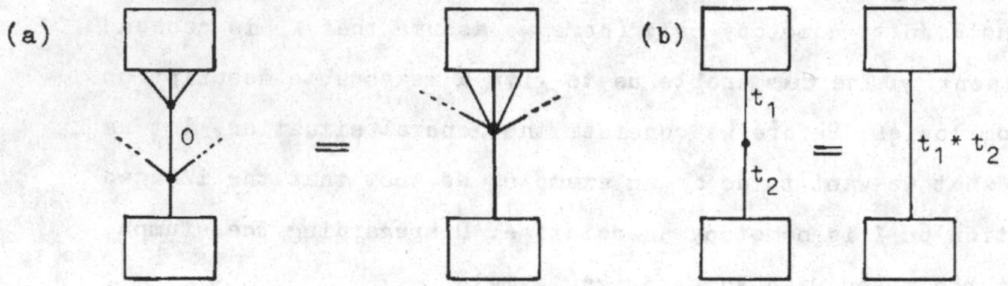

Each point in $W\mathfrak{A}(n,1)$ is represented by a pair (τ,x) where τ is a tree with n twigs and x a point in a k-cube $C(\tau)$ where k is the number of internal edges of τ. We give $W\mathfrak{A}(n,1)$ the obvious quotient topology from the disjoint union of the cubes $C(\tau)$.

The associating homotopy is therefore given by two 1-cubes $C(\tau_1)$ and $C(\tau_2)$ where τ_1 and τ_2 are the two outside trees of the previous example. These two cubes have common lower faces which are identified with the 0-cube $C(\tau_3)$ where τ_3 is the middle tree of the previous example.

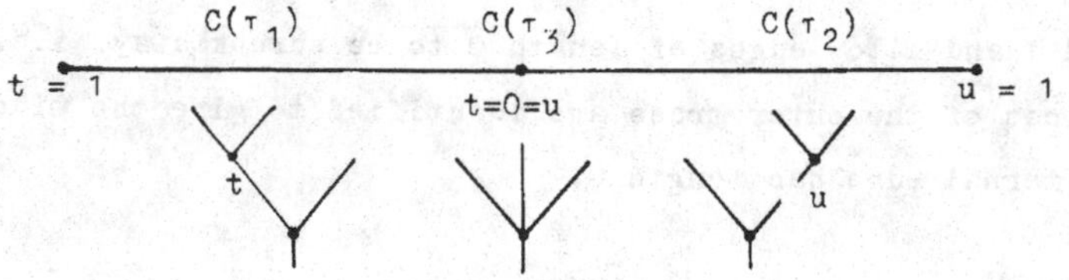

Let us give a pictorial description of what we mean by a $W\mathfrak{A}$-structure on a space, a different and, may be, clearer definition will be given in the next chapter.

Definition 1.20: A *cherry tree* on a space X is a tree (with lengths) in some space $W\mathfrak{A}(n,1)$ to each twig of which is assigned a point of X, which we call a *cherry*, that is, a cherry tree is a point of some space $W\mathfrak{A}(n,1) \times X^n$.

Examples:

(i)

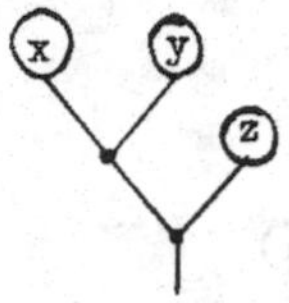

(ii)

(iii)

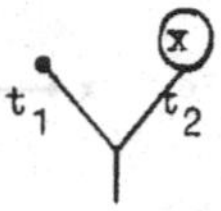

Definiton 1.21: A $W\mathfrak{A}$-*structure* on a space X is a continuous map

$$F : \bigcup_{n=0}^{\infty} W\mathfrak{A}(n,1) \times X^n \longrightarrow X$$

from the space of all cherry trees on X to the space X, subject to two conditions,

(c) we can cut down fully grown cherry trees without affecting their values;

(d) the value of the trivial cherry tree with cherry x is x (see Expl. (ii)).

Relation (c) demands some explanation: We say a cherry tree is *fully grown*, if some internal edge has length 1. To *cut it down* we replace the subcherry tree sitting on that edge by its value under F, regarded as a cherry in X and the cut branch becomes a twig.

Example:

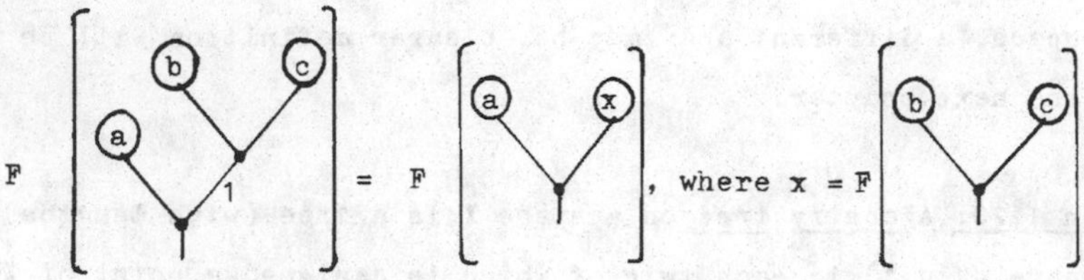

We now want to connect our definition of a $W\mathfrak{A}$-space, i.e. a space with a $W\mathfrak{A}$-structure, with Stasheff's definition of an A_∞-space. We have pointed out in the beginning of Section 1 that we work with homotopy units rather than strict ones. They are represented by stumps, because we have a 1-cube

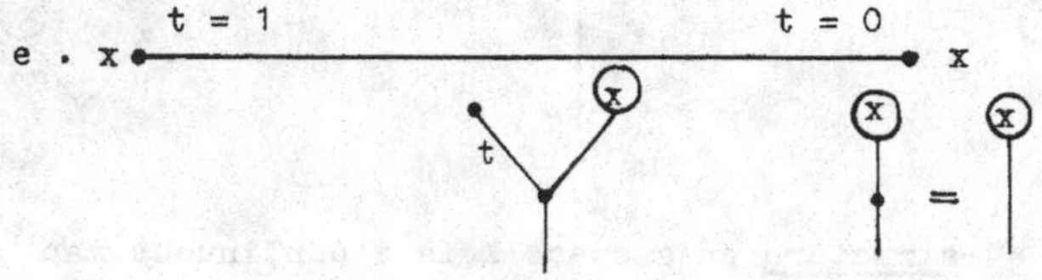

Therefore we can compare the two definitions only if we disregard stumps and units. In this case Stasheff's definition reduces to those structures to which Proposition 1.11 applies. We leave it to the reader to check that such a structure coincides with a $W\mathfrak{A}$-structure if we disregard stumps. Let us make this statement precise.

Proposition 1.22: Let $S\mathfrak{A}(n,1)$ be the subspace of $W\mathfrak{A}(n,1)$ of all trees which do not have stumps. Then the following are equivalent.

(i) There exists a continuous map

$$F : \bigcup_{n=0}^{\infty} S\mathfrak{A}(n,1) \times X^n \longrightarrow X$$

satisfying 1.21 (c) and (d).

(ii) There exist maps $M_n : K_n \times X^n \longrightarrow X$ for $n \geq 0$ satisfying 1.8(ii).

In fact, one can show that $S\mathfrak{A}(n,1)$ is just a subdivided copy of K_n, and relation 1.21(c) corresponds to 1.8(ii).

Example: $S\mathfrak{A}(4,1)$ is K_4 subdivided into five 2-cubes

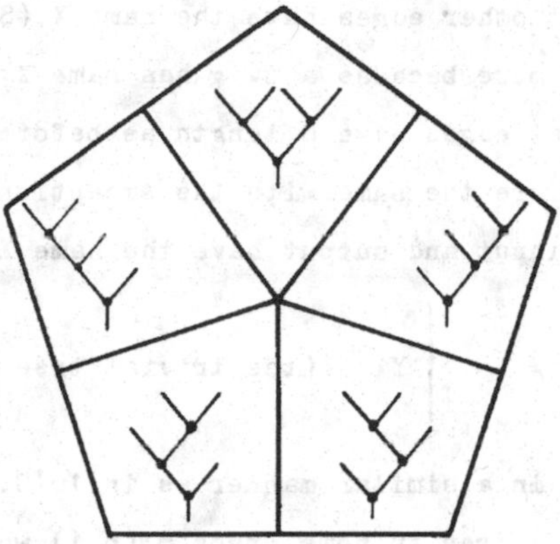

Before we give a proof of Proposition 1.11, let us illustrate how to define A_∞-maps with the aid of trees. We then can formulate Proposition 1.11 in its full strength.

At this stage it would lead too far to define A_∞-maps between A_∞-spaces, or better $W\mathfrak{A}$-spaces. This will be done in one of the coming chapters. Let us be content with the definition of an A_∞-map from a $W\mathfrak{A}$-space to a monoid.

Let $f : X \longrightarrow Y$ be a map from a $W\mathfrak{A}$-space to a monoid. In order to fit f into our tree description, we consider it as an electrical box with one input and one output

Is x the value of the input, then the value of the output is f(x). We compose this electrical box with the trees of the operations in X and Y as before by wiring together. To be able to decide where the operation takes place, in X or in Y, we give each edge a name X or Y, which we later call the colour of the edge. Because of the relations of a monoid structure, we have exactly one operation $Y^n \longrightarrow Y$ for any n. We make use of this in the definition of the models $M\mathfrak{A}(n,1)$ for such an A_∞-map.

Definition 1.23: A point of $M\mathfrak{A}(n,1)$ is a tree with n twigs, the root has the name Y while all other edges have the name X.(So the trivial tree does not occur any more because a twig has name X while a root has name Y.) The internal edges have a length as before and the relations 1.19(a) and (b) are the same with the exception that 1.19(b) may be applied only of input and output have the name X, i.e.

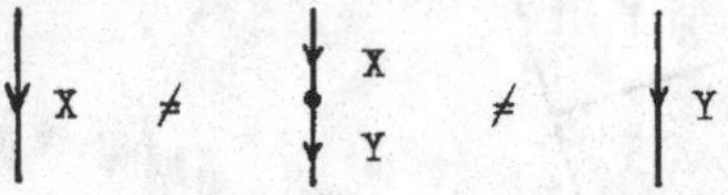

$$\text{\textdownarrow}\, X \;\neq\; \begin{array}{c}\text{\textdownarrow}\,X\\ \text{\textdownarrow}\,Y\end{array} \;\neq\; \text{\textdownarrow}\,Y \qquad \text{(the trivial tree does not occur)}$$

The topology is defined in a similar manner as in 1.19. A planted cherry tree on (X,Y) is a tree in some space $M\mathfrak{A}(n,1)$ which has a point of X assigned to each twig.

Definition 1.24: Let X be a W$\mathfrak{A}$-space whose structure is given by $F : \bigcup_n W\mathfrak{A}(n,1) \times X^n \longrightarrow X$ and Y be a monoid. A W$\mathfrak{A}$-structure on a map $f : X \longrightarrow Y$ is a continuous map

$$G : \bigcup_{n=0}^{\infty} M\mathfrak{A}(n,1) \times X^n \longrightarrow Y$$

from the space of all planted cherry trees on (X,Y) to Y such that

(c*) we can cut down fully grown trees without affecting their values under G.

(d*) $\qquad G \left[\begin{array}{c} \boxed{x} \\ \text{\textdownarrow}\,X \\ \text{\textdownarrow}\,Y \end{array} \right] = f(x)$

Again an explanation of relation (c*): To cut down a fully grown tree we replace the subtree sitting on an internal edge of length 1 by its value under F, regarded as a cherry in X. The cut branch becomes a twig. (Note that all edges of the subtree have name X.)

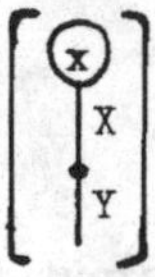

Example:

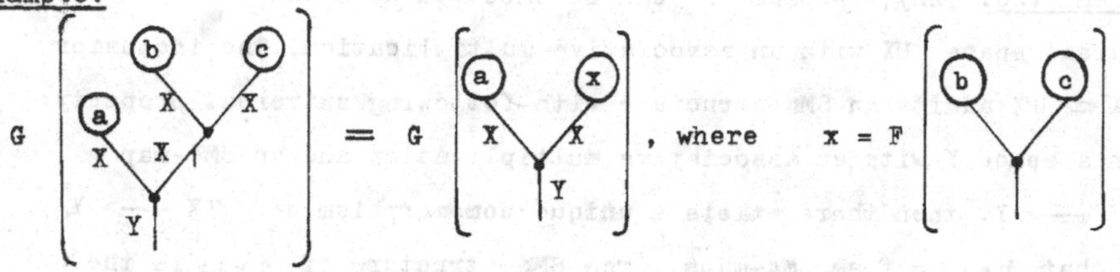

If we want to put our definition in relation to the one of Stasheff, we again run into the trouble that we have homotopy units instead of the strict units of Stasheff's structures. So to make precise statements, we have to neglect units. Let $SM\mathfrak{A}(n,1)$ be the subspace of $M\mathfrak{A}(n,1)$ consisting of the trees without stumps. We leave it to the reader to check the following result.

Proposition 1.25: Let X be an $S\mathfrak{A}$-space (see 1.22) and hence an A_∞-space with the exception that $M_2 : X^2 \longrightarrow X$ need not have a unit. Let $f : X \longrightarrow Y$ be a map from X to a space Y with an associative multiplication. Then f admits the structure of an A_∞-map (see 1.13) iff it admits an $SM\mathfrak{A}$-structure.

Example: The model $SM\mathfrak{A}(3,1)$

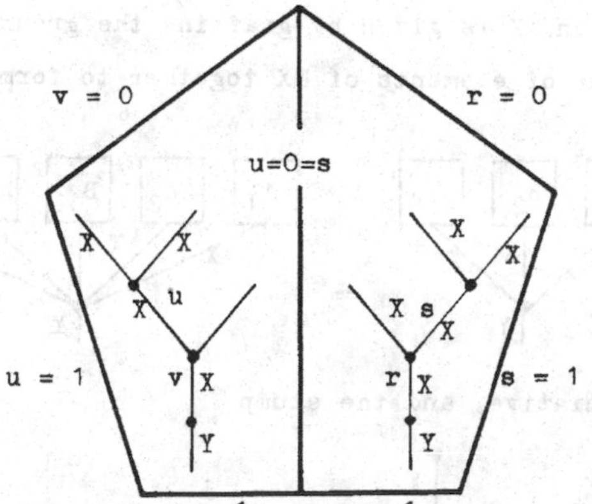

We now can formulate and prove a stronger version of Proposition 1.11.

Theorem 1.26: Any $S\underline{A}$-space X can be embedded as a SDR in a universal space UX with an associative multiplication. The inclusion $i : X \subset UX$ admits an $SM\mathfrak{A}$-structure with following universal property: Given a space Y with an associative multiplication and an $SM\mathfrak{A}$-map $f : X \longrightarrow Y$, then there exists a unique homomorphism $h : UX \longrightarrow Y$ such that $h \circ i = f$ as $SM\mathfrak{A}$-maps. (The $SM\mathfrak{A}$-structure of $h \circ i$ is the obvious one.)

There is a similar result involving structures with homotopy units.

Theorem 1.27: For any $W\mathfrak{A}$-space X there exists a monoid MX containing X as SDR. The inclusion $i : X \subset MX$ admits a $W\mathfrak{A}$-structure with following universal property: If $g : X \longrightarrow Y$ is a $W\mathfrak{A}$-map into a monoid Y, then there exists a unique homomorphism $h : MX \longrightarrow Y$ such that $h \circ i = g$ as $W\mathfrak{A}$-maps, where $h \circ i$ has the obvious $W\mathfrak{A}$-structure from i.

Since the proof of Theorem 1.26 is essentially the same as the one of Theorem 1.27 we only prove the latter.

Proof of 1.27: The monoid MX is obtained from the space $\bigcup_{n=0}^{\infty} M\mathfrak{A}(n,1) \times X^n$ by factoring out the relation 1.24(c*), i.e. a fully grown planted cherry tree is equivalent to the cut down one. The monoid structure on MX is given by grafting the ground vertices of two representatives of elements of MX together to form a new ground vertex

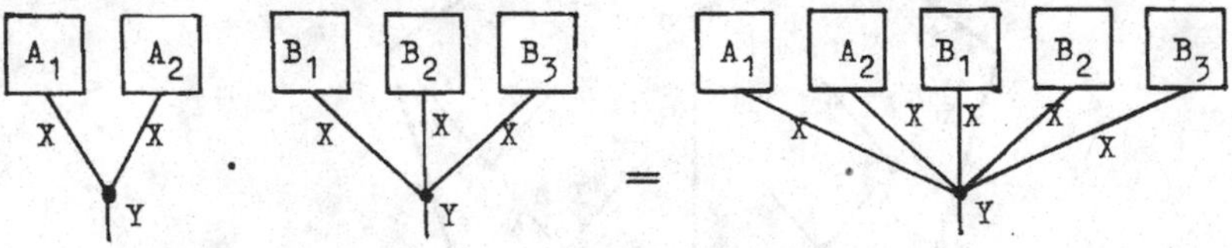

This is clearly associative, and the stump

serves as unit. So MX is a monoid. We define the inclusion $i : X \longrightarrow MX$

by

$i(x) =$ 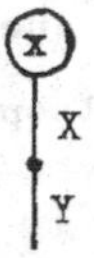

and the quotient map $G : \bigcup_{n=0}^{\infty} M\mathfrak{A}(n,1) \times X^n \longrightarrow MX$ endows i with a $W\mathfrak{A}$-structure. The deforming homotopy $H_t : MX \longrightarrow MX$ is given by

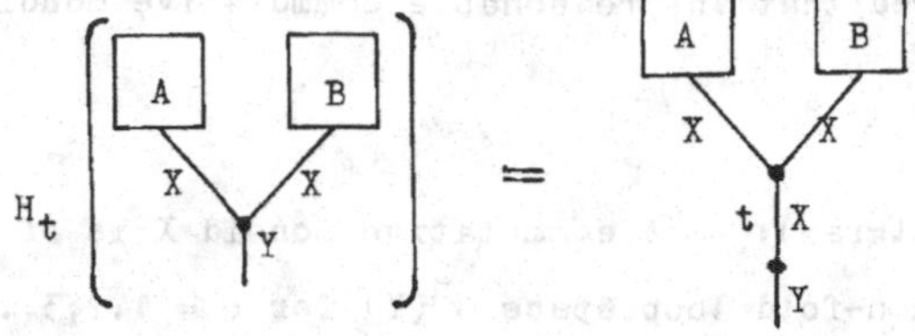

For t = 0, we can shrink away the additional edge of the right hand side tree, hence H_0 is the identity. At t = 1, we can cut down completely to obtain a tree representing an element in i(X). Because of relation 1.19(b), the homotopy H_t leaves i(X) pointwise fixed. ∎

We have pointed out at the end of previous section that A_∞-maps are difficult to handle even if the spaces involved are monoids. Our models are a little more complicated than the cubes in the definition of Sugawara. So, a priori, there is no reason to assume that our approach makes life easier. Nevertheless, this is the case because our $W\mathfrak{A}$-structures are universal in some sense which we do not want to elaborate on at this stage. Let us say so much as that it often is possible to replace a naturally occuring structure on a space by a $W\mathfrak{A}$-structure because of this universality.

5. SOME REMARKS ON COMMUTATIVITY

So far we have investigated spaces of the homotopy type of a monoid and hence of a loop space. We are interested in a more general question; we want to find conditions on the H-space structure of a space X under which X is of the homotopy type of an n-fold loop space.

An interesting special part of this question is to find conditions under which a space is an infinite loop space.

Definition 1.28: A space X is called an infinite loops space if there exist spaces X_n, $n = 0,1,2,\ldots$ such that $X_o = X$ and $X_n \simeq \Omega X_{n+1}$.

Milgram [37] showed that any reasonable commutative monoid is an infinite loop space.

Propsition 1.29 (Milgram): A commutative monoid X is of the weak homotopy type of an n-fold loop space $\Omega^n(Y)$ for $n = 1,2,3,\ldots$. Moreover, there is an H-structure on $\Omega^n(Y)$ such that the weak homotopy equivalence preserves the multiplication up to homotopy.

As in the associative case, the structure of a commutative monoid is a bad one from the view point of homotopy theory for the same reasons as there. So one is interested in structures which are not quite commutative and which live in homotopy theory. For example, we could search for the weakest structure on a space X such that X is a double loop space. Attempts on this line have been made by Sugawara[53]. He looked for conditions on a space X such that X has a classifying space which is an H-space. This is somewhat different from our question because we want the classifying space to be a loop space.

Definition 1.30 (Sugawara): A monoid X with unit e is called strongly homotopy-commutative if there exist maps

$$C_n : (I^n \times X^{2n}, I^n \times e^{2n}) \longrightarrow (X,e) \qquad n = 1,2,3,\ldots$$

such that

$$C_1(0,x,y) = xy \qquad\qquad C_1(1,x,y) = yx$$

and

$$C_n(t_1,\dots,t_n,x_1,\dots,x_n,y_1,\dots,y_n) = \begin{cases} x_1C_{n-1}(t_2,\dots,t_n,x_2,\dots,x_n,y_1y_2,\dots,y_n) & t_1 = 0 \\ C_{n-1}(t_1,\dots,\hat{t_i},\dots,t_n,x_1,\dots,x_{i-1}x_i,\dots,x_n,y_1,\dots, \\ \qquad y_iy_{i+1},\dots,y_n) & t_i = 0,\ 1 < i < n \\ C_{n-1}(t_1,\dots,t_{n-1},x_1,\dots,x_{n-1}x_n,y_1,\dots,y_{n-1})y_n & t_n = 0 \\ C_{i-1}(t_1,\dots,t_{i-1},x_1,\dots,x_{i-1},y_1,\dots,y_{i-1}) \\ \qquad y_ix_iC_{n-i}(t_{i+1},\dots,t_n,x_{i+1},\dots,x_n,y_{i+1},y_n) & t_i = 1 \end{cases}$$

Examples:

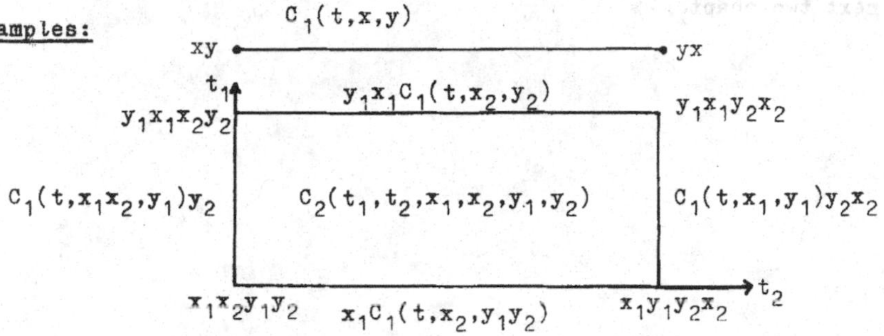

Proposition 1.31 (Sugawara): (a) The loop space ΩX of a countable CW-complex X is strongly homotopy-commutative iff X is an H-space.
(b) The Milnor-classifying space of a countable CW-group G is an H-space iff G is strongly homotopy-commutative.

We see that Sugawara requires that the H-spaces are at least associative besides being strongly homotopy-commutative. This is a condition which does not do for us. To find what we think is the correct structure to work with we again use our tree language, but we have to make some changes. For the definition of a $W\mathfrak{A}$-structure we took the operations of a monoid structure and replace the relations by homotopies. If we try to do the same for a commutative monoid we cannot restrict our attention to the operations

$$\lambda_n : (x_1,\dots,x_n) \longrightarrow x_1 + \dots + x_n$$

because permutations come in in the definition of commutativity which

reads

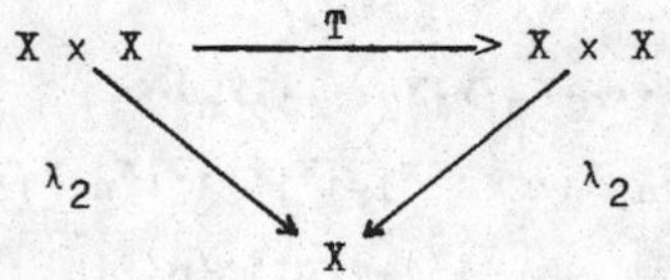

commutes, where T interchanges the factors. So we have to add permutations as operations to our trees and we must impose an additional relation. The details of this more general construction will be given in the next two chapters.

II. CHAPTER

TOPOLOGICAL-ALGEBRAIC THEORIES

In this chapter we introduce the notions from categorical algebra we need. For the sake of topologists our proofs are less formal than a category theorist would like them to be. So we construct adjoint functors explicitly and are even willing to use elements instead of restricting ourselves to formal concepts. Our treatment of categorical algebra is broader than absolutely necessary for the understanding of the following chapters. We wanted to give a self-contained exposition of categorical algebra involving more than just one underlying object. Bénabou [4] has investigated such algebras in the category of sets; we work with topological spaces as underlying objects. Readers who are only interested in the topological aspect of these notes can skip over most of this chapter. It suffices to read section 1, the proof of Proposition 2.5, sections 3, 5, and 6. We want to remind that we work in the category of compactly generated spaces, which we denote by $\mathfrak{Top}$.

1. DEFINITIONS

Lawvere [25] has formalized the concept of an algebraic theory given by operations and laws without existential quantifiers. As examples we have the theories of monoids, groups, rings etc. (whose axioms can be put into the required form), but not the theory of fields. He considers the category of <u>all</u> operations that can be written down in the theory, instead of selecting certain operations that generate the rest.

Each theory contains a distinguished collection of operations, the <u>set</u> <u>operations</u>: Let

$$\sigma : \{1,2,\ldots,m\} \longrightarrow \{1,2,\ldots,n\}$$

be a function and X a topological space. Then there is an operation

$$(2.1) \qquad \sigma^* : X^n \longrightarrow X^m$$

given by $\sigma^*(x_1,\ldots,x_n) = (x_{\sigma 1},\ldots,x_{\sigma m})$. Let $\mathfrak{S}$ be the category consisting of one finite set n for each cardinal n less than infinity and all functions between them.

<u>Definition 2.2</u>: A (finitary) <u>algebraic</u> <u>theory</u> is a category Θ with objects 0,1,2,... together with a faithful functor $\mathfrak{S}^{op} \longrightarrow \Theta$ preserving objects and products. Call Θ <u>topological-algebraic</u> if each set $\Theta(m,n)$ is topologized and if composition and products are continuous (the latter means that $\Theta(m,n) \cong \Theta(m,1)^n$ is a homeomorphism).
A Θ-<u>space</u> is a continuous functor $\Theta \longrightarrow \mathfrak{Top}$, such that $\mathfrak{S}^{op} \longrightarrow \Theta \longrightarrow \mathfrak{Top}$ preserves products, the image of 1 is called the <u>underlying</u> <u>space</u>.
A <u>homomorphism</u> between Θ-spaces is a natural transformation between such functors.
If Θ_1 and Θ_2 are theories, a <u>theory</u> <u>functor</u> is a continuous functor $\Theta_1 \longrightarrow \Theta_2$ such that the following diagram commutes

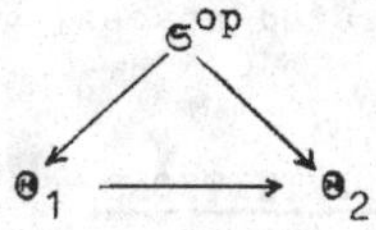

The results of sections 2,3,4 stay true if we omit the condition that $\mathfrak{S}^{op} \longrightarrow \Theta$ be faithful (i.e. monic on morphism sets), but in view of the definition of a Θ-space the case where $\mathfrak{S}^{op} \longrightarrow \Theta$ is faithful is the only interesting one.

The image of $\mathfrak{S}^{op}$ in Θ consists of all set operations as described in (2.1). We call the elements of $\Theta(m,1)$ the m-<u>ary</u> <u>operations</u> of Θ. In abuse of language we often identify a Θ-space with its underlying space X and say "X admits a Θ-structure" or "Θ acts on X" or "X is a Θ-space".

Essentially for the study of maps we need a generalization, namely theories on several objects. Let K be a set and $\mathfrak{S}_K$ the category $\mathfrak{S}$ over K, whose objects are functions $\underline{i} : [n] = \{1,\dots,n\} \longrightarrow K$ and whose morphisms from $\underline{i}$ to $\underline{j}$ are all functions f making

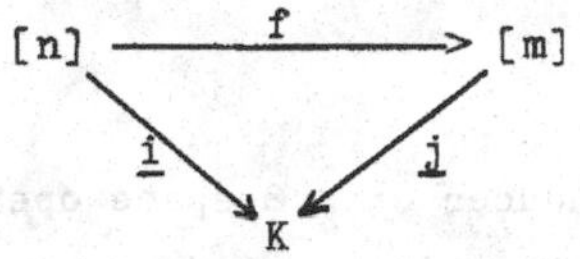

commute. We often use an alternative description of $\mathfrak{S}_K$: Its objects are ordered collections $\{k_1,\dots,k_n\}$ of elements of K and its morphisms are functions $f : \{k_1,\dots,k_n\} \longrightarrow \{k'_1,\dots,k'_m\}$ such that $f^{-1}(k'_r)$ is either empty or consists only of elements equal to k'_r. An object $\underline{i} : [1] \longrightarrow K$ is often denoted by its image $k = \underline{i}(1)$. We call such an object basic. A function $r : K \longrightarrow L$ induces a $r_* : \mathfrak{S}_K^{op} \longrightarrow \mathfrak{S}_L^{op}$ given by $r_*(\underline{i}) = r \circ \underline{i}$ and $r_*(f : [n] \longrightarrow [m]) = f$.

Definition 2.3: A (finitary) K-coloured algebraic theory is a category Θ with the same objects as $\mathfrak{S}_K$ together with a faithful functor $\mathfrak{S}_K^{op} \longrightarrow \Theta$ preserving objects and products. Call Θ topological-algebraic if each set $\Theta(\underline{i},\underline{j})$ is topologized and if composition and products are continuous (the latter means $\Theta(\underline{i},\underline{j}) \cong \Theta(\underline{i},\underline{j}(1)) \times \dots \times \Theta(\underline{i},\underline{j}(m))$ is a homeomorphism).

A Θ-space is a continuous functor $\Theta \longrightarrow \mathfrak{Top}$ such that $\mathfrak{S}_K^{op} \longrightarrow \Theta \longrightarrow \mathfrak{Top}$ preserves products; the images of the basic objects form the collection of its underlying spaces; we have one for each $k \in K$.

A homomorphism between Θ-spaces is a natural transformation between such functors.

A theory functor from a K-coloured theory Θ_1 to an L-coloured theory Θ_2 is a continuous functor $F : \Theta_1 \longrightarrow \Theta_2$ together with a function $f : K \longrightarrow L$ such that

$$\begin{array}{ccc} \mathfrak{S}_K^{op} & \xrightarrow{\ f_*\ } & \mathfrak{S}_L^{op} \\ \downarrow & & \downarrow \\ \Theta_1 & \xrightarrow{\ F\ } & \Theta_2 \end{array}$$

commutes.

Remark: Among category theorists a coloured theory is better known by the name "sorted theory".

The image of $\mathfrak{S}_K^{op}$ in Θ induces on a Θ-space operations given again by formula (2.1). We therefore call its elements <u>set operations</u>.

If we do not specify the function $f : K \longrightarrow L$ of a theory functor, we assume that it is the identity.

In any type of theory the set operations have the useful property that we can push them to the right. Given $\sigma \in \mathfrak{S}_K(\underline{i},\underline{j})$ with $\underline{i} : [m] \longrightarrow K$, $\underline{j} : [n] \longrightarrow K$ and $a = (a_1,\ldots,a_n) \in \Theta(\underline{k},\underline{j})$ with $a_r \in \Theta(\underline{k},\underline{j}(r))$ and $b_r \in \Theta(\underline{k}_r,\underline{j}(r))$, $k_r : [p_r] \longrightarrow K$. Denoting the product bifunctor $\Theta \times \Theta \longrightarrow \Theta$ by $\oplus$ we have the following formulae

$$(2.4) \qquad \begin{array}{l} \sigma^* \circ a = (a_{\sigma 1},\ldots,a_{\sigma m}) \\ \sigma^* \circ (b_1 \oplus \ldots \oplus b_n) = (b_{\sigma 1} \oplus \ldots \oplus b_{\sigma m}) \circ \sigma(p_1,\ldots,p_n)^* \end{array}$$

Here $\sigma(p_1,\ldots,p_n) \in \mathfrak{S}_K(\bigcup_i \underline{k}_{\sigma i}, \bigcup_i \underline{k}_i)$ is given by

$$\sigma(p_1,\ldots,p_n)\,(r) = p_1 + p_2 + \ldots + p_{\sigma(u)-1} + v$$

if $r = p_{\sigma 1} + \ldots + p_{\sigma(u-1)} + v$ and $0 < v \le p_{\sigma(u)}$. Loosely speaking, we consider $\Sigma\, p_{\sigma i}$ as m blocks of $p_{\sigma 1},\ldots,p_{\sigma m}$ elements and $\Sigma\, p_i$ as n blocks of $p_1,\ldots,p_n$ elements, and $\sigma(p_1,\ldots,p_n)$ maps the u-th block $p_{\sigma u}$ of $\Sigma\, p_{\sigma i}$ identically onto the $\sigma(u)$-th block $p_{\sigma u}$ of $\Sigma\, p_i$. We call $\sigma(p_1,\ldots,p_n)$ a <u>block function</u>. Note that $\sigma(p_1,\ldots,p_n)$ is a permutation whenever σ is one.

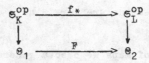

2. FREE THEORIES

In this section we show how to obtain a theory from a set of operations and relations between them.

Define categories $\mathfrak{Theories}$, $\mathfrak{Gr\ spaces}$, $\mathfrak{Gr\ spaces}^o$, and $\mathfrak{Eq\ spaces}$ as follows:

$\mathfrak{Theories}$ is the category of coloured theories and theory functors.

$\mathfrak{Gr\ spaces}$ has $(\mathrm{ob}\,\mathfrak{S}_K \times K)$-graded spaces $\{X_{\underline{i},k}\}$, K some set, as objects. A morphism from an $(\mathrm{ob}\,\mathfrak{S}_K \times K)$-graded space $\{X_{\underline{i},k}\}$ to an $(\mathrm{ob}\,\mathfrak{S}_L \times L)$-graded space $\{Y_{\underline{j},l}\}$ is a pair (g,f) where $f : K \longrightarrow L$ is a function and $g : \{X_{\underline{i},k}\} \longrightarrow \{Y_{\underline{j},l}\}$ a continuous graded map sending $X_{\underline{i},k}$ to $Y_{f\circ\underline{i},f(k)}$.

$\mathfrak{Gr\ spaces}^o$ has $(\mathrm{ob}\,\mathfrak{S}_K \times K)$-graded spaces $(X_{\underline{i},k})$ as objects but each $X_{k,k}$, $k \in K$, is based (recall that $k \in \mathrm{ob}\,\mathfrak{S}_K$ is the basic object determined by k). The morphisms are defined as for $\mathfrak{Gr\ spaces}$ but are supposed to preserve base points.

$\mathfrak{Eq\ spaces}$ is the category of objects in $\mathfrak{Gr\ spaces}^o$ with an $\mathfrak{S}_K$-action and equivariant morphisms in $\mathfrak{Gr\ spaces}^o$. An $\mathfrak{S}_K$-action on an $(\mathrm{ob}\,\mathfrak{S}_K \times K)$-graded space $\{X_{\underline{i},k}\}$ is a colloection of maps

$$a_{\underline{i},\underline{j}} : X_{\underline{i},k} \times \mathfrak{S}_K(\underline{i},\underline{j}) \longrightarrow X_{\underline{j},k}$$

such that $a_{\underline{i},\underline{i}}(x,\mathrm{id}_{\underline{i}}) = x$ and the following diagram commutes

$$\begin{array}{ccc}
X_{\underline{i},k} \times \mathfrak{S}_K(\underline{i},\underline{j}) \times \mathfrak{S}_K(\underline{j},\underline{l}) & \xrightarrow[a_{\underline{i},\underline{j}}\times \mathrm{id}]{} & X_{\underline{j},k} \times \mathfrak{S}_K(\underline{j},\underline{l}) \\
\Big\downarrow {\scriptstyle \mathrm{id}\times\mathrm{composition}} & & \Big\downarrow {\scriptstyle a_{\underline{j},\underline{l}}} \\
X_{\underline{i},k} \times \mathfrak{S}_K(\underline{i},\underline{l}) & \xrightarrow[a_{\underline{i},\underline{l}}]{} & X_{\underline{l},k}
\end{array}$$

A morphism (g,f) from an $\mathfrak{S}_K$-space $\{X_{\underline{i},k}\}$ to an $\mathfrak{S}_L$-space $\{X_{\underline{p},l}\}$ in $\mathfrak{Gr\ spaces}^o$ is called equivariant if

$$\begin{array}{ccc}
X_{\underline{i},k} \times \mathfrak{S}_K(\underline{i},\underline{j}) & \xrightarrow{\text{action}} & X_{\underline{j},k} \\
\Big\downarrow {\scriptstyle g\times f_*} & & \Big\downarrow {\scriptstyle g} \\
Y_{f\circ\underline{i},f(k)} \times \mathfrak{S}_L(f\circ\underline{i},f\circ\underline{j}) & \xrightarrow{\text{action}} & Y_{f\circ\underline{j},f(k)}
\end{array}$$

commutes.

We call the (ob $\mathfrak{S}_K \times K$)-graded spaces of the last three categories its K-<u>coloured</u> <u>objects</u>.

We also use a more conceptual definition of the last three categories: Let $\mathfrak{D}\mathfrak{S}_K$ be the subcategory of $\mathfrak{S}_K$ consisting of all objects and the identity morphisms. An object of $\mathfrak{Gr}$ $\mathfrak{spaces}$ can be considered as a K-indexed collection of functors $R_k : \mathfrak{D}\mathfrak{S}_K \longrightarrow \mathfrak{Top}$. A morphism from $\{R_k | k \in K\}$ to $\{Q_l | l \in L\}$ is a pair (α, f) consisting of a function $f : K \longrightarrow L$ and a K-indexed collection α of natural transformations $\alpha_k : R_k \longrightarrow Q_{f(k)} \circ f_*$:

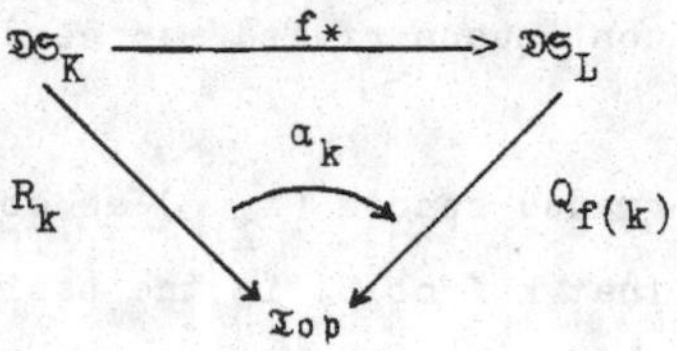

$\mathfrak{Gr}$ $\mathfrak{spaces}^o$ can be defined similarly with the exception that we have to introduce a base point e_k in $R_k(k)$ and that α_k has to preserve this base point. Finally, an object of $\mathfrak{Eq}$ $\mathfrak{spaces}$ can be considered as a K-indexed collection of functors $R_k : \mathfrak{S}_K \longrightarrow \mathfrak{Top}$ such that $R_k(k)$ is based. A morphism from $\{R_k | k \in K\}$ to $\{Q_l | l \in L\}$ consists of a function $f : K \longrightarrow L$ and a collection α of base point preserving natural transformations α_k

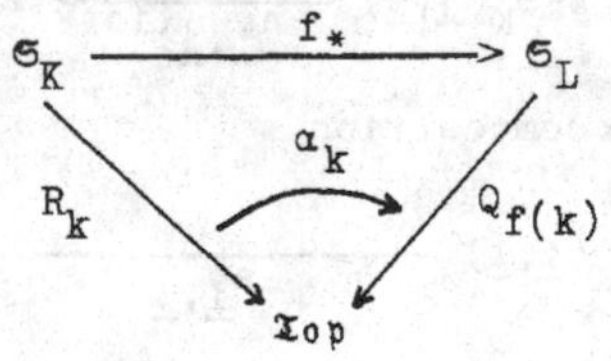

We have forgetful functors

$$\mathfrak{Theories} \xrightarrow{U_1} \mathfrak{Eq}\ \mathfrak{spaces} \xrightarrow{U_2} \mathfrak{Gr}\ \mathfrak{spaces}^o \xrightarrow{U_3} \mathfrak{Gr}\ \mathfrak{spaces}$$

where U_2 and U_3 are the obvious ones while $U_1\Theta = \{R_k | k \in K\}$, Θ a K-coloured theory, is given by $R_k(\underline{i}) = \Theta(\underline{i}, k)$ and $R_k(\sigma)$ is composition on the right by σ^*. The base points are $id_k \in \Theta(k,k)$.

Proposition 2.5: Each functor U_i has a left adjoint F_i.

Proof: F_3 is given by adjoining an extra point which becomes the base point to each space $X_{k,k}$ of $(X_{\underline{i},k})$.

Let $(\alpha,f) : \{R_k | k\in K\} \longrightarrow \{Q_l | l\in L\}$ be a morphism in $\mathfrak{Gr}$ spaces^{o}. Define

$$F_2\{R_k | k\in K\} = \{ \bigcup_{\underline{j}\in\mathfrak{S}_K} R_k(\underline{j}) \times \mathfrak{S}_K(\underline{j},-) | k\in K\}$$

and $F_2(\alpha,f) = (\beta,f)$ with

$$\beta_k(\underline{i}) : \bigcup_{\underline{j}\in\mathfrak{S}_K} R_k(\underline{j})\times \mathfrak{S}_K(\underline{j},\underline{i}) \xrightarrow{\cup\alpha_k(\underline{j})\times f_*} \bigcup_{\underline{j}\in\mathfrak{S}_K} Q_{f(k)}(f\circ\underline{j})\times\mathfrak{S}_L(f\circ\underline{j},f\circ\underline{i}) \subset F_2\{Q_l | l\in L\}$$

Since $R_k(k) \times \mathfrak{S}_K(k,k) \cong R_k(k)$, we can take the base points of the $R_k(k)$ as base points of $F_2\{R_k | k\in K\}$.

The construction of F_1 is more lengthy, though straightforward. For a K-coloured object $\{X_{\underline{i},k}\}$ of $\mathfrak{Gq}$ spaces we want to construct a K-coloured theory $\Theta = F_1\{X_{\underline{i},k}\}$. We consider the points of $X_{\underline{i},k}$ as the indecomposable operations from $\underline{i}$ to k. A general operation is a composite of products (formally written $\oplus$) of such indecomposable operations. The set operations of Θ stand, of course, in some relation to the $\mathfrak{S}_K$-action on $\{X_{\underline{i},k}\}$. More explicitly, a letter from $\underline{i}$ to $\underline{j} = \{\underline{j}(1),\ldots,\underline{j}(p)\}$ is either a formal product $x_1 \oplus \ldots \oplus x_p$ with $x_q \in X_{\underline{i}_q,\underline{j}(q)}$, $\underline{i} = \underline{i}_1 \oplus \ldots \oplus \underline{i}_p$, or an element $\sigma^* \in \mathfrak{S}_K^{op}(\underline{i},\underline{j})$. A morphism in Θ from $\underline{i}$ to $\underline{j}$ is an equivalence class of words $[a_1|\ldots|a_n]$ in letters a_i such that source a_i = target a_{i+1}, source $a_n = \underline{i}$, target $a_1 = \underline{j}$. The equivalence relation is generated by

(i) $[id_{\underline{i}}] = [\ \ ] = [e_{\underline{i}(1)} \oplus \ldots \oplus e_{\underline{i}(n)}]$, where $\underline{i} : [n] \longrightarrow K$ in $\mathfrak{S}_K$ and $e_k \in X_{k,k}$ the base point

(ii) $[x \oplus e_1 | e_2 \oplus y] = [e_2' \oplus y | x \oplus e_1'] = [x \oplus y]$ for appropriate formal products x and y and formal products e_1, e_2, e_1', e_2' of base points

(iii) $[\sigma^* | x_1 \oplus \ldots \oplus x_n] = [x_{\sigma 1} \oplus \ldots \oplus x_{\sigma m} | \sigma(p_1,\ldots,p_n)^*]$ for $\sigma^* \in \text{mor}\, \mathfrak{S}_K^{op}$, $x_r \in X_{\underline{i}_r,k_r}$, $\underline{i}_r : [p_r] \longrightarrow K$ (compare 2.4)

(iv) $[\sigma^* | \pi^*] = [(\pi\circ\sigma)^*]$

(v) $[x \cdot \sigma] = [x|\sigma^*]$ where $x \cdot \sigma$ is the image of (x,σ) under the $\mathfrak{S}_K$-action on $\{X_{\underline{i},k}\}$.

Composition of morphisms is induced by juxtaposition of words. Relation (i) makes the base points to units, relation (ii) allows us to define a bifunctor $\oplus$, (i), (iii), and (iv) assure that the set operations behave correctly, and (v) makes them compatible with the $\mathfrak{S}_K$-action on $\{X_{\underline{i},k}\}$.

For later purposes we want to have a better grip on the morphisms of Θ. We now give an alternative description of representatives using the tree language developed in Chapter I. This enables us to do without the relations (ii), (iii), and (iv).

<u>Definition 2.6</u>: A <u>tree</u> from $\underline{j} : [n] \longrightarrow K$ to the basic object k associated with $\{X_{\underline{i},k}\}$ consists of

(a) a finite directed planar tree as known from graph theory except that the edges need not have vertices on both ends. There is exactly one edge called the <u>root</u> which has no end vertex, but there may be arbitrarily many edges called <u>twigs</u> with no beginning vertex; all other edges are called <u>internal</u>. Each vertex has exactly one outgoing edge (though not necessarily an incoming one).

(b) a function assigning to each edge an element of K called its <u>colour</u>. The colours of the twigs are elements from $\underline{j}([n]) \subset K$, the colour of the root is k. This function together with the underlying graph is called the <u>shape</u> of the tree.

(c) a function assigning to each vertex, whose incoming edges have the colours $i_1,..,i_n$ in clockwise order and whose outgoing edge has the colour l, a point in $X_{\underline{i},l}$ called its <u>label</u>, where $\underline{i}=\{i_1,..,i_n\}$.

(d) a function assigning to each twig a label $\in \{1,...,n\}$ such that a twig with label r has the colour $\underline{j}(r)$.

We allow <u>trivial trees</u> whose underlying graphs consist of one edge and no vertex. Hence we have one trivial tree from $\underline{j}$ to k for each label $r \in \{1,\ldots,n\}$ with $\underline{j}(r) = k$. A tree consisting of one vertex with no incoming edge is called a <u>stump</u>.

A tree of a given shape is specified by its vertex labels and twig labels and can therefore be considered as a point in some product space $(\prod X_{\underline{i},l}) \times \mathfrak{S}_K(\underline{k},\underline{j})$ where $\underline{k} = \{k_1,\ldots,k_r\}$ is the collection of twig colours taken in clockwise order. The disjoint union of these product spaces defines a topology on the set of all trees from $\underline{j}$ to k associated with $\{X_{\underline{i},k}\}$.

As in the previous chapter the trees are inspired by the attempt to obtain a general composite operation from a collection of indecomposable operations. Starting with a K-indexed family of topological spaces Z_k and topologized collections $X_{\underline{i},l}$ of operations $a : Z_{\underline{i}(1)} \times \ldots \times Z_{\underline{i}(r)} \longrightarrow Z_l$, a tree from $\underline{j} = \{k_1,\ldots,k_n\}$ to k represents a composite of such operations, composed with a set operation. Its source is $Z_{k_1} \times \ldots \times Z_{k_n}$, its target Z_k. The value of this composite on an element $(z_1,\ldots,z_n)$ of the source can be computed as follows: We assign to each twig with label $r \in \{1,\ldots,n\}$ the value z_r. Inductively we give the outgoing edge of a vertex with label a the value $a(y_1,\ldots,y_m)$ if the values of the incoming edges are $y_1,\ldots,y_m$ in clockwise order. The value of the root is the value of the tree operation on $(z_1,\ldots,z_n)$.

<u>Example</u>: (The direction of the graph will always be given by "gravity", i.e. from top to bottom). Let K = {X,Y}. Given operations $a : X\times Y \longrightarrow Y$ and $b : X\times X \longrightarrow X$, the tree

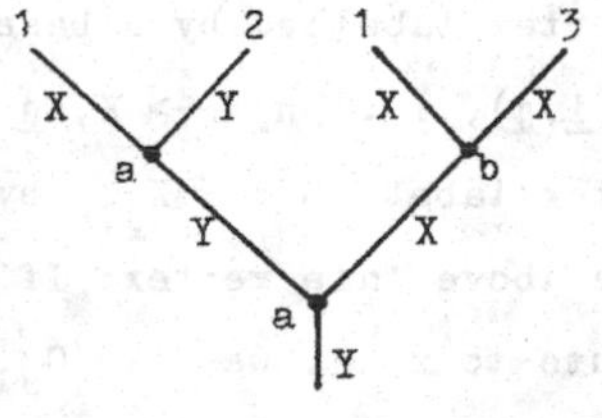

represents the operation $c : X \times Y \times X \longrightarrow Y$ given by $c(x_1,y,x_2)=a(a(x_1,y), b(x_1,x_2))$.

While a tree represents an operation into a single space, i.e. its target is a basic object, a general operation is represented by a copse:

Definition 2.7: A copse with source $\underline{i} : [n] \longrightarrow K$ and target $\underline{j} : [m] \rightarrow K$, associated with $\{X_{\underline{i},k}\}$, is an ordered collection of m trees whose sources are $\underline{i}$ and whose targets are $\underline{j}(1),\ldots,\underline{j}(m)$. The copses with a given source and target inherit a topology from their trees.

We can compose two copses $A_1 : \underline{i} \longrightarrow \underline{j}$ and $A_2 : \underline{j} \longrightarrow \underline{l}$ to form a copse $A_2 \circ A_1 : \underline{i} \longrightarrow \underline{l}$ by grafting the r-th tree of the copse A_1 to each twig of A_2 with label $r \in \{1,\ldots,m\}$. This defines a continuous associative composition of copses with the copse consisting of m trivial trees with twig labels $1,\ldots,m$ acting as identity of $\underline{j}$. Hence the collection of copses associated with $\{X_{\underline{i},k}\}$ forms a category Ψ. In fact, Ψ is a K-coloured theory. The functor

$$\mathfrak{S}_K^{op} \longrightarrow \Psi$$

sends $\sigma \in \mathfrak{S}_K(\underline{i},\underline{j})$, $\underline{i} : [n] \longrightarrow K$, $\underline{j} : [m] \longrightarrow K$, to the copse consisting of n trivial trees with colours $\underline{i}(r)$ and labels $\sigma(r)$, $r=1,\ldots,n$.

It is evident, that the tree description takes care of the relations (ii), (iii), (iv). We still have to account for (i) and (v). The theory Θ, we look for, is the quotient (with the quotient topology) of Ψ under the equivalence relation generated by

(2.8) (a) we may remove any vertex labelled by a base point

(b) if $x \in X_{\underline{i},k}$, $\sigma \in \mathfrak{S}_K(\underline{i},\underline{j})$, $\underline{i} : [n] \longrightarrow K$, $\underline{j} = [m] \longrightarrow K$, we may replace the vertex label $x \cdot \sigma \in X_{\underline{j},k}$ by x by changing the part of the tree above this vertex: If $C_1,\ldots,C_m$ are the subtrees on the inputs to $x \cdot \sigma$, we take $C_{\sigma 1},\ldots,C_{\sigma n}$ as subtrees over x.

Θ is a K-coloured theory, the functor $\mathfrak{S}_K^{op} \longrightarrow \Theta$ is induced by the one for Ψ. The space $X_{\underline{i},k}$ can be considered as subspace of $\Theta(\underline{i},k)$ by identifying $x \in X_{\underline{i},k}$, $\underline{i} : [n] \longrightarrow K$, with the tree consisting of one vertex only, whose label is x. The labels of the n twigs are 1,...,n

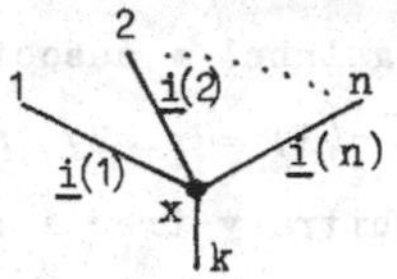

Let Ξ be an L-coloured theory. A theory functor $(P,f) : \Theta \longrightarrow \Xi$ induces a morphism $(g,f) : \{X_{\underline{i},k}\} \longrightarrow U_1\, \Xi$ in $\mathfrak{Eq\ spaces}$ by $g(x) = P(x)$. Conversely, since a theory functor $\Theta \longrightarrow \Xi$ is completely determined by its values on the indecomposable elements, a morphism $(g,f) : \{X_{\underline{i},k}\} \longrightarrow U_1\, \Xi$ induces a theory functor $(P,f) : \Theta \longrightarrow \Xi$ by $P(x) = g(x)$. The correspondence $(g,f) \longleftrightarrow (P,f)$ yields a natural bijection

$$\mathfrak{Theories}\ (F_1\{X_{\underline{i},k}\},\Xi) \cong \mathfrak{Eq\ spaces}\ (\{X_{\underline{i},k}\},U_1\Xi)$$

This completes the proof of Proposition 2.5. ∎

<u>Remark</u>: If $\{X_{\underline{i},k}\} \in \mathfrak{Or\ spaces}$, then $F_1 \circ F_2 \circ F_3\{X_{\underline{i},k}\}$ is the category Ψ of copses as constructed above. If $\{X_{\underline{i},k}\} \in \mathfrak{Or\ spaces}^o$ then $F_1 \circ F_2\{X_{\underline{i},k}\}$ is Ψ modulo the relation (2.8 a).

<u>Definition 2.9</u>: Let $\{X_{\underline{i},k}\} \in \mathfrak{Or\ spaces}$, $\mathfrak{Or\ spaces}^o$, or $\mathfrak{Eq\ spaces}$. The <u>free theory</u> on $\{X_{\underline{i},k}\}$ is its image under $F_1 \circ F_2 \circ F_3$, $F_1 \circ F_2$, or F_1 respectively.

If there is no chance of confusion we denote any of the three free functors into $\mathfrak{Theories}$ by F and its adjoint by U.

We want to describe the front and back adjunctions

(2.10) $\eta : \mathrm{Id} \longrightarrow UF$ $\quad\quad \varepsilon : FU \longrightarrow \mathrm{Id}$

The adjunction map η is induced by the inclusion $X_{\underline{i},k} \subset \Upsilon(\underline{i},k)$. The adjunction map $\varepsilon : FU\Theta \longrightarrow \Theta$ for any theory Θ is given by taking the composite operation in Θ.

Definition 2.11: The <u>composite operation</u> in Θ represented by the trivial tree $A : \underline{i} \longrightarrow k$ with twig label r associated with $U\Theta$ is the set operation $\sigma^* \in \Theta(\underline{i},k)$ given by $\sigma(1) = r$. By induction, the composite operation represented by an arbitrary tree $A : \underline{i} \longrightarrow k$ associated with $U\Theta$ is $x\circ(B_1 \ \ldots \ B_n)$, where x is the label of the root vertex and $B_1,\ldots,B_n$ the previously defined composite operations represented by the n subtrees above the inputs of x. In other words, a composite operation is obtained by composing the vertex labels (which are elements in Θ) and the set operations given by the twig labels.

Let $\{X_{\underline{i},k}\}$ and $\{R_{\underline{i},k}\}$ be K-coloured objects and $q_1,q_2 : \{R_{\underline{i},k}\} \longrightarrow UF\{X_{\underline{i},k}\}$ be morphisms in $\mathfrak{Gr}\ \mathfrak{spaces}$, $\mathfrak{Gr}\ \mathfrak{spaces}^o$, or $\mathfrak{Eq}\ \mathfrak{spaces}$. Passing to the adjoints we obtain morphisms of theories

$$F\{R_{\underline{i},k}\} \underset{p_2}{\overset{p_1}{\rightrightarrows}} F\{X_{\underline{i},k}\} \tag{2.12}$$

The difference cokernel $r : F\{X_{\underline{i},k}\} \longrightarrow \Xi$ of p_1 and p_2 exists in $\mathfrak{Theories}$, and we say that Ξ is <u>generated by</u> $\{X_{\underline{i},k}\}$ <u>with the relations</u> $(q_1\{R_{\underline{i},k}\},\ q_2\{R_{\underline{i},k}\})$. We call the diagram (2.12) a <u>presentation</u> of Ξ.

<u>Examples</u>: The relations are given by pairs of trees, the maps q_1,q_2 are the projections onto the first respectively the second factor. Instead of writing pairs (A_1,A_2) we use $A_1 = A_2$.

(1) <u>A presentation of the theory of abelian monoids</u>.

Since the theory is monochrome it suffices to specify the generating n-ary operations and the relations between them. We have a binary operation + and a 0-ary operation, the unit e. These satisfy the relations

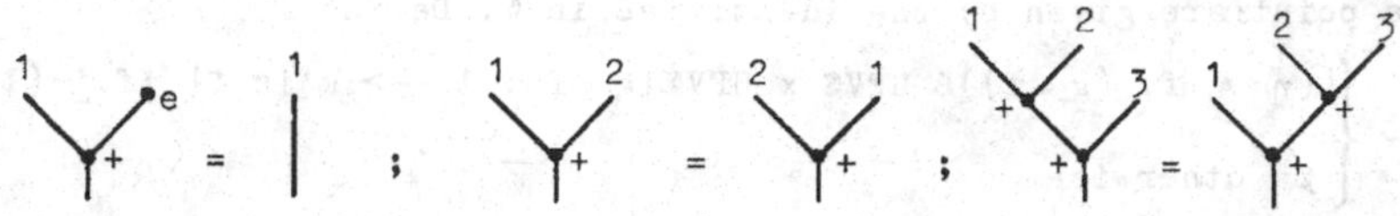

(2) A presentation of the theory of groups.

The theory of groups is generated by three operations: a binary operation x, a unitary operation i and a constant operation e, satisfying the relations

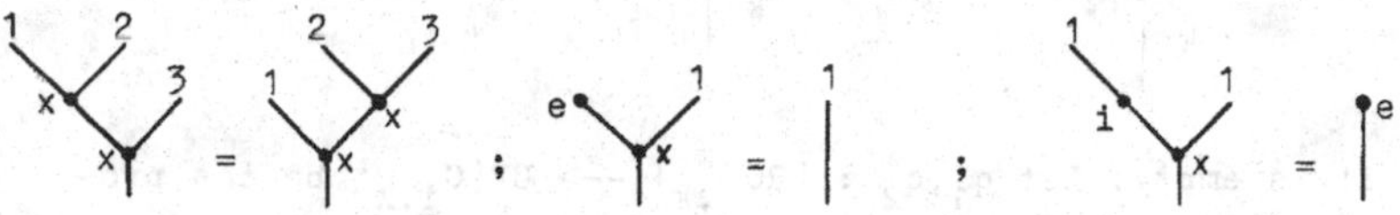

(ab)c = a(bc) ea = a $a^{-1}a = e$

Proposition 2.13: Each K-coloured theory can be presented as the difference cokernel of free K-coloured theories.

Proof: Given a theory Θ, let $R_{\underline{i},k} = \{(x,y) \in FU\Theta(\underline{i},k) \times FU\Theta(\underline{i},k) \mid \varepsilon(x) = \varepsilon(y)\}$ where $\varepsilon : FU\Theta \longrightarrow \Theta$ is the back adjunction. Then

$$F\{R_{\underline{i},k}\} \overset{p_1}{\underset{p_2}{\rightrightarrows}} FU\Theta \xrightarrow{\ \varepsilon\ } \Theta$$

is a difference cokernel diagram. ∎

Let $\mathfrak{Cat}$ denote the category of small topological categories and continuous functor. A category is called topological if its morphism sets are topologized and composition is continuous. A functor is called continuous if the induced function of the morphism spaces is continuous. There is an obvious functor

$$V : \mathfrak{Cat} \longrightarrow \mathfrak{Gr}\ \mathfrak{Spaces}^{O}$$

sending $\mathfrak{C} \in \mathfrak{Cat}$ to $\{C_{\underline{i},k}\}$ given by

$$C_{\underline{i},k} = \begin{cases} \mathfrak{C}(l,k) & \text{if } \underline{i} = (l) \\ \emptyset & \text{otherwise} \end{cases}$$

The base pointsare given by the identities in $\mathfrak{C}$. Define

$$RC_{\underline{i},k} = \begin{cases} \{(\eta g \circ \eta f, \eta(g \circ f)) \in UFV\mathfrak{C} \times UFV\mathfrak{C} \mid g \circ f : l \longrightarrow k \text{ in } \mathfrak{C}\} & \text{if } \underline{i} = (l) \\ \emptyset & \text{otherwise} \end{cases}$$

where η is the front adjunction (2.10). Hence if $\underline{i}$ is a basic object, l say, $RC_{\underline{i},k}$ consists of pairs of trees

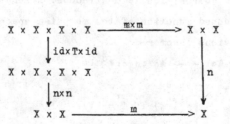

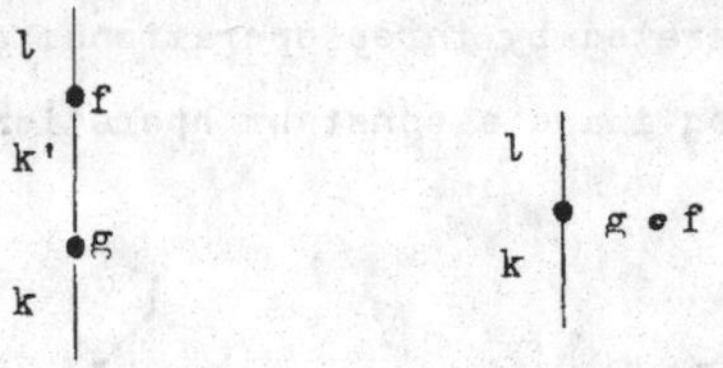

otherwise it is empty. Let $q_1, q_2 : \{RC_{\underline{i},k}\} \longrightarrow UF\{C_{\underline{i},k}\}$ be the projections. Let $T\mathfrak{C}$ be the theory generated by $\{C_{\underline{i},k}\}$ with the relations $\{RC_{\underline{i},k}\}$. Thus we obtain an embedding functor

$$T : \mathfrak{Cat} \longrightarrow \mathfrak{Theories}$$

which enables us to consider small topological categories as theories. Note that $T\mathfrak{C}$ is (ob $\mathfrak{C}$)-coloured.

3. INTERCHANGE

It has been known for some time that the concept of "interchange" of two structures on a space is fundamental in the study of H-spaces. Take, for example, a space X with two monoid structures m and n. Let $T : X \times X \longrightarrow X \times X$ be the interchange factors. If m and n interchange, i.e. if

$$\begin{array}{ccc} X \times X \times X \times X & \xrightarrow{\;m \times m\;} & X \times X \\ \downarrow{\scriptstyle id \times T \times id} & & \\ X \times X \times X \times X & & \downarrow{\scriptstyle n} \\ \downarrow{\scriptstyle n \times n} & & \\ X \times X & \xrightarrow{\;\;m\;\;} & X \end{array}$$

commutes, then m and n agree and define a commutative monoid structure on X.

We formalize this concept.

Let Θ be a K-coloured theory and $F_l : \Theta \longrightarrow \mathfrak{Top}$ an L-indexed family of Θ-spaces with underlying spaces $\{X_{l,k} | k \in K\}$. Then there is a canonical Θ-space $F : \Theta \longrightarrow \mathfrak{Top}$, called the <u>product Θ-space</u> of the F_l, whose collection of underlying spaces is $\{\prod_{l\in L} X_{l,k} | k \in K\}$. Explicitly, F is given on objects $\underline{i} : [n] \longrightarrow K$ by

$$F(\underline{i}) = \prod_{p=1}^{n} \left(\prod_{l\in L} X_{l,\underline{i}(p)}\right)$$

(for $\underline{i} : \emptyset \longrightarrow K$ put $F(\underline{i})$ = single point) and on morphisms $a : \underline{i} \longrightarrow \underline{j}$, $j : [m] \longrightarrow K$ by

$$\prod_{p=1}^{n}\left(\prod_{l\in L} X_{l,\underline{i}(p)}\right) \cong \prod_{l\in L} F_l(\underline{i}) \xrightarrow{\prod F_l(a)} \prod_{l\in L} F_l(\underline{j}) \cong \prod_{q=1}^{m}\left(\prod_{l\in L} X_{l,\underline{j}(q)}\right)$$

<u>Definition 2.14</u>: Let Θ_1 be a K-coloured and Θ_2 an L-coloured theory. Let $\{X_{k,l}\}$ be a K×L-indexed family of topological spaces such that each subfamily $\{X_{k,l} | k \in K\}$ is the collection of underlying spaces of a Θ_1-space F_l and each subfamily $\{X_{k,l} | l \in L\}$ the collection of underlying spaces of a Θ_2-space G_k. We say the Θ_1- and Θ_2-action on $\{X_{i,k}\}$ interchange if each $a : \underline{i} \longrightarrow \underline{j}$, $\underline{i} : [n] \longrightarrow K$, $\underline{j} : [m] \longrightarrow K$, in the Θ_1-action induces a homomorphism from the product Θ_2-space of $G_{\underline{i}(1)}, \dots, G_{\underline{i}(n)}$ to the product Θ_2-space of $G_{\underline{j}(1)}, \dots, G_{\underline{j}(m)}$. Or, equivalently, if each $b : \underline{u} \longrightarrow \underline{v}$, $\underline{u} : [p] \longrightarrow L$, $\underline{v} : [q] \longrightarrow L$, induces a homomorphism from the product Θ_1-space of $F_{\underline{u}(1)}, \dots, F_{\underline{u}(p)}$ to the product Θ_1-space of $F_{\underline{v}(1)}, \dots, F_{\underline{v}(q)}$.

Going back to the definition of product Θ-spaces, the actions interchange if following "shuffle" diagram commutes for all $a \in \Theta_1$ and all $b \in \Theta_2$

(2.15)

$$\begin{array}{ccc}
\prod\limits_{t=1}^{p}\Big(\prod\limits_{s=1}^{n} X_{\underline{i}(s),\underline{u}(t)}\Big) \cong \prod\limits_{s=1}^{n} G_{\underline{i}(s)}(\underline{u}) & \xrightarrow{\ \prod G_{\underline{i}(s)}(b)\ } & \prod\limits_{s=1}^{n} G_{\underline{i}(s)}(\underline{v}) \cong \prod\limits_{t=1}^{q}\Big(\prod\limits_{s=1}^{n} X_{\underline{i}(s),\underline{v}(t)}\Big) \\
\Vert & & \Vert \\
\prod\limits_{t=1}^{p} F_{\underline{u}(t)}(\underline{i}) & & \prod\limits_{t=1}^{q} F_{\underline{v}(t)}(\underline{i}) \\
\Big\downarrow {\scriptstyle \prod F_{\underline{u}(t)}(a)} & & {\scriptstyle \prod F_{\underline{v}(t)}(a)} \Big\downarrow \\
\prod\limits_{t=1}^{p} F_{\underline{u}(t)}(\underline{j}) & & \prod\limits_{t=1}^{q} F_{\underline{v}(t)}(\underline{j}) \\
\Vert & & \Vert \\
\prod\limits_{t=1}^{p}\Big(\prod\limits_{s=1}^{m} X_{\underline{j}(s),\underline{u}(t)}\Big) \cong \prod\limits_{s=1}^{m} G_{\underline{j}(s)}(\underline{u}) & \xrightarrow{\ \prod G_{\underline{j}(s)}(b)\ } & \prod\limits_{s=1}^{m} G_{\underline{j}(s)}(\underline{v}) \cong \prod\limits_{t=1}^{q}\Big(\prod\limits_{s=1}^{m} X_{\underline{j}(s),\underline{v}(t)}\Big)
\end{array}$$

The horizontal lines are the operations of $b \in \Theta_2$ on the product Θ_2-spaces of the $G_{\underline{i}(s)}$ respectively the $G_{\underline{j}(s)}$, the vertical lines give the homomorphism induced by a.

In the case where Θ_1 and Θ_2 are monochrome theories with underlying space X, the shuffle diagram becomes more transparent. Let $a : X^n \longrightarrow X^m$ be an operation from Θ_1 and $b : X^p \longrightarrow X^q$ one from Θ_2. Then (2.15) reads

$$\begin{array}{ccc}
(X^n)^p \cong (X^p)^n & \xrightarrow{\ \ b^n\ \ } & (X^q)^n \cong (X^n)^q \\
\Big\downarrow {\scriptstyle a^p} & & \Big\downarrow {\scriptstyle a^q} \\
(X^m)^p \cong (X^p)^m & \xrightarrow{\ \ b^m\ \ } & (X^q)^m \cong (X^m)^q
\end{array}$$

If we have two theories Θ_1 and Θ_2, whose actions on $\{X_{k,l}\}$ interchange as defined in (2.14), the actions of Θ_1 and Θ_2 together with the shuffle relations (2.15) induce an action of a K×L-coloured theory $\Theta_1 \otimes \Theta_2$, which we are going to describe, on $\{X_{k,l}\}$, and each action of $\Theta_1 \otimes \Theta_2$ comes from interchanging actions of Θ_1 and Θ_2. Let $\{Y_{\underline{i},(k,l)}\} \in \mathfrak{Or}$ *spaces* be the following K×L-coloured object:

Let $\underline{i} = \{(k_1,l_1),\dots,(k_n,l_n)\}$, $\underline{i}_K = \{k_1,\dots,k_n\}$, $\underline{i}_L = \{l_1,\dots,l_n\}$.

(a) If $k_1 = k_2 = \dots = k_n = k$ and $l_1 = l_2 = \dots = l_n = l$, then

$Y_{\underline{i},(k,l)} = \Theta_1(\underline{i}_K,k) \cup \Theta_2(\underline{i}_L,l)$

(b) If $l_1 = \ldots = l_n = l$ and (a) does not apply, then $Y_{\underline{i},(k,l)} = \Theta_1(\underline{i}_K,k)$

(c) If $k_1 = k_2 = \ldots = k_n = k$ and (a) does not apply, then

$Y_{\underline{i},(k,l)} = \Theta_2(\underline{i}_L,l)$

(d) If neither of (a),(b),(c) apply, then $Y_{\underline{i},(k,l)} = \emptyset$.

The elements of $\{Y_{\underline{i},(k,l)}\}$ are the generators of $\Theta_1 \otimes \Theta_2$. They are uniquely determined by a pair $(a,r) \in (\bigcup_{\underline{i},k} \Theta_1(\underline{i},k)\times L) \cup (\bigcup_{\underline{j},l} \Theta_2(\underline{j},l)\times K)$.
The source of $(a,r) \in \Theta_1(\underline{i},k)\times L$ is
$\{(k_1,r),\ldots,(k_n,r)\}$ if $\underline{i}=\{k_1,\ldots,k_n\}$, and its target (k,r); similarly for $(a,r) \in \Theta_2(\underline{j},l)\times K$.

(2.15) We have the following relations(between the trees of $F\{Y_{\underline{i},(k,l)}\}$:

(i) The same as (2.8a): We may remove any vertex labelled by an identity in Θ_1 or Θ_2.

(ii) The same as (2.8b): If $a \in \Theta_1(\underline{i},k)$ respectively $\Theta_2(\underline{p},l)$ and $\sigma \in \mathfrak{S}_K(\underline{i},\underline{j})$ respectively $\mathfrak{S}_L(\underline{p},\underline{q})$, we may replace the vertex label $a \circ \sigma^*$ by a changing the part of the tree above this vertex: If $C_1,\ldots,C_m$ are the subtrees above the inputs of $a \circ \sigma^*$ and σ is a function from $[n]$ to $[m]$, we take $C_{\sigma 1},\ldots,C_{\sigma n}$ as subtrees over a.

(iii) The relations of Θ_1 and Θ_2: Any edge joining two vertices whose labels are both in Θ_1 or both in Θ_2 can be removed. We unite the two vertices to form a new vertex, whose label is the tree composite in Θ_1 or Θ_2 (see 2.11) of the tree consisting of these two vertices and their incoming and outgoing edges (see example below).

(iv) The shuffle relations: Given $a \in \Theta_1(\underline{i},k)$ and $b \in \Theta_2(\underline{j},l)$, $\underline{i} = \{k_1,\ldots,k_n\}$, $\underline{j} = \{l_1,\ldots,l_m\}$. Let $\varphi_{n,m} : [n]\times[m] \longrightarrow [n\cdot m]$ be the bijection $(p,q) \longmapsto (q-1)n+p$. Let A be the tree with root operation (a,l), the operation (b,k_r) on the r-th input to (a,l), and label $\varphi_{n,m}(p,q)$ for the twig with colour (k_p,l_q). Let B be the tree with root operation (b,k), the operation (a,l_r) on the r-th input to (b,k), and label $\varphi_{n,m}(p,q)$ for the twig with colour (k_p,l_q). Then $A = B$. (See example below).

Illustrations:

(iii) Let $a : \{k_1,k_2,k_3\} \longrightarrow k$ and $b : \{k_4,k_5\} \longrightarrow k_2$ be morphisms in Θ_1. Then

$$c = a \circ (id_{k_1} \oplus b \oplus id_{k_3})$$

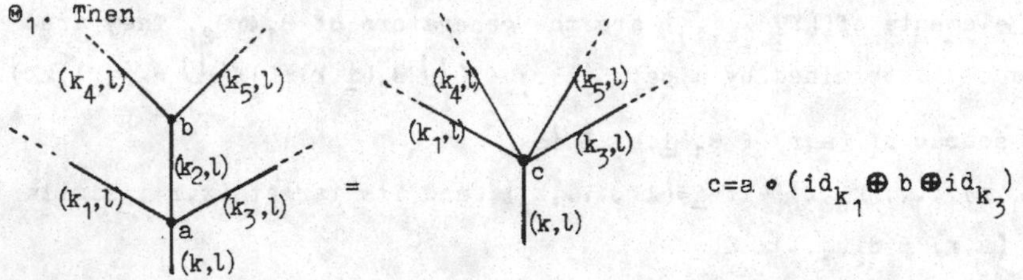

(iv) Given $a : \{k_1,k_2\} \longrightarrow k$ in Θ_1 and $b : \{l_1,l_2\} \longrightarrow l$ in Θ_2. Then

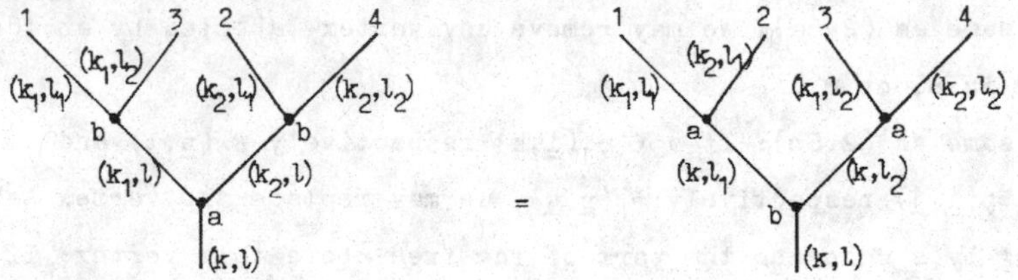

Let $a \otimes b$ denote the element represented by these trees. Then the generators (a,l) and (b,k) can be identified with $a \otimes id_l$ and $id_k \otimes b$.

Convention: If we do not specify the twig labels of a tree with n twigs, we assume they are $1,\ldots,n$ in clockwise order.

Examples:

(1) $\mathfrak{S}_K^{op} \otimes \mathfrak{S}_L^{op} \cong \mathfrak{S}_{K \times L}^{op}$ as theories

(2) If Θ_m is the theory of monoids and Θ_{cm} the one of abelian monoids, then

$$\Theta_{cm} \cong \Theta_m \otimes \ldots \otimes \Theta_m \qquad \text{n times, } n \geq 2$$

A proof in theory language of this classical result can be found in Pareigis [41;p.113,114].

From the relations and the tree calculus we obtain the following identities, (for clarification use the shuffle diagram and note that the horizontal rows are the operations $id \otimes b$, the vertical ones the operations $a \otimes id$)

(2.16)
$$a \otimes b = (a \otimes id)\circ(id \otimes b) = (id \otimes b)\circ(a \otimes id)$$
$$(a \otimes id)\circ(\bar{a} \otimes id) = (a \circ \bar{a}) \otimes id$$
$$(id \otimes b)\circ(id \otimes \bar{b}) = id \otimes (b \circ \bar{b})$$
$$(a_1,\dots,a_m)\otimes(b_1,\dots,b_q) = (a_1 \otimes b_1,\dots,a_m \otimes b_1,\dots,a_1 \otimes b_q,\dots,a_m \otimes b_q)$$

where $a,\bar{a},a_i \in \Theta_1$, $b,\bar{b},b_j \in \Theta_2$ and $(a_1,\dots,a_m)$ is the morphism into $\{k_1,\dots,k_m\}$ induced by the $a_i \in \Theta_1(\underline{p},k_i)$. Hence the correspondences $a \longmapsto a \otimes id_{\underline{u}}$ and $b \longmapsto id_{\underline{i}} \otimes b$ induce for each $\underline{i} \in \mathrm{ob}\,\Theta_1$ and each $\underline{u} \in \mathrm{ob}\,\Theta_2$ functors

(2.17)
$$\Theta_1 \xrightarrow{R_{\underline{u}}} \Theta_1 \otimes \Theta_2 \xleftarrow{Q_{\underline{i}}} \Theta_2$$

__Proposition 2.18__: The tensor product of theories is commutative and associative up to isomorphism.

__Proof__: The correspondence $a \otimes id \longrightarrow id \otimes a$, $id \otimes b \longrightarrow b \otimes id$ induces an isomorphism $T : \Theta_1 \otimes \Theta_2 \cong \Theta_2 \otimes \Theta_1$. From (2.16) it is clear that the correspondence $a \otimes (b \otimes c) \longrightarrow (a \otimes b) \otimes c$ induces an isomorphism, too. ∎

Given a $(K \times L)$-indexed family $\{X_{k,l}\}$ of spaces such that each $\{X_{k,l} \mid k \in K\}$ is the collection of underlying spaces of a Θ_1-space F_l and each $\{X_{k,l} \mid l \in L\}$ one of a Θ_2-space G_k. We said that the actions of Θ_1 and Θ_2 interchange, if each operation b of the Θ_2-action induces a homomorphism of appropriate product Θ_1-spaces obtained from the F_l. A different way of expressing this is saying that the $\{F_l \mid l \in L\}$ form a collection of underlying spaces of a Θ_2-space in the category of Θ_1-spaces. Let us explain this in more detail:

Denote the category of Θ_1-spaces and homomorphisms by $\mathfrak{Top}^{\Theta_1}$. Since a homomorphism $f : F_1 \longrightarrow F_2$ between two Θ_1-spaces is uniquely determined by the maps $f(k) : F_1(k) \longrightarrow F_2(k)$ of the underlying spaces, we can regard $\mathfrak{Top}^{\Theta_1}(F_1,F_2)$ as subspace of $\prod_{k\in K} \mathfrak{Top}(F_1(k),F_2(k))$. This makes $\mathfrak{Top}^{\Theta_1}$ into a topological category, and we can define Θ_2-spaces in $\mathfrak{Top}^{\Theta_1}$ in the same manner as in (2.3) by replacing $\mathfrak{Top}$ by $\mathfrak{Top}^{\Theta_1}$.

<u>Proposition 2.19</u>: $\mathfrak{Top}^{\Theta_1\otimes\Theta_2} \cong (\mathfrak{Top}^{\Theta_1})^{\Theta_2}$ as topological categories, i.e. the functors induce homeomorphisms on the morphism spaces.

<u>Proof</u>: The bijection $\mathfrak{Funct}(\Theta_1\times\Theta_2,\mathfrak{Top}) \cong \mathfrak{Funct}(\Theta_2,\mathfrak{Funct}(\Theta_2,\mathfrak{Top}))$ induces a bijection

$$\mathfrak{Funct}_{p,p}(\Theta_1\times\Theta_2,\mathfrak{Top}) \cong \mathfrak{Funct}_p(\Theta_2,\mathfrak{Funct}_p(\Theta_1,\mathfrak{Top})) = (\mathfrak{Top}^{\Theta_1})^{\Theta_2} \quad (*)$$

from the category of bifunctors $\Theta_1\times\Theta_2 \longrightarrow \mathfrak{Top}$ which preserve products in each argument to the category of product preserving functors from Θ_2 to the category of product preserving functors $\Theta_1 \longrightarrow \mathfrak{Top}$. Since a natural transformation between two objects F,G of $\mathfrak{Funct}_{p,p}(\Theta_1\times\Theta_2,\mathfrak{Top})$ is uniquely determined by its values on the basic objects $(k,l)\in\Theta_1\times\Theta_2$, we can regard $\mathfrak{Funct}_{p,p}(\Theta_1\times\Theta_2,\mathfrak{Top})(F,G)$ as subspace of $\prod_{(k,l)\in K\times L} \mathfrak{Top}(F(k,l),G(k,l))$. With this topology (*) becomes an isomorphism of topological categories. We further establish an isomorphism of topological categories

$$\mathfrak{Top}^{\Theta_1\otimes\Theta_2} \cong \mathfrak{Funct}_{p,p}(\Theta_1\times\Theta_2,\mathfrak{Top})$$

The functor $P : \Theta_1 \times \Theta_2 \longrightarrow \Theta_1\otimes\Theta_2$, sending $(\underline{i},\underline{j})$, $\underline{i} : [n] \longrightarrow K$, $\underline{j} : [m] \longrightarrow L$, to the object $(\underline{i}\times\underline{j})\circ\varphi^{-1}_{n,m} : [n\cdot m] \longrightarrow K\times L$ (see 2.15 (iv)) and the morphism (a,b) to $a\otimes b$, induces a continuous functor

$$P^* : \mathfrak{Top}^{\Theta_1\otimes\Theta_2} \longrightarrow \mathfrak{Funct}_{p,p}(\Theta_1\times\Theta_2,\mathfrak{Top})$$

Since $F \in \mathfrak{Top}^{\Theta_1\otimes\Theta_2}$ and $G \in \mathfrak{Funct}_{p,p}(\Theta_1\times\Theta_2,\mathfrak{Top})$ are uniquely determined on objects by their values on the basic objects (k,l) and on morphisms by their values on $a\otimes\mathrm{id}$, $\mathrm{id}\otimes b$ respectively (a, id), (id,b), the functor P^* is a bijection on the objects. Furthermore, we consider $\mathfrak{Top}^{\Theta_1\otimes\Theta_2}(F_1,F_2)$ and $\mathfrak{Funct}_{p,p}(\Theta_1\times\Theta_2,\mathfrak{Top})(F_1\circ P,F_2\circ P)$ as the same subspaces of $\prod_{(k,l)\in K\times L}(F_1(k,l),F_2(k,l))$ so that P* induces homeomorphisms of the morphism spaces. ∎

In general, the structure of $\Theta_1\otimes\Theta_2$ is far from clear because the shuffle relations are difficult to handle. Given a morphism $a : \underline{i} \to \underline{j}$, $\underline{i} : [n] \to K$, $\underline{j} : [m] \to K$ in Θ_1 and $b : \underline{u} \to \underline{v}$, $\underline{u} : [p] \to L$, $\underline{v} : [q] \to L$ in Θ_2. Let $\pi_{n,p}$ be the permutation

$$\pi_{n,p} = \varphi_{p,n}\cdot T\cdot\varphi_{n,p}^{-1} : [n\cdot p] \to [n]\times[p] \to [p]\times[n] \to [n\cdot p]$$

where T interchanges the factors. Then the shuffle relation reads

$$\begin{aligned}(2.20)\quad a\bullet b &= \pi_{q,m}^*\circ[(id_{\underline{j}1}\otimes b)\oplus\ldots\oplus(id_{\underline{j}m}\otimes b)]\circ\pi_{m,p}^*\circ[(a\otimes id_{\underline{u}1})\oplus\ldots\oplus(a\otimes id_{\underline{u}p})]\\ &= [(a\otimes id_{\underline{v}1})\oplus\ldots\oplus(a\otimes id_{\underline{v}q})]\circ\pi_{q,n}^*\circ[(id_{\underline{i}1}\otimes b)\oplus\ldots\oplus(id_{\underline{i}n}\otimes b)]\circ\pi_{n,p}^*\end{aligned}$$

So each morphism of $\Theta_1\otimes\Theta_2$ can be written as

$$(2.21)\qquad a_1\circ b_1\circ\ldots\circ a_k\circ b_k\circ\xi^*$$

where a_i is of the form $(c_1\otimes id_{l_1})\oplus\ldots\oplus(c_r\otimes id_{l_r})$ and b_i of the form $(id_{k_1}\otimes d_1)\oplus\ldots\oplus(id_{k_r}\otimes d_r)$, the c and d are morphisms into basic objects in Θ_1 respectively Θ_2 and different from set operations.

The permutations $\pi_{n,p}$ cause the difficulties in the attempt to determine the structure of $\Theta_1\otimes\Theta_2$. In case Θ_2 has 1-ary operations only, i.e. $\Theta_2 = \mathfrak{C}$ is a small topological category considered as theory, we can handle the shuffle relations. If b in (2.20) is a 1-ary operation the four permutations are identities. So any morphism of $\Theta_1\otimes\mathfrak{C}$ into a basic object can be written as $(a\otimes id_l)\circ[(id_{l_1}\otimes b_1)\oplus\ldots\oplus(id_{k_r}\otimes b_r)]\circ\xi^*$ and hence by (2.15 (ii)) as

$$(a\otimes id_l)\circ[(id_{l_1}\otimes b_1)\oplus\ldots\oplus(id_{k_r}\otimes b_r)]$$

We obtain

Lemma 2.22: Let Θ be a K-coloured theory and $\mathfrak{C}$ a small topological category. Let $\underline{i}=\{(k_1,l_1),\ldots,(k_n,l_n)\}\in ob(\Theta\otimes\mathfrak{C})$ and $\underline{j}=\{k_1,\ldots,k_n\}\in ob\,\Theta$. Then there is a natural homeomorphism

$$(\Theta\otimes\mathfrak{C})(\underline{i},(k,l)) \cong \Theta(\underline{j},k)\times\mathfrak{C}(l_1,l)\times\ldots\times\mathfrak{C}(l_n,l) \qquad ∎$$

The most interesting example for us is the structure of $\Theta \otimes \Omega_n$ where Ω_n is the "linear" category whose objects are $0,1,\ldots,n$, with one morphism $i \longrightarrow j$ if $i \le j$ and none otherwise. A $(\Theta \otimes \Omega_n)$-space is determined by a sequence of Θ-spaces and homomorphisms (see 2.19)

$$F_0 \longrightarrow F_1 \longrightarrow F_2 \longrightarrow \ldots \longrightarrow F_n$$

We have one useful general result which helps us to get rid of constant operations. Let 0 denote the unique object $\emptyset \longrightarrow K$ in any K-coloured theory.

Lemma 2.23: Let Θ_1 be a K-coloured and Θ_2 an L-coloured theory such that $\Theta_1(0,k) \neq \emptyset \neq \Theta_2(0,l)$ for all $k \in K, l \in L$. Then $(\Theta_1 \otimes \Theta_2)(0,(k,l))$ contains exactly one element. Moreover if Θ_1' and Θ_2' denote the subtheories of Θ_1 and Θ_2 without the constant operations, then $(\Theta_1 \otimes \Theta_2)(\underline{i},(k,l))$ is a quotient of $(\Theta_1' \otimes \Theta_2')(\underline{i},(k,l))$ for $\underline{i} \neq 0$.

Proof: Let $c \in \Theta_1(0,k)$ and $d \in \Theta_2(0,l)$. From the shuffle relation we obtain that $c \otimes d = c \otimes id_l = id_k \otimes d$. Given any tree in $\Theta_1 \otimes \Theta_2$, we can prune away all stumps one by one: If an edge joins a vertex labelled c to a vertex labelled $b \in \Theta_2$, we replace c by d and compose in Θ_2 according to 2.15 (iii) to remove that edge. We end up with either a tree without stumps or a tree consisting of a stump only. Since the relations between the trees of $\Theta_1' \otimes \Theta_2'$ also hold between the trees of $\Theta_1 \otimes \Theta_2$ the result follows. ∎

4. FREE Θ-SPACES AND TRIPLES

Instead of theories many people prefer to work with triples (e.g. see [2]). Since K-coloured triples up to date have not been investigated (to the authors' knowledge), we include this chapter for the sake of completeness and to put our further constructions into a wider frame work.

Let $\mathfrak{Top}_K$ denote the category $\mathfrak{Top}$ over K, i.e. its objects are continuous maps $X \longrightarrow K$, $X \in \mathrm{ob}\,\mathfrak{Top}$, and its morphisms from $g : X \longrightarrow K$ to $h : Y \longrightarrow K$ are continuous maps $f : X \longrightarrow Y$ such that

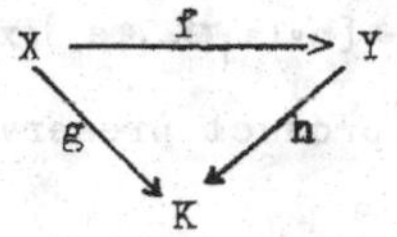

commutes. We can identify $\mathfrak{Top}_K$ with the category of K-graded spaces $X = \{X_k | k \in K\}$ and grading preserving maps by putting $X_k = g^{-1}(k)$ for an object $g : X \longrightarrow K$ of $\mathfrak{Top}_K$. Topologize $\mathfrak{Top}_K$ by $\mathfrak{Top}_K(X,Y) = \prod_{k \in K} \mathfrak{Top}(X_k, Y_k)$.

A function $r : K \longrightarrow L$ induces a functor

$$r_* : \mathfrak{Top}_K \longrightarrow \mathfrak{Top}_L$$

given on objects $g : X \longrightarrow K$ by $r_*(g) = r \circ g$ and on morphisms $f : g \longrightarrow h$ by $r_*(f) = f$. It has a right adjoint

$$r^* : \mathfrak{Top}_L \longrightarrow \mathfrak{Top}_K$$

sending $\{X_l | l \in L\}$ to $\{X_{r(k)} | k \in K\}$.

For any K-coloured theory Θ we have a continuous forgetful functor

$$U : \mathfrak{Top}^{\Theta} \longrightarrow \mathfrak{Top}_K$$

mapping the Θ-space $G : \Theta \longrightarrow \mathfrak{Top}$ to the graded space $\{G(k) | k \in K\}$.

<u>Defineorem 2.24</u>: The functor U has a continuous left adjoint

$$F : \mathfrak{Top}_K \longrightarrow \mathfrak{Top}^{\Theta}$$

The image FX of X is called the <u>free</u> <u>Θ-space</u> on X. Moreover, the natural bijection

$$\mathfrak{Top}^{\Theta}(FX,G) \cong \mathfrak{Top}_K(X,UG)$$

is a homeomorphism.

<u>Proof</u>: Given $X = \{X_k | k \in K\} \in \mathfrak{Top}_K$. For $\underline{i} : [n] \longrightarrow K$ denote $X_{\underline{i}1} \times \ldots \times X_{\underline{i}n}$ by $X_{\underline{i}}$ and $f_{\underline{i}1} \times \ldots \times f_{\underline{i}n} : X_{\underline{i}1} \times \ldots \times X_{\underline{i}n} \longrightarrow Y_{\underline{i}1} \times \ldots \times Y_{\underline{i}n}$ by $f_{\underline{i}}$. Define

$$FX(k) = \bigcup_{\underline{i} \in \Theta} \Theta(\underline{i},k) \times X_{\underline{i}} / \sim$$

with the identification

$$(a \circ \sigma^*;\, x_1,\ldots,x_n) = (a;\, \sigma^*(x_1,\ldots,x_n)),$$

σ^* a set operation (see 2.1). For $b \in \Theta(\underline{j}, l)$ define

$FX(b) : FX(\underline{j}1) \times \ldots \times FX(\underline{j}m) \longrightarrow FX(l)$ by

$$[(a_1; x_1^1,\ldots,x_{n_1}^1),\ldots,(a_m; x_1^m,\ldots,x_{n_m}^m)] \longmapsto [b\circ(a_1 \oplus\ldots\oplus a_m); x_1^1,\ldots,x_{n_1}^1,\ldots,x_1^m,\ldots,x_{n_m}^m]$$

These data determine a continuous product preserving functor $FX : \Theta \longrightarrow \mathfrak{Top}$.

Given a morphism $f : X \longrightarrow Y$ in $\mathfrak{Top}_K$ the maps $id \times f_{\underline{i}} : \Theta(\underline{i},k)\times X_{\underline{i}} \to \Theta(\underline{i},k)\times Y_{\underline{i}}$ induce a map $Ff(k) : FX(k) \longrightarrow FY(k)$, which is continuous in f. The collection $\{Ff(k) \mid k \in K\}$ determines a natural transformation $Ff : FX \to FY$ continuous in f.

Define a continuous natural map

$$\rho : \mathfrak{Top}^{\Theta}(FX,G) \longrightarrow \mathfrak{Top}_K(X,UG)$$

by $\rho(g)_k(x) = g(k)(id_k;x)$, $x \in X_k$, $g : FX \longrightarrow G$. Given $f : X \longrightarrow UG$ in $\mathfrak{Top}_K$, let $\varkappa(f) : FX \longrightarrow G$ be the homomorphism induced on the basic object k by the maps

$$\begin{array}{rcl} \Theta(\underline{i},k) \times X_{\underline{i}} & \longrightarrow & G(k) \\ (a; x_1,\ldots,x_n) & \longmapsto & G(a)(f_{\underline{i}}(x_1,\ldots,x_n)) \end{array}$$

Then $\varkappa : \mathfrak{Top}_K(X,UG) \longrightarrow \mathfrak{Top}^{\Theta}(FX,G)$ is a continuous inverse of ρ. ∎

The front and back adjunction $\eta : Id \longrightarrow UF$ and $\varepsilon : FU \longrightarrow Id$ of any adjoint pair F,U satisfy [41; p.45]

$$U\varepsilon \circ \eta U = id_U$$

$$\varepsilon F \circ F\eta = id_F$$

Setting $T = UF$ and $\mu = U\varepsilon F$ we therefore have commutative diagrams

(2.25)

$$\text{(a)}\quad \begin{array}{ccccc} T & \xrightarrow{T\eta} & T\circ T & \xleftarrow{\eta T} & T \\ & \diagdown\!\!\diagdown & \downarrow{\scriptstyle\mu} & /\!\!/ & \\ & & T & & \end{array} \qquad \text{(b)}\quad \begin{array}{ccc} T\circ T\circ T & \xrightarrow{T\mu} & T\circ T \\ \downarrow{\scriptstyle\mu T} & & \downarrow{\scriptstyle\mu} \\ T\circ T & \xrightarrow{\mu} & T \end{array}$$

<u>Definition 2.26</u>: A continuous endofunctor $T : \mathfrak{C} \longrightarrow \mathfrak{C}$ of a topological

category $\mathfrak{C}$ together with natural transformations $\eta : Id_{\mathfrak{C}} \longrightarrow T$ and $\mu : T \circ T \longrightarrow T$ satisfying (2.25 a,b) is called a (continuous) <u>triple</u> on $\mathfrak{C}$.

So any K-coloured theory Θ determines a triple (T,η,μ) on $\mathfrak{Top}_K$, which associates to each $X \in$ ob $\mathfrak{Top}_K$ the collection of underlying spaces of the free Θ-space on X. The natural transformations $\eta X : X \to TX$ and $\mu X : TTX \longrightarrow TX$ are induced by $x \longmapsto (id_k; x)$, $x \in X_k$, respectively $[b;(a_1; y_1),\ldots,(a_m,y_m)] \longmapsto [b \circ (a_1 \oplus \ldots \oplus a_m);y_1,\ldots,y_m]$ where y_r stands for some n_r-tuple $x_1^r,\ldots,x_{n_r}^r$ of elements of X.

<u>Definition 2.27</u>: A <u>triple morphism</u> $(T,\eta,\mu) \longrightarrow (T',\eta',\mu')$ from a triple on $\mathfrak{Top}_K$ to a triple on $\mathfrak{Top}_L$ is a pair (τ,f) consisting of a function $f : K \longrightarrow L$ and a natural transformation $\tau : f_* \circ T \longrightarrow T' \circ f_*$ such that

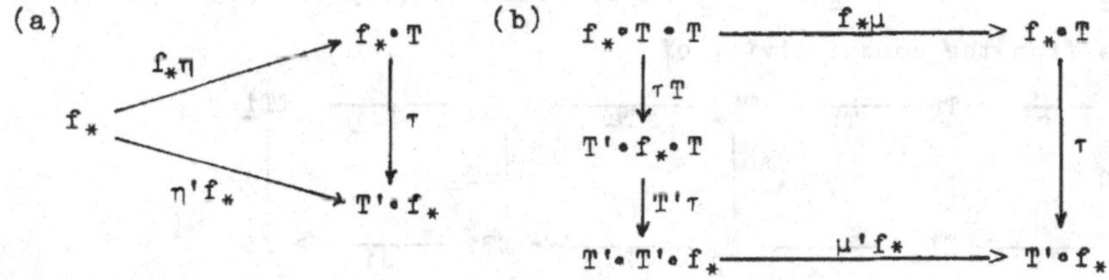

commute. Composition of triple morphisms is defined by $(\rho,g) \circ (\tau,f) = (\rho f_* \circ g_*\tau, g \circ f)$. Let $\mathfrak{Triples}$ denote the category of such triples and triple morphisms.

A theory functor $(P,f) : \Theta \longrightarrow \Theta'$ determines a triple morphism $(T,\eta,\mu) \longrightarrow (T',\eta',\mu')$ of the associated triples as follows: Let $(a;y) \in \Theta(\underline{i},k) \times X_{\underline{i}}$ be a representative of $(f_*TX)_l$ (this implies $f(k) = l$). Then $\tau X : f_*TX \longrightarrow T'f_*X$ is induced by

$$(a;y) \longmapsto (Pa;y) \in \Theta'(f \circ \underline{i}, f(k)) \times (f_*X)_{f \circ \underline{i}}$$

(recall that $(f_*X)_{f\circ \underline{i}} = [\amalg X_k | f(k) = f(\underline{i}1)] \times \ldots \times [\amalg X_k | f(k) = f(\underline{i}n)]$). Since $\tau \circ f_*\eta$ maps a representative $x \in X_k \subset (f_*X)_{f(k)}$ to $(P id_l;x) = (id_{f(k)};x) = \eta' f_*(x)$, diagram (2.27 a) commutes. The commutativity of

(2.27 b) follows from

$$\begin{array}{ccc}
[b;(a_1;y_1),\ldots,(a_m;y_m)] & \xrightarrow[f_*\mu]{} & [b\circ(a_1\oplus\ldots\oplus a_m);y_1,\ldots,y_m] \\
\downarrow{\scriptstyle \tau T} & & \downarrow{\scriptstyle \tau} \\
[Pb;(a_1;y_1),\ldots,(a_m;y_m)] & & [P(b\circ(a_1\oplus\ldots\oplus a_m));y_1,\ldots,y_m] \\
\downarrow{\scriptstyle T'\tau} & & \| \\
[Pb;(Pa_1;y_1),\ldots,(Pa_m;y_m)] & \xrightarrow[\mu f_*]{} & [Pb\circ(Pa_1\oplus\ldots\oplus Pa_m);y_1,\ldots,y_m]
\end{array}$$

This defines us a functor $\mathfrak{Theories} \longrightarrow \mathfrak{Triples}$.

Conversely, we can construct a functor $\mathfrak{Triples} \longrightarrow \mathfrak{Theories}$. Given any triple (T,η,μ) on $\mathfrak{Top}_K$, we obtain a K-coloured theory Θ by putting $\Theta(\underline{i},\underline{j}) = \mathfrak{Top}_K(\underline{j},T\underline{i})$. The composite of $a \in \Theta(\underline{i},\underline{j})$, $a : \underline{j} \longrightarrow T\underline{i}$ with $b \in \Theta(\underline{j},\underline{p})$, $b : \underline{p} \longrightarrow T\underline{j}$ is defined as

$$b\circ a : \underline{p} \xrightarrow[b]{} T\underline{j} \xrightarrow[Ta]{} TT\underline{i} \xrightarrow[\mu\underline{i}]{} T\underline{i}$$

Given $\underline{i} \xrightarrow{a} \underline{j} \xrightarrow{b} \underline{p} \xrightarrow{c} \underline{q}$ in Θ, associativity of the composition follows from the commutativity of

$$\begin{array}{ccccccccc}
\underline{q} & \xrightarrow[c]{} & T\underline{p} & \xrightarrow[Tb]{} & TT\underline{j} & \xrightarrow[TTa]{} & TTT\underline{i} & \xrightarrow[T\mu\underline{i}]{} & TT\underline{i} \\
\downarrow{\scriptstyle c} & & & & \downarrow{\scriptstyle \mu\underline{j}} & & \downarrow{\scriptstyle \mu T\underline{i}} & & \downarrow{\scriptstyle \mu\underline{i}} \\
T\underline{p} & \xrightarrow[Tb]{} & TT\underline{j} & \xrightarrow[\mu\underline{j}]{} & T\underline{j} & \xrightarrow[Ta]{} & TT\underline{i} & \xrightarrow[\mu\underline{i}]{} & T\underline{i}
\end{array}$$

The commutativity of

$$\begin{array}{ccccc}
\underline{j} & \xrightarrow[a]{} & T\underline{i} & & \\
\downarrow{\scriptstyle \eta\underline{j}} & & \downarrow{\scriptstyle \eta T\underline{i}} & \searrow\!\!\!\!\searrow & \\
T\underline{j} & \xrightarrow[Ta]{} & TT\underline{i} & \xrightarrow[\mu\underline{i}]{} & T\underline{i}
\end{array}
\qquad
\begin{array}{ccc}
\underline{p} & \xrightarrow[b]{} & T\underline{j} \\
\downarrow{\scriptstyle b} & \nearrow\!\!\!\!\nearrow & \uparrow{\scriptstyle \mu\underline{j}} \\
T\underline{j} & \xrightarrow[T\eta\underline{j}]{} & TT\underline{j}
\end{array}$$

shows that $\eta\underline{j} : \underline{j} \longrightarrow T\underline{j}$ acts as the identity of $\underline{j}$ in Θ. Finally, the set operation corresponding to $\sigma \in \mathfrak{S}_K(\underline{i},\underline{j})$ is given by the composite

$$\underline{i} \xrightarrow[\sigma]{} \underline{j} \xrightarrow[\eta\underline{j}]{} T\underline{j}$$

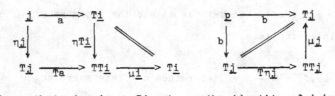

A triple morphism $(\tau,f) : (T,\eta,\mu) \longrightarrow (T',\eta',\mu')$ induces a theory functor $(P,f) : \Theta \longrightarrow \Theta'$ of the associated theories. It maps $a \in \Theta(\underline{i},\underline{j})$, $a : \underline{j} \longrightarrow T\underline{i}$ in $\mathfrak{Top}_K$, to the morphism of $\Theta'(f\circ\underline{i},f\circ\underline{j})$ given by the composite

$$f_*\underline{j} \xrightarrow{f_*a} f_*T\underline{i} \xrightarrow{\tau\underline{i}} T'f_*\underline{i}$$

The upper sequence in the following commutative diagram represents $P(b \circ a)$ while the lower one represents $Pb \circ Pa$. Hence P preserves composition

$$\begin{array}{ccccccccc}
f_*\underline{p} & \xrightarrow{f_*b} & f_*T\underline{j} & \xrightarrow{f_*Ta} & f_*TT\underline{i} & \xrightarrow{f_*\mu\underline{i}} & f_*T\underline{i} & \xrightarrow{\tau\underline{i}} & T'f_*\underline{i} \\
 & & \downarrow{\scriptstyle \tau\underline{j}} & & \downarrow{\scriptstyle \tau T\underline{i}} & & & & \uparrow{\scriptstyle \mu f_*\underline{i}} \\
 & & T'f_*\underline{j} & \xrightarrow{T'f_*a} & T'f_*T\underline{i} & \xrightarrow{T'\tau\underline{i}} & T'T'f_*\underline{i} & \longrightarrow & T'f_*\underline{i}
\end{array}$$

Finally the commutativity of

$$\begin{array}{ccccc}
 & & & & f_*T\underline{j} \\
f_*\underline{i} & \xrightarrow{f_*\sigma} & f_*\underline{j} & \begin{array}{c}\nearrow^{f_*\eta\underline{j}} \\ \searrow_{\eta' f_*\underline{j}}\end{array} & \downarrow{\scriptstyle \tau\underline{j}} \\
 & & & & T'f_*\underline{j}
\end{array} \qquad \sigma \in \mathfrak{S}_K(\underline{i},\underline{j})$$

ensures that P preserves set operations and, in particular, identities.

<u>Proposition 2.28</u>: The dual Θ^{op} of a K-coloured theory Θ is isomorphic as topological category to the full subcategory of $\mathfrak{Top}^\Theta$ of the free Θ-spaces $F\underline{i}$, $\underline{i} \in \text{ob}\,\mathfrak{S}_K$.

<u>Proof</u>: The correspondence $\underline{i} \longmapsto \Theta(\underline{i},-)$ and $a \longmapsto \Theta(a,-)$ defines a full embedding $\Theta^{op} \subset \mathfrak{Funct}[\Theta,\mathfrak{Top}]$ by the Yoneda lemma [41; p.37]. Since $\Theta(\underline{i},-)$ preserves products, this embedding factors through $\mathfrak{Top}^\Theta$, and it is easy to check that it is a homeomorphism on the morphism spaces. We have natural homeomorphisms

$$\mathfrak{Top}^\Theta(F\underline{i},G) \underset{(1)}{\cong} \mathfrak{Top}_K(\underline{i},UG) \underset{(2)}{\cong} G(\underline{i}) \underset{(3)}{\cong} \mathfrak{Top}^\Theta(\Theta(\underline{i},-),G)$$

(1) holds by (2.24), (2) is obvious, and (3) follows from the Yoneda lemma. The isomorphism $\Theta(\underline{i},-) \longrightarrow F\underline{i}$ is given on the basic object k by

$$a \longmapsto (a;\, k_1,\ldots,k_n)$$

if $\underline{i} = (k_1,\ldots,k_n)$. ∎

Corollary 2.29: The composite functor $\mathfrak{Theories} \longrightarrow \mathfrak{Triples} \longrightarrow \mathfrak{Theories}$ is naturally equivalent to the identity.

Proof: Given a theory functor $(P,f) : \Theta \longrightarrow \Theta'$, let $\tau : f_*T \longrightarrow T'f_*$ be the corresponding triple morphism, and $(\hat{P},f) : \hat{\Theta} \longrightarrow \hat{\Theta}'$ the theory functor induces by (τ,f). The corollary follows from the commutative diagram

$$\begin{array}{ccccccc}
\hat{\Theta}(\underline{i},\underline{j}) & = & \mathfrak{Top}_K(\underline{j},T\underline{i}) & \cong & \mathfrak{Top}^{\Theta}(F\underline{j},F\underline{i}) & \cong & \Theta(\underline{i},\underline{j}) \\
 & & \big\downarrow f_* & & & & \\
\Big\downarrow \hat{P} & & \mathfrak{Top}_L(f_*\underline{j},f_*T\underline{i}) & & & & \Big\downarrow P \\
 & & \big\downarrow (\tau\underline{i})_* & & & & \\
\hat{\Theta}'(f_*\underline{i},f_*\underline{j}) & = & \mathfrak{Top}_L(f_*\underline{j},T'f_*\underline{i}) & \cong & \mathfrak{Top}^{\Theta'}(F'f_*\underline{j},F'f_*\underline{i}) & \cong & \Theta'(f_*\underline{i},f_*\underline{j})
\end{array} \quad \blacksquare$$

Call a triple finitary if it lies in the image of $\mathfrak{Theories} \rightarrow \mathfrak{Triples}$. Each triple T has an associated finitary triple T_{fin}, namely the image of T under $\mathfrak{Triples} \longrightarrow \mathfrak{Theories} \longrightarrow \mathfrak{Triples}$. Note that

$$(T_{fin}\, X)_k = \bigcup_{\underline{i}} \mathfrak{Top}_K(k,T\underline{i}) \times X_{\underline{i}} / \sim$$

and we can identify $X_{\underline{i}}$ with $\mathfrak{Top}_K(\underline{i},X)$. The maps

$$\begin{array}{ccc}
\mathfrak{Top}_K(k,T\underline{i}) \times \mathfrak{Top}_K(\underline{i},X) & \longrightarrow & \mathfrak{Top}_K(k,TX) = (TX)_k \\
(f,g) & \longmapsto & (Tg \circ f)
\end{array}$$

hence induce a natural transformation $T_{fin} \longrightarrow T$, which is an isomorphism if T is finitary, but not in general. We obtain

Proposition 2.30: The category $\mathfrak{Theories}$ is naturally equivalent to the full subcategory of finitary triples in $\mathfrak{Triples}$. $\blacksquare$

If we enlarge $\mathfrak{Theories}$ by adding infinitary theories, it becomes naturally equivalent to the whole category $\mathfrak{Triples}$. To define infinitary theories we need the notion of a continuous product. In $\mathfrak{Top}$, an

X-indexed product of copies of the space Y is defined to be $\mathfrak{Top}(X,Y)$. The exponential law

$$\mathfrak{Top}(Z,\mathfrak{Top}(X,Y)) \cong \mathfrak{Top}(X,\mathfrak{Top}(Z,Y))$$

generalizes the usual functorial equation of a product $\prod_{\alpha} Y_{\alpha}$. In our case, we have to define continuous products in $\mathfrak{Top}_K$. The object $K \in \mathfrak{Top}_K$ given by $id : K \longrightarrow K$ substitutes the point object of $\mathfrak{Top}$. Since

$$\begin{aligned}\mathfrak{Top}_K^{op}(X,Y)=\mathfrak{Top}_K(Y,X)\cong \prod_{k\in K} \mathfrak{Top}(Y_k,X_k)\cong \prod_{k\in K} \mathfrak{Top}(Y_k,\mathfrak{Top}(X_k,k))&\cong\mathfrak{Top}_K(Y,\mathfrak{Top}_K(K,X))\\ &=\mathfrak{Top}_K(Y,\mathfrak{Top}_K^{op}(X,K))\end{aligned}$$

we call Y a Y-indexed product of K in $\mathfrak{Top}_K$.

<u>Definition 2.31</u>: An <u>infinitary</u> K-<u>coloured</u> <u>topological</u> <u>theory</u> is a topological category Θ with ob Θ = ob $\mathfrak{Top}_K$ together with a continuous functor $P : \mathfrak{Top}_K^{op} \longrightarrow \Theta$ preserving objects and products, i.e. the diagram

$$\begin{array}{ccc}\mathfrak{Top}_K^{op}(X,Y) & \cong & \mathfrak{Top}_K(Y,\mathfrak{Top}_K^{op}(X,K)) \\ \downarrow P & & \downarrow P_* \\ \Theta(X,Y) & \cong & \mathfrak{Top}_K(Y,\Theta(X,K))\end{array}$$

commutes.

A Θ-<u>space</u> is a product preserving functor $G : \Theta \longrightarrow \mathfrak{Top}_K$. In particular, $G(X) \cong \mathfrak{Top}_K(X,G(K))$.

It follows directly from the Yoneda lemma that Θ^{op} is equivalent to the category of free Θ-spaces. Hence $\mathfrak{Triples}$ and the enlarged category $\mathfrak{Theories}$ are equivalent by the same argument as above. The equivalence is given by

$$\begin{array}{ccccc}\mathfrak{Theories} & \longrightarrow & \mathfrak{Triples} & \longrightarrow & \mathfrak{Theories} \\ \Theta & \longmapsto & T \qquad T' & \longmapsto & \Theta'\end{array}$$

where $T(X) = \Theta(X,-)$ and $\Theta'(X,Y) = \mathfrak{Top}_K(Y,T'X)$.

Because of the strong connection between theories and triples it is no surprise that we can define the category of T-spaces for a triple T and that it is connected with the category of Θ-spaces of a theory Θ.

<u>Definition 2.32</u>: Let T be a triple on $\mathfrak{Top}_K$. A <u>T-space</u> consists of an object $X \in \mathfrak{Top}_K$ and a morphism $\xi : TX \longrightarrow X$ such that

$$\begin{array}{ccc} X & \xrightarrow{\eta X} & TX \\ & \diagdown\!\!\diagdown & \downarrow \xi \\ & & X \end{array} \qquad\qquad \begin{array}{ccc} TTX & \xrightarrow{\mu X} & TX \\ T\xi \downarrow & & \downarrow \xi \\ TX & \xrightarrow{\xi} & X \end{array}$$

commute. X is called the <u>underlying space</u> of (X,ξ).
A <u>homomorphism</u> $(X,\xi) \longrightarrow (Y,\xi)$ of T-spaces is a morphism $f : X \longrightarrow Y$ such that

$$\begin{array}{ccc} TX & \xrightarrow{Tf} & TY \\ \xi \downarrow & & \downarrow \xi \\ X & \xrightarrow{f} & Y \end{array}$$

commutes. Let $\mathfrak{Top}^T$ denote the topological category of T-spaces and homomorphisms ($\mathfrak{Top}^T((X,\xi),(Y,\xi))$ is topologized as subspace of $\mathfrak{Top}_K(X,Y)$).

<u>Proposition 2.33</u>: Let T be the associated triple of a (finitary) K-coloured theory Θ. Then there exists an isomorphism $R : \mathfrak{Top}^\Theta \longrightarrow \mathfrak{Top}^T$ of topologized categories such that

$$\begin{array}{ccc} \mathfrak{Top}^\Theta & \xrightarrow{R} & \mathfrak{Top}^T \\ & {\scriptstyle U}\searrow \quad \swarrow{\scriptstyle V} & \\ & \mathfrak{Top}_K & \end{array} \qquad U,V \text{ underlying space functors}$$

commutes.

<u>Proof</u>: For a Θ-space G define $R(G) = (UG, U\varepsilon)$ with the back adjunction $\varepsilon : FU \longrightarrow Id$ (recall that $T = UF$). The inverse of R maps the T-space (X,ξ) to the Θ-space G for which $G(\underline{i}) = X_{\underline{i}}$ and $G(a), a \in \Theta(\underline{i},\underline{j})$,

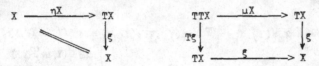

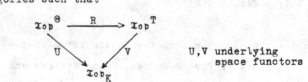

is given by the composite

$$X_{\underline{i}} \xrightarrow{(\eta X)_{\underline{i}}} (TX)_{\underline{i}} \xrightarrow{U(FX(a))} (TX)_{\underline{i}} \xrightarrow{\xi_{\underline{i}}} X_{\underline{i}} \quad \blacksquare$$

As corollary we obtain a generalization of the classical result that each group is the epimorphic image of a free group.

<u>Corollary 2.34</u>: An object $Z \in \mathfrak{Top}_K$ is a Θ-space iff the injection $\eta Z : Z \longrightarrow UFZ$ admits a retraction $\xi : UFZ \longrightarrow Z$ which makes

$$\begin{array}{ccc} UFUFZ & \xrightarrow{U\varepsilon FZ} & UFZ \\ \downarrow{\scriptstyle UF\xi} & & \downarrow{\scriptstyle \xi} \\ UFZ & \xrightarrow[\xi]{} & Z \end{array}$$

commute ($U : \mathfrak{Top}^{\Theta} \longrightarrow \mathfrak{Top}_K$ is the underlying space functor, F its left adjoint and η and ε the adjunction maps). $\blacksquare$

5. SPINES

We do not know how to handle general theories. So we restrict attention to those kinds of theories that interest us most and in which we can work satisfactorily. It is clear from the interpretation of Θ^{op} as the category of free Θ-spaces on the elements of $\mathfrak{S}_K$, that theories tend to be inconveniently large.

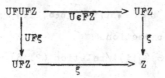

Let Θ be a K-coloured theory given by generators $\{X_{\underline{i},k}\}$ and relations $\{R_{\underline{i},k}\}$ in $\mathfrak{G}r$ spaces. The elements of $\{R_{\underline{i},k}\}$ are pairs of trees with vertex labels in $\{X_{\underline{i},k}\}$. A tree from $\underline{i}$ to k with m twigs of colours $j_1,\ldots,j_m$ and labels $n_1,\ldots,n_m \in [n]$ determines a morphism in $\mathfrak{S}_K$ from $\underline{j} = \{j_1,\ldots,j_m\}$ to $\underline{i}$ induced by $j_r \longrightarrow n_r \in [n]$. Let $\mathfrak{A}$ be the subcategory of $\mathfrak{S}_K$ generated under composition and disjoint union $\oplus$ by the morphisms determined by the trees of $\{R_{\underline{i},k}\}$, and let $\mathfrak{B}$ be the subcategory of Θ generated under composition and the product bifunctor $\oplus$ by the elements of $\{X_{\underline{i},k}\}$ and $\mathfrak{S}^{op}$. (We assume that all these categories have objects ob $\mathfrak{S}_K$).

Definition 2.35: The subcategory $\mathfrak{B}$ of Θ is called a $\mathfrak{S}$-spine of Θ.

An element a of Θ can be written $a = b \circ \sigma^*$ with $b \in \mathfrak{B}$. If $\sigma \in \mathfrak{S}$ implies that all block functions (see 2.4) associated with σ are in $\mathfrak{S}$, this decomposition is unique up to the equivalence

(2.36) $$(b \circ \mu^*) \circ \sigma^* = b \circ (\sigma \circ \mu)^* \qquad b \in \mathfrak{B},\ \mu \in \mathfrak{S},\ \sigma \in \mathfrak{S}_K.$$

Hence there is a continuous bijection

(2.37) $$\Big(\bigcup_{\underline{j}} \mathfrak{B}(\underline{j},\underline{r}) \times \mathfrak{S}_K(\underline{j},\underline{i})\Big)/\text{relation } (2.36) \longrightarrow \Theta(\underline{i},\underline{r})$$

If this is a homeomorphism we can recover Θ from $\mathfrak{B}$, its topology included. In this case we call $\mathfrak{B}$ a proper $\mathfrak{S}$-spine of Θ.

The products in Θ are no longer products in $\mathfrak{B}$ unless $\mathfrak{S} = \mathfrak{S}_K$. Instead we have an associative "product" functor

$$\oplus : \mathfrak{B} \times \mathfrak{B} \longrightarrow \mathfrak{B}$$

sending the object $(\underline{i},\underline{j})$ to $\oplus(\underline{i},\underline{j}) = \underline{i} \oplus \underline{j}$, the sum of $\underline{i}$ and $\underline{j}$ in $\mathfrak{S}_K$. The correspondence $(a_1,\ldots,a_n,\tau) \longmapsto (a_1 \oplus \ldots \oplus a_n) \circ \tau^*$ defines a homeomorphism

(2.38) $$h : \bigcup_{\underline{j}} \bigcup_{\oplus \underline{j}_q = \underline{j}} \Theta(\underline{j}_1,\underline{r}(1)) \times \ldots \times \Theta(\underline{j}_n,\underline{r}(n)) \times \mathfrak{S}_K(\underline{j},\underline{i})/\sim\ \cong \Theta(\underline{i},\underline{r})$$

with $r : [n] \longrightarrow K$ and the relation

(2.39) $$(a_1 \circ \sigma_1^*,\ldots,a_n \circ \sigma_n^*,\tau) \sim (a_1,\ldots,a_n,\tau \circ (\sigma_1 \oplus \ldots \oplus \sigma_n))$$

If $\mathfrak{B}$ is a proper spine, (2.37) and (2.38) determine a homeomorphism

$$\bigcup_{\underline{p}} \mathfrak{B}(\underline{p},\underline{r}) \times \mathfrak{S}_K(\underline{p},\underline{i})/\sim\ \cong \bigcup_{\underline{j}} \bigcup_{\oplus \underline{j}_q = \underline{j}} \mathfrak{B}(\underline{j}_1,\underline{r}(1)) \times \ldots \times \mathfrak{B}(\underline{j}_n,\underline{r}(n)) \times \mathfrak{S}_K(\underline{j},\underline{i})/\sim$$

with the relation (2.36) on the left and the relation (2.39) on the right (then, of course, each $\sigma_i \in \mathfrak{S}$). At least in the cases we want to consider, namely

(A) $\mathfrak{S}$ contains only the identities of $\mathfrak{S}_K$

(B) $\mathfrak{S}$ contains all isomorphisms of $\mathfrak{S}_K$

this homeomorphism induces a homeomorphism

(2.40) $\mathfrak{B}(\underline{p},\underline{r}) \cong \bigcup_{\underline{i}} \bigcup_{\oplus \underline{j}_q = \underline{i}} \mathfrak{B}(\underline{j}_1,\underline{r}(1)) \times \ldots \times \mathfrak{B}(\underline{j}_n,\underline{r}(n)) \times \mathfrak{S}(\underline{j},\underline{p}) / \text{relation}\,(2.39)$

Given a proper $\mathfrak{S}$-spine $\mathfrak{B}$ of Θ, a Θ-space is completely determined by a continuous functor $R : \mathfrak{B} \longrightarrow \mathfrak{Top}$ preserving the set operations of $\mathfrak{S}$ and the product functor $\oplus$, i.e. the diagram

$$\begin{array}{ccc} \mathfrak{B} \times \mathfrak{B} & \xrightarrow{\;\oplus\;} & \mathfrak{B} \\ \downarrow{\scriptstyle R\times R} & & \downarrow{\scriptstyle R} \\ \mathfrak{Top} \times \mathfrak{Top} & \xrightarrow{\;\times\;} & \mathfrak{Top} \end{array}$$

commutes. We call such a functor R a $\mathfrak{B}$-<u>space</u> and a natural transformation between such functors a <u>homomorphism</u> of $\mathfrak{B}$-spaces. The free Θ-space FX on $X \in \mathfrak{Top}_K$ is given by

$$FX(k) = \bigcup_{\underline{i}} \mathfrak{B}(\underline{i},k) \times X_{\underline{i}}/\sim$$

with $(a \circ \mu^*; x_1,\ldots,x_n) \sim (a; \mu^*(x_1,\ldots,x_n))$, $\mu \in \mathfrak{S}$.

<u>Spines of type (A)</u>: We investigate proper $\mathfrak{S}$-spines for which $\mathfrak{S}$ consists of identities only. A simple example of a theory with such a spine is the monochrome theory Θ_m of monoids. We pay particular attention to the elements $\lambda_n \in \Theta_m(n,1)$ corresponding to $z_1 z_2 \ldots z_n \in F[n]$, the free monoid on n generators $z_1,\ldots,z_n$ (under the isomorphism (2.28)), i.e. λ_n represents the operation $(z_1,z_2,\ldots,z_n) \longrightarrow z_1 z_2 \ldots z_n$ (here n denotes the unique object $[n] \longrightarrow K = \{*\}$). The subcategory $\mathfrak{A}$ of Θ_m generated under $\oplus$ and composition by the λ_n is the required $\mathfrak{S}$-spine. In view of (2.40), any morphism of $\mathfrak{A}$ has uniquely the form $\lambda_{n_1} \oplus \ldots \oplus \lambda_{n_r}$. Since relation (2.36) is trivial, any morphism of Θ_m is uniquely expressible as $\lambda \circ \sigma^*$ with $\lambda \in \mathfrak{A}$.

Θ_m serves as sort of a terminal object for theories with proper spines of type (A). Let $\mathfrak{Is}_K$ denote the category which has the elements of K as objects and exactly one morphism between any two objects. The subcategory $\mathfrak{A} \otimes \mathfrak{Is}_K$ of $\Theta_m \otimes \mathfrak{Is}_K$ has exactly one morphism from any object $\underline{i}$ to a basic object k (compare 2.22). Hence given a K-coloured theory Θ with proper Θ-spine $\mathfrak{B}$, there exists a unique object-preserving theory functor $P : \Theta \longrightarrow \Theta_m \otimes \mathfrak{Is}_K$ such that $P^{-1}(\mathfrak{A} \otimes \mathfrak{Is}_K) = \mathfrak{B}$.

The above considerations give a characterization of proper Θ-spines of type (A).

Lemma 2.41: A topological category $\mathfrak{B}$ is a proper Θ-spine of type (A) of a K-coloured theory iff ob $\mathfrak{B}$ = ob $\mathfrak{S}_K$ and there is a strictly associative bifunctor $\oplus : \mathfrak{B} \times \mathfrak{B} \longrightarrow \mathfrak{B}$ such that

(a) $\oplus(\underline{i},\underline{j}) = \underline{i} \oplus \underline{j}$, the sum in $\mathfrak{S}_K$

(b) the correspondence $(a_1,\ldots,a_n) \longmapsto a_1 \oplus \ldots \oplus a_n$ yields a homeomorphism

$$\bigcup_{\oplus \underline{i}_j = \underline{i}} \mathfrak{B}(\underline{i}_1,\underline{r}(1)) \times \ldots \times \mathfrak{B}(\underline{i}_n,\underline{r}(n)) \cong \mathfrak{B}(\underline{i},\underline{r}) \qquad \blacksquare$$

Definition 2.42: We call a category $\mathfrak{B}$ satisfying (2.41) a K-coloured PRO (for product functor) and a K-coloured theory Θ having a PRO as proper spine a split theory over Θ_m. If the spaces $\mathfrak{B}(\underline{i},\underline{r})$ are CW-complexes and the homeomorphisms (2.41 (b)) and composition are skeletal, we call $\mathfrak{B}$ a CW-PRO.

Note that the morphism spaces $\mathfrak{B}(\underline{i},k)$ of morphisms into a basic object and the composition maps $[\mathfrak{B}(\underline{r}_1,\underline{i}(1)) \times \ldots \times \mathfrak{B}(\underline{r}_n,\underline{i}(n)] \times \mathfrak{B}(\underline{i},k) \longrightarrow \mathfrak{B}(\underline{r}_1 \oplus \ldots \oplus \underline{r}_n,k)$ sending $(a_1,\ldots,a_n,b)$ to $b \circ (a_1 \oplus \ldots \oplus a_n)$ completely determine the PRO $\mathfrak{B}$.

If Θ is a K-coloured split theory over Θ_m with PRO $\mathfrak{B}$, the free Θ-space FX on $X \in \mathfrak{Top}_K$ is given by

$$FX(k) = \bigcup_{\underline{i} \in \Theta} \mathfrak{B}(\underline{i},k) \times X_{\underline{i}}$$

Spines of type (B): Here the theory of commutative monoids Θ_{cm} takes the place of Θ_m. Generating morphisms are again the morphisms λ_n representing the operations $(z_1,\ldots,z_n) \longrightarrow z_1 + \ldots + z_n$. Since $\lambda_n \circ \pi^* = \lambda_n$ for all permutations $\pi \in S_n$, we also have to include the permutations. It is easily verified that the resulting spine may be identified with the category $\mathfrak{S}$ of finite sets as defined in section 1 (not to be confused with the set operations from $\mathfrak{S}^{op} \subset \Theta_{cm}$). An explicit description of the inclusion functor $\mathfrak{S} \subset \Theta_{cm}$ is given as follows: The isomorphism (2.28) identifies $a \in \mathfrak{S}(m,n) \subset \Theta(m,1)^n$ with $(y_1,\ldots,y_n) \in (F[m])^n$, where $y_r = z_{i_1} + \ldots + z_{i_q} \in F[m]$ if $a^{-1}(r) = \{i_1,\ldots,i_q\} \subset [m]$. In particular, for a permutation $\pi \in \mathfrak{S}(m,m)$ we have $\pi^{-1} = \pi^*$.

The abelianization $\Theta_m \longrightarrow \Theta_{cm}$ identifies $\mathfrak{A} \subset \Theta_m$ with the subcategory of $\mathfrak{S}$ of order preserving maps.

As in the non-commutative case, for any K-coloured theory Θ with proper $\mathfrak{G}$-spine $\mathfrak{B}$ there exists a unique object-preserving theory functor $P : \Theta \longrightarrow \Theta_{cm} \otimes \mathfrak{Js}_K$ such that $P^{-1}(\mathfrak{S} \otimes \mathfrak{Js}_K) = \mathfrak{B}$.

A morphism $\pi : \underline{i} \longrightarrow \underline{j}$ of $\mathfrak{S}_K$

$$\begin{array}{ccc} [n] & \xrightarrow{\pi} & [m] \\ & {\scriptstyle \underline{i}}\searrow \quad \swarrow{\scriptstyle \underline{j}} & \\ & K & \end{array}$$

is in $\mathfrak{G}$ iff n=m and π is a permutation. Hence a morphism of $\mathfrak{G}$ is given by its source or target and a permutation. If $K = \{*\}$ it is given by a permutation alone. If source or target are clear from the context, we therefore often write $\pi \in S_n$ instead of $\pi \in \mathfrak{G}(\underline{i},\underline{j})$. A $\mathfrak{G}$-spine of type (B) has more structure than one of type (A) because of the permutations. The analogue of Lemma 2.41 is

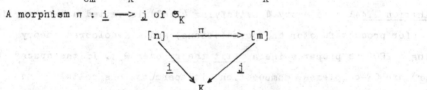

Lemma 2.43: A topological category $\mathfrak{B}$ is a proper $\mathfrak{G}$-spine of type (B) of a K-coloured theory iff ob $\mathfrak{B}$ = ob $\mathfrak{S}_K$ and we have a strictly as-

sociative bifunctor $\oplus : \mathfrak{B} \times \mathfrak{B} \longrightarrow \mathfrak{B}$ and an inclusion functor $\mathfrak{S} \subset \mathfrak{B}$ such that

(a) $\oplus(\underline{i},\underline{j}) = \underline{i} \oplus \underline{j}$, the sum in $\mathfrak{S}_K$

(b) The correspondence $(b_1,\ldots,b_n,\pi) \longmapsto (b_1 \oplus \ldots \oplus b_n) \circ \pi$ yields a homeomorphism

$$\Big(\bigcup_{\oplus \underline{j}_q = \underline{j}} \mathfrak{B}(\underline{j}_1,\underline{r}(1)) \times \ldots \times \mathfrak{B}(\underline{j}_n,\underline{r}(n)) \times \mathfrak{S}(\underline{i},\underline{j})\Big)/\sim\ \cong \mathfrak{B}(\underline{i},\underline{r})$$

where $(b_1 \circ \pi_1,\ldots,b_n \circ \pi_n,\pi) \sim (b_1,\ldots,b_n,(\pi_1 \oplus \ldots \oplus \pi_n) \circ \pi)$, $\pi_i,\pi \in \mathfrak{S}$

(c) $\pi_1 \oplus \pi_2 \in \mathfrak{S}$ is the sum (in $\mathfrak{S}_K$) of π_1 and π_2

(d) Given r morphisms $b_q : \underline{i}_q \longrightarrow \underline{j}_q$, $\underline{i}_q : [m_q] \longrightarrow K$, $\underline{j}_q : [n_q] \longrightarrow K$, and $\pi \in \mathfrak{S}$. Then

$$\pi(n_1,\ldots,n_r) \circ (b_1 \oplus \ldots \oplus b_r) = (b_{\pi^{-1}(1)} \oplus \ldots \oplus b_{\pi^{-1}(r)}) \circ \pi(m_1,\ldots,m_r)$$

(see (2.4) for the block permutations) ∎

If the reader is disturbed by the π^{-1} in (d) he should note that the inclusion functor $\mathfrak{S} \subset \mathfrak{B}$ is given by $\pi \longrightarrow (\pi^{-1})^*$.

Definition 2.44: A category $\mathfrak{B}$ satisfying (2.43) is called a K-coloured PROP (for product functor and permutations) and a K-coloured theory having a PROP as proper spine a split theory over Θ_{cm}. If the spaces $\mathfrak{B}(\underline{i},\underline{r})$ are CW-complexes, composition with permutations cellular and the homeomorphisms (2.43 (b)) and the composition in $\mathfrak{B}$ skeletal, we call $\mathfrak{B}$ a CW-PROP.

Note that a PROP $\mathfrak{B}$ is completely determined by its morphism spaces $\mathfrak{B}(\underline{i},k)$ of morphisms into a basic object, by the composition of $a \in \mathfrak{B}(\underline{i},k)$ with a permutation on the right, and by the composition maps

$$[\mathfrak{B}(\underline{r}_1,\underline{i}(1)) \times \ldots \times \mathfrak{B}(\underline{r}_n,\underline{i}(n))] \times \mathfrak{B}(\underline{i},k) \longrightarrow \mathfrak{B}(\underline{r}_1 \oplus \ldots \oplus \underline{r}_n,k)$$

sending $(a_1,\ldots,a_n,b)$ to $b \circ (a_1 \oplus \ldots \oplus a_n)$.

A PROP is a more general concept than a PRO because we can add all isomorphisms of $\mathfrak{S}_K$ to make a spine $\mathfrak{B}$ of type (A) into a spine $\mathfrak{B}'$ of type (B), and we have an inclusion functor $\mathfrak{B} \subset \mathfrak{B}'$.

<u>Definition 2.45</u>: A functor $F : \mathfrak{B} \longrightarrow \mathfrak{B}'$ of PROPs is called a <u>PROP-functor</u> if it is continuous, carries basic objects to basic objects, and preserves the product functor $\oplus$ and the permutations. Analogously for PRO-<u>functors</u>.

Obviously, a PROP-functor is the restriction of a theory functor P and completely determines P.

Our principal concern will be E-spaces.

<u>Definition 2.46</u>: A K-coloured PROP $\mathfrak{B}$ is called a <u>K-coloured E-theory</u> if each space $\mathfrak{B}(\underline{i},k)$, $k \in K$, is contractible, in other words, if $P : \mathfrak{B} \longrightarrow \mathfrak{S} \otimes \mathfrak{J}\mathfrak{s}_K$ is topologically a homotopy equivalence. An object $X \in \mathfrak{Top}_K$ is called an <u>E-space</u> if it allows a $\mathfrak{B}$-action for some K-coloured E-theory $\mathfrak{B}$. (The monochrome E-spaces are identical with the homotopy-everything H-spaces of [8]).

<u>Remark</u>: Our whole theory developed in part from the theory of PROPs and PACTs propounded by Adams and MacLane [29]. Their PROPs are essentially the algebraic analogue of ours, and a PACT is the analogue for chain complexes. A Steenrod PACT then corresponds to an E-space.

Spines inherit the notion of interchange from their envelopping theories. One observation is of importance. Given spines $\mathfrak{B}, \mathfrak{B}'$, each a PRO or a PROP, then $\mathfrak{B} \otimes \mathfrak{B}'$ is a PROP because $\Theta_m \otimes \Theta_m \cong \Theta_m \otimes \Theta_{cm} \cong \Theta_{cm} \otimes \Theta_{cm} \cong \Theta_{cm}$ (see Example 2, section 3).

6. EXAMPLES OF PROs AND PROPs

(2.47) The categories $\mathfrak{A}$ and $\mathfrak{S}$ are examples of a CW-PRO and a CW-PROP, which we already have discussed. The $\mathfrak{A}$-spaces are exactly the topological monoids and the $\mathfrak{S}$-spaces exactly the commutative topological monoids.

(2.48) Trivial examples of PROs can be obtained in the following way. If $\mathfrak{C}$ is a topological category, we obtain an (ob $\mathfrak{C}$)-coloured PRO by setting

$$\mathfrak{B}(\underline{i},l) = \begin{cases} \emptyset & \text{if } \underline{i} \text{ is not a basic object} \\ \mathfrak{C}(k,l) & \text{if } \underline{i} \text{ is the basic object } k \in \text{ob } \mathfrak{C} \end{cases} \qquad l \in \text{ob } \mathfrak{C}$$

Composition is given by the composition in $\mathfrak{C}$. Note that $\mathfrak{B}$ is a spine of $\mathfrak{C}$ considered as theory. If $\mathfrak{C}$ has the discrete topology, a $\mathfrak{B}$-space is just a $\mathfrak{C}$-diagram of topological spaces. In general, a $\mathfrak{B}$-space is a $\mathfrak{C}$-diagram with a topology on the morphisms. If there is no chance of confusion we denote $\mathfrak{B}$ again by $\mathfrak{C}$.

(2.49) For each $n \geq 1$ we define a monochrome PROP $\mathfrak{Q}_n$, the n-th <u>little-cube category</u>, which operates on the n-th loop space $X = \Omega^n Y$, the space of all maps $(I^n, \partial I^n) \longrightarrow (Y,*)$, where I^n is the standard n-cube, ∂I^n its bounadry, and * the base point of Y. As before, denote the unique object $[m] \longrightarrow K = \{*\}$ by m. A point $a \in \mathfrak{Q}_n(m,1)$ is an ordered collection of m n-cubes I_i^n, linearly embedded in I^n, with disjoint interiors, and with axes parallel to those of I^n. Such an embedding is uniquely determined by the images in I^n of the lowest vertex $(0,0,\ldots,0)$ and the upper vertex $(1,1,\ldots,1)$. Hence a is given by a 2m-tuple $(x_1,y_1,\ldots,x_m,y_m)$ of points in I^n, where x_i is the lowest vertex and y_i the upper vertex of I_i^n. We topologize $\mathfrak{Q}_n(m,1)$ as subspace of I^{2mn}. Composition of a with a permutation $\pi \in \mathfrak{S}_m$ is given by $a \circ \pi = (x_{\pi 1}, y_{\pi 1}, \ldots, x_{\pi m}, y_{\pi m})$. Let $b_i \in \mathfrak{Q}_n(r_i,1)$, $i=1,\ldots,m$, let

$b_{ij} : I^n \subset I^n$ be the linear embedding of the j-th cube of b_i and let $a_i : I^n \subset I^n$ be the linear embedding of the i-th cube of a. Then the $r_1+\ldots+r_m$ linear embeddings of I^n into I^n which correspond to $a\circ(b_1\oplus\ldots\oplus b_m)$ are given by $a_1\circ b_{11},\ldots,a_1\circ b_{1r_1},a_2\circ b_{21},\ldots,a_2\circ b_{2r_2},\ldots,a_m\circ b_{m1},\ldots,a_m\circ b_{mr_m}$. This defines a continuous composition in $\mathfrak{D}_n$.

$\mathfrak{D}_n$ acts on X as follows: Given $(f_1,f_2,\ldots,f_m)\in X^m$, the map $a(f_1,\ldots,f_m) : I^n \longrightarrow I$ is given by f_i on the embedded cube I^n_i and zero elsewhere.

Let $a \in \mathfrak{D}_n(m,1)$, $a = (x_1,y_1,\ldots,x_m,y_m)$ with $x_i = (x_{i1},\ldots,x_{in})$, $y_i = (y_{i1},\ldots,y_{in}) \in I^n$. The correspondence $a \longmapsto a'=(x_1',y_1',\ldots,x_m',y_m')$ with $x_i' = (x_{i1},\ldots,x_{in},0)$ and $y_i' = (y_{i1},\ldots,y_{in},1)$ defines an inclusion of PROPs $\mathfrak{D}_n \subset \mathfrak{D}_{n+1}$. Let $\mathfrak{D}_\infty$ be the PROP with $\mathfrak{D}_\infty(m,1) = \bigcup_{n=1}^{\infty} \mathfrak{D}_n(m,1)$ with the direct limit topology.

<u>Lemma 2.50</u>: $\mathfrak{D}_\infty(m,1)$ is contractible for all m; hence $\mathfrak{D}_\infty$ is an E-category.

<u>Proof</u>: First observe that $\mathfrak{D}_n(m,1)$ has a very nice product neighbourhood N in $\mathfrak{D}_{n+1}(m,1)$, namely the set of all points $(x_1,y_1,\ldots,x_m,y_m)\in(I^{n+1})^{2m}$ with $x_{i,n+1} < \frac{1}{2}$ and $y_{i,n+1} > \frac{1}{2}$ for all i. Then $N=\mathfrak{D}_n(m,1)\times[0,\frac{1}{2})^m\times(\frac{1}{2},1]^m$. It follows that $\mathfrak{D}_n(m,1)\subset \mathfrak{D}_{n+1}(m,1)$ is a S_m-NDR (see Appendix II). By Lemma A 4.10 (the prefix A refers to the appendix), it suffices to show that $\mathfrak{D}_n(m,1)$ is contractible in $\mathfrak{D}_{n+1}(m,1)$. The contracting homotopy $H : \mathfrak{D}_n(m,1) \times I \longrightarrow \mathfrak{D}_{n+1}(m,1)$ is given by

$$H_t : (x_1,y_1,\ldots,x_m,y_m) = (x_1(t),y_1(t),\ldots,x_m(t),y_m(t))$$

where

$$x_i(t) = \begin{cases}(x_{i1},\ldots,x_{in},2t(i-1)/m) & 0\le t\le\frac{1}{2}\\ ((2-2t)x_{i1},\ldots,(2-2t)x_{in},(i-1)/m) & \frac{1}{2}\le t\le 1\end{cases}$$

$$y_i(t) = \begin{cases}(y_{i1},\ldots,y_{in},1-2t(1-i/m)) & 0\le t\le\frac{1}{2}\\ (2t-1+(2-2t)y_{i1},\ldots,2t-1+(2-2t)y_{in},i/m) & \frac{1}{2}\le t\le 1\end{cases}$$ ∎

Originally, we proved that $\mathfrak{D}_n(m,1)$ is (n-2)-connected for all m (as indicated in [8]) using results of Fadell and Neuwirth on configuration spaces. Since we only need that $\mathfrak{D}_\infty(m,1)$ is contractible, we prefer the present more direct and shorter proof. Our original version can be found in [34;chapter 4].

The E-category $\mathfrak{D}_\infty$ is quite important because it acts on strict infinite loop spaces.

Definition 2.51: A space Z is called a strict infinite loop space if there exists a sequence of based spaces Z_i and based homeomorphisms $\omega_i : Z_i \cong \Omega Z_{i+1}$, i=0,1,2,... such that $Z \cong Z_o$.

The exponential law $\mathfrak{Top}((I^n,\partial I^n),\mathfrak{Top}((I,\partial I),Z))\cong\mathfrak{Top}((I^n,\partial I^n)\times(I,\partial I),Z)$ defines homeomorphisms $q_n : \Omega^n(\Omega Z_{n+1}) \cong \Omega^{n+1}Z_{n+1}$. The ω_i and q_i combine to maps

$$r_n = q_{n-1}\circ\Omega^{n-1}\omega_{n-1}\circ\ldots\circ q_i\circ\Omega^i\omega_i\circ\ldots\circ q_1\circ\Omega^1\omega_1\circ\omega_o : Z \cong \Omega^n Z_n$$

The action of $\mathfrak{D}_\infty$ on Z is now defined as follows: Given $\overline{a}\in\mathfrak{D}_\infty(m,1)$ and $(z_1,\ldots,z_m)\in Z^m$. Let $a\in\mathfrak{D}_n(m,1)$ be a representative of $\overline{a}$. We define

$$\overline{a}(z_1,\ldots,z_m) = r_n^{-1}(a(r_n z_1,\ldots,r_n z_m))$$

To show that this definition is independent of the choice of the representative a, we have to verify that

$$r_n^{-1}(a(r_n z_1,\ldots,r_n z_m)) = r_{n+1}^{-1}(a'(r_{n+1}z_1,\ldots,r_{n+1}z_m))$$

where a' is the image of a in $\mathfrak{D}_{n+1}(m,1)$. Since $r_{n+1}=q_n\circ\Omega^n\omega_n\circ r_n$, this amounts to showing that

$$q_n[\Omega^n\omega_n(a(r_n z_1,\ldots,r_n z_m))] = a'(r_{n+1}z_1,\ldots,r_{n+1}z_m),$$

which is easily verified. We thus obtain

Proposition 2.52: A strict infinite loop space is an E-space. ∎

(2.53) We next define a PRO $\mathfrak{D}$ which acts on loop spaces ΩY. A point

$a \in \mathfrak{D}(m,1)$ is a 2m-tuple $(x_1,y_1,\dots,x_m,y_m)$ of points in I such that $0 \le x_1 < y_1 \le \dots \le x_i < y_i \le \dots \le x_n < y_n \le 1$. We topologize $\mathfrak{D}(m,1)$ as subspace of I^{2m}. Considering $(x_1,y_1,\dots,x_m,y_m)$ as element of $\mathfrak{D}_1(m,1)$, we obtain an inclusion $\mathfrak{D}(m,1) \subset \mathfrak{D}_1(m,1)$. Composition in $\mathfrak{D}$ is induced by the composition in $\mathfrak{D}_1$ so that we have a functor $\mathfrak{D} \longrightarrow \mathfrak{D}_1$. Since $\mathfrak{D}_1$ acts on ΩY, so does $\mathfrak{D}$.

<u>Lemma 2.54</u>: $\mathfrak{D}(m,1)$ is contractible for all m.

<u>Proof</u>: The contracting homotopy $\mathfrak{D}(m,1)\times I \longrightarrow \mathfrak{D}(m,1)$ is given by

$$[(x_1,y_1,\dots,x_m,y_m),t] \longrightarrow (x_1(t),y_1(t),\dots,x_m(t),y_m(t))$$

with $x_i(t) = (1-t)x_i + t(i-1)/m$ and $y_i(t) = (1-t)y_i + t\cdot i/m$. ∎

(2.54) Our next PRO $\mathfrak{A}_\infty$ acts on A_∞-spaces (see 1.8). We use the models K_i (see 1.7) for its definition: $\mathfrak{A}_\infty(m,1) = K_m$, m=0,1,2,..., with $K_o = K_1 = *$. In (1.7) we have already defined boundary maps $\partial_l(r,s) : K_r \times K_s \longrightarrow K_i$, $r+s = i+1$, which correspond to the copy of $K_r \times K_s$ in the boundary of K_i indexed by $1\,2\dots(l\;l+1\dots l+s-1)\dots i$. According to [46; §6], one can inductively construct degeneracy maps $s_j : K_i \longrightarrow K_{i-1}$, $i\ge1$, for $1\le j\le i$, satisfying

$$\partial_j(r,s+t-1)\circ(1\times\partial_k(s,t)) = \partial_{j+k-1}(r+s-1,t)\circ(\partial_j(r,s)\times 1)$$

$$\partial_{j+s-1}(r+s-1,t)\circ(\partial_k(r,s)\times 1) = \partial_k(r+t-1,s)\circ(\partial_j(r,t)\times 1)\circ(1\times \text{twist})$$

$$s_j\circ s_k = s_k\circ s_{j+1} \qquad \text{for } k\le j$$

$$s_j\partial_k(r,s) = \begin{cases} \partial_{k-1}(r-1,s)\circ(s_j\times 1) & \text{for } j<k \text{ and } r>2 \\ \partial_k(r-1,s)\circ(s_{j-s+1}\times 1) & \text{for } k+s\le j \\ \partial_k(r,s-1)\circ(1\times s_{j-k+1}) & \text{for } s>2,\ k\le j<k+s \end{cases}$$

$$s_j\partial_k(i-1,2) = pr_1 \qquad \text{for } 1<j = k<i \text{ and } 1<j = k+1\le i$$

$$s_1\partial_2(2,i-1) = s_i\partial_1(2,i-1) = pr_2$$

where pr_i is the projection onto the i-th factor.

We obtain a composition in $\mathfrak{A}_\infty$ if we specify the composites

$$c = a \cdot (\underbrace{1 \oplus \ldots \oplus 1}_{k-1} \oplus b \oplus \underbrace{1 \oplus \ldots \oplus 1}_{n-k})$$

with $a \in \mathfrak{A}_\infty(n,1)$ and $b \in \mathfrak{A}_\infty(m,1)$. We define

$$\begin{aligned} c &= s_k(a) \in \mathfrak{A}_\infty(n-1,1) && \text{if } m=0 \\ &= a && \text{if } m=1 \\ &= \partial_k(n,m)(a,b) \in \mathfrak{A}_\infty(n+m-1,1) && \text{if } m>1 \end{aligned}$$

The identities listed above imply the associativity of the composition.

Stasheff shows [46; Thm 5, Lemma 7] that an A_∞-space X admits maps $M_i : K_i \times X^i \longrightarrow X$ for i=2,3,4,... such that

(a) $M_2(*,e,x) = M_2(*,x,e) = x$ for $x \in X$, $*=K_2$, e a distinguished point of X

(b) for $(k_1,k_2) \in K_r \times K_s$, $r+s = i+1$, we have

$$M_i(\partial_k(r,s)(k_1,k_2),x_1,\ldots,x_i) = M_r(k_1,x_1,\ldots,x_{k-1},M_s(k_2,x_k,\ldots,x_{k+s-1}),x_{k+s},\ldots,x_i)$$

(c) for $k \in K_i$ and $i>2$, we have

$$M_i(k,x_1,\ldots,x_{j-1},e,x_{j+1},\ldots,x_i) = M_{i-1}(s_j(k),x_1,\ldots,x_{j-1},x_{j+1},\ldots,x_i)$$

The adjoints $K_i \longrightarrow \mathfrak{Top}(X^i,X)$ of the M_i define an action of $\mathfrak{A}_\infty$ on X.

If X is an A_n-space, we have such maps M_i for $2 \le i \le n$. Hence the sub-PRO $\mathfrak{A}_n$ of $\mathfrak{A}_\infty$ generated under composition and $\oplus$ by the morphisms in $\mathfrak{A}_\infty(m,1)$, $0 \le m \le n$, acts on an A_n-space.

For future use we note [46; Prop.3] that $\mathfrak{A}_\infty(m,1)$ is contractible for all m and $\mathfrak{A}_n(m,1)$ for all $m \le n$.

7. PARTLY HOMOGENEOUS THEORIES

We introduced coloured theories mainly for the purpose of studying maps between monochrome Θ-spaces. Given two monoids X and Y and a homomorphism $f : X \longrightarrow Y$, we can construct an $\mathfrak{A} \otimes \mathfrak{L}_1$-space $F : \mathfrak{A} \otimes \mathfrak{L}_1 \longrightarrow \mathfrak{Top}$ such that $F|\mathfrak{A} \otimes \{0\}$ defines the monoid structure on X and $F|\mathfrak{A} \otimes \{1\}$ the monoid structure on Y (Recall that $\mathfrak{L}_1$ is the category with objects 0,1 and one morphism from 0 to 1. We denote the subcategory consisting of the object i by $\{i\}$). Now $\mathfrak{A} \circledast \mathfrak{B}_1$ is $\{0,1\}$-coloured, and we have exactly one morphism $\{0,1\} \longrightarrow \{1\}$ (see (2.22)). It is mapped by F to $g : X \times Y \longrightarrow Y$ given by $g(x,y) = f(x) \cdot y$. Although such "mixed" maps occur naturally, it sometimes seems desirable to allow only operations $X^n \longrightarrow X$, $X^n \longrightarrow Y$, and $Y^n \longrightarrow Y$ for the study of maps from X to Y. In the literature this restriction always has been made. For this purpose we define partly homogeneous theories.

Let $H_L\mathfrak{S}_{K\times L}$ be the full subcategory of $\mathfrak{S}_{K\times L}$ consisting of all objects $\underline{i} : [n] \longrightarrow K\times L$ such that (projection) $\circ\, \underline{i} : [n] \longrightarrow K\times L \longrightarrow L$ is constant.

<u>Definition 2.55</u>: A (finitary) L-<u>homogeneous</u> (K×L)-<u>coloured</u> <u>theory</u> is a topological category Θ with ob Θ = ob $H_L\mathfrak{S}_{K\times L}$ together with an object and product preserving functor $H_L\mathfrak{S}^{op}_{K\times L} \longrightarrow \Theta$. Again we assume that $\Theta(\underline{i},\underline{j}_1 \oplus \ldots \oplus \underline{j}_n)$ is homeomorphic to $\Theta(\underline{i},\underline{j}_1) \times \ldots \times \Theta(\underline{i},\underline{j}_n)$. The definitions for a Θ-<u>space</u> and a <u>homomorphism</u> of Θ-spaces are analogue to those of (2.3). A <u>theory</u> <u>functor</u> from a L-homogeneous (K×L)-coloured theory Θ_1 to a N-homogeneous (M×N)-coloured theory Θ_2 consists of functions $f : K \longrightarrow M$, $g : L \longrightarrow N$, and a continuous functor $F: \Theta_1 \longrightarrow \Theta_2$ such that

$$\begin{array}{ccc} H_L\mathfrak{S}_{K\times L} & \xrightarrow{(f\times g)_*} & H_N\mathfrak{S}_{M\times N} \\ \downarrow & & \downarrow \\ \Theta_1 & \xrightarrow{\quad F \quad} & \Theta_2 \end{array}$$

commutes.

Here again we denote the product by $\oplus$. In contrary to inhomogeneous theories, $\oplus$ is not a bifunctor on a partly homogeneous theory because it is not everywhere defined, but it behaves like a bifunctor where it is defined.

Note that (2.55) is more general than (2.3) because it contains the inhomogeneous theories, just put $L = \{*\}$. If $K = \{*\}$, we call a

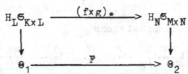

L-homogeneous theory completely homogeneous.

Any (K×L)-coloured theory Θ has a L-homogeneous part, denoted by $H_L\Theta$, namely the full subcategory of all objects coming from $H_L\mathfrak{S}^{op}_{K\times L}$. If $\mathfrak{C}$ is a topological category, we frequently denote the (ob $\mathfrak{C}$)-homogeneous part of $\Theta\otimes\mathfrak{C}$ by $H_{\mathfrak{C}}(\Theta\otimes\mathfrak{C})$.

Interchange, free Θ-spaces, PROs and PROPs can be defined for the partly homogeneous version in an analogous manner and similar results hold. We just want to mention one fact. The bifunctor $\oplus$ for PROs and PROPs coming from the product bifunctor of the enveloping theories is not everywhere defined for partly homogeneous PROs and PROPs. On objects, $\underline{i}\oplus\underline{j}$ exists whenever it exists in $H_L\mathfrak{S}_{K\times L}$, and $f\oplus g$ exists for $f\in\mathfrak{B}(\underline{i},\underline{p})$, $g\in\mathfrak{B}(\underline{j},\underline{q})$ whenever $\underline{i}\oplus\underline{j}$ and $\underline{p}\oplus\underline{q}$ exist. If $\oplus$ is defined, it behaves in the same manner as for ordinary PROs and PROPs.

(2.56) Example: Let $K=\{*\}$ and $L=\{0,1\}$. We construct a L-homogeneous (K×L)-coloured PRO $\mathfrak{B}$ which defines A_∞-maps between monoids (see (1.14)). There are exactly two objects $[n]\longrightarrow K\times L$ in $\mathfrak{B}$ for $n>0$, namely one for each object in L. Denote the one corresponding to 0 by n^0 and the one corresponding to 1 by n^1. Define $\mathfrak{B}(m^0,1^0)=\{\lambda_m\}$, $\mathfrak{B}(m^1,1^1)=\{\mu_m\}$, $\mathfrak{B}(m^0,1^1)=I^{m-1}$. So the full subcategories of $\mathfrak{B}$ consisting of all objects $0,1^0,2^0,\ldots$ and $0,1^1,2^1,\ldots$ are copies of the PRO $\mathfrak{A}$. It remains to define the compositions

$$\mu_m\circ[(t_1^1,\ldots,t_{r_1}^1)\oplus\ldots\oplus(t_1^m,\ldots,t_{r_m}^m)]=(t_1^1,\ldots,t_{r_1}^1,1,t_1^2,\ldots,t_{r_2}^2,1,t_1^3,\ldots,1,t_1^m,\ldots,t_{r_m}^m)$$

$$(t_1,\ldots,t_m)\circ(\lambda_{r_1}\oplus\ldots\oplus\lambda_{r_{m+1}})=(\underbrace{0,\ldots,0}_{r_1-1},t_1,\underbrace{0,\ldots,0}_{r_2-1},t_2,0,\ldots,t_m,\underbrace{0,\ldots,0}_{r_{m+1}-1})$$

According to (1.14), a map $f:X\longrightarrow Y$ between monoids is an A_∞-map if there are maps $F_i: I^{i-1}\times X^i\longrightarrow Y$ such that $F_1=f$ and

$$F_i(t_1,\ldots,t_{i-1},\ldots,x_1,\ldots,x_i)=\begin{cases}F_{i-1}(t_1,\ldots,\hat{t}_j,\ldots,t_{i-1},x_1,\ldots,x_jx_{j+1},\ldots,x_i) & \text{if } t_j=0\\ F_j(t_1,\ldots,t_{j-1},x_1,\ldots,x_j)F_{i-j}(t_{j+1},\ldots,t_i,x_{j+1},\ldots,x_i) & \text{if } t_j=1\end{cases}$$

Hence the adjoints of F_i define an action H of $\mathfrak{B}$ such that $H(\mathfrak{B}(1^0,1^1))=\{f\}$, and vice versa. If $f : X \longrightarrow Y$ is only an A_n-map between monoids, we can define an action of a subcategory $\mathfrak{B}'$ of $\mathfrak{B}$ on (X,Y) extending f and the monoid structures. $\mathfrak{B}'$ is generated under composition and $\oplus$ by the two copies of $\mathfrak{A}$ in $\mathfrak{B}$ and the morphisms of $\mathfrak{B}(m^o,1^1)$ for $m \leq n$.

We note that the morphism spaces of $\mathfrak{B}$ are contractible and so are the morphism spaces of $\mathfrak{B}'$ into basic objects with exception of $\mathfrak{B}'(m^o,1^1)$ for $m > n$.

III CHAPTER

THE BAR CONSTRUCTION FOR THEORIES

In Chapter I we defined a structure W𝔄 which is a monoid structure up to coherent homotopies. In this chapter we generalize the process 𝔄 ⟼ W𝔄 to general theories Θ. We need results from homotopy theory, which are not directly connected with the development of our theory and therefore proved in an appendix. Recall that we refer to the appendix with the prefix A.

1. THE THEORY WΘ

Let Θ be a K-coloured theory, let U : 𝔗𝔥𝔢𝔬𝔯𝔦𝔢𝔰 ⟶ 𝔊𝔯 𝔰𝔭𝔞𝔠𝔢𝔰 be the forgetful functor, and F : 𝔊𝔯 𝔰𝔭𝔞𝔠𝔢𝔰 ⟶ 𝔗𝔥𝔢𝔬𝔯𝔦𝔢𝔰 the free functor. Starting point for the construction of WΘ is the category FUΘ of copses associated with UΘ (see II; §2). To each internal edge of a tree of FUΘ we associate a real number in I = [0,1], called its <u>length</u>. A tree of a given shape λ can be considered as a point of a topological space $(\prod \Theta(\underline{i},k))\times \mathfrak{S}_K(\underline{i},\underline{j})$ (see Def. 2.6 ff); and if λ has r internal edges, a tree of shape λ with lengths can be considered as a point of the topological space

$$M_\lambda = (\prod \Theta(\underline{i},k))\times \mathfrak{S}_K(\underline{i},\underline{j})\times I^r$$

We impose three kinds of relations on the space of trees with lengths:

(3.1 a) <u>We may remove any vertex labelled by an identity of Θ</u>: We give the resulting edge the length $t_1 * t_2 = t_1 + t_2 - t_1 t_2$, where t_1 and

t_2 are the lengths of the edges below and above this vertex (By convention, the roots and twigs have lengths 1)

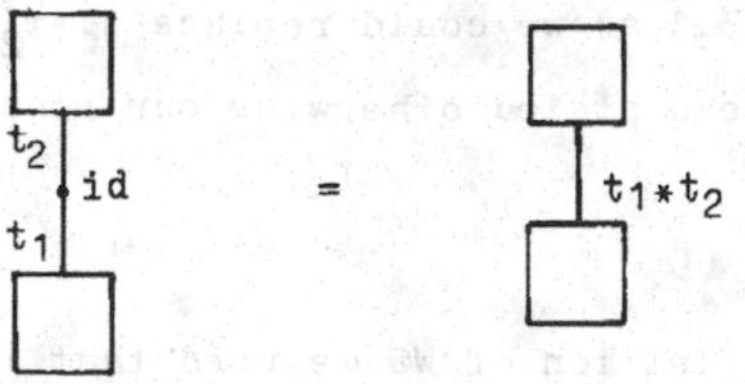

(3.1 b) *We may replace any vertex label* $a \cdot \sigma^*$ *by* a, *by changing the part of the tree above this vertex* as in (2.8 (b)), but for trees with lengths.

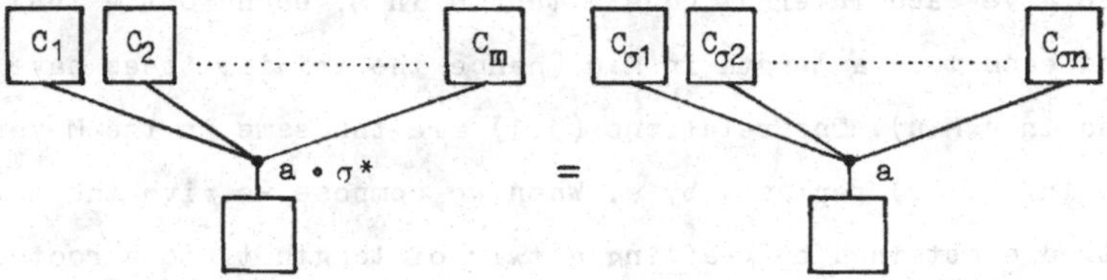

(3.1 c) *We may remove any edge of length* 0: We unite the vertices at the two ends to form a new vertex, whose label is the tree composite in Θ of the tree consisting of the two vertices and their incoming and outgoing edges (compare I,§4)

where $c = a \cdot (id_{k_1} \oplus \ldots \oplus id_{k_{i-1}} \oplus b \oplus id_{k_{i+1}} \oplus \ldots \oplus id_{k_n})$ if the incoming edges of a have the colours $k_1,\ldots,k_n$ and b sits on the i-th edge.

$W\Theta(\underline{i},\underline{j})$ is the space of all copses on $U\Theta$ with lengths modulo these three relations. We compose two copses with lengths by taking their composite in $FU\Theta$ (see 2.7ff) and giving the new internal edges obtained by grafting the roots of the right copse to the twigs of the left one the length 1. This makes $W\Theta$ into a theory.

(3.1 a*) In relation (3.1 a) we could replace $t_1*t_2=t_1+t_2-t_1t_2$ by $t_1*t_2=\max(t_1,t_2)$. Unless stated otherwise our results hold for both definitions of t_1*t_2.

Remark 3.2: For the definition of $W\Theta$ we used that the unit internal I with multiplication $*$ is a monoid. We can make the same construction for an arbitrary monoid M with multiplication $*$ and unit e having an idempotent $u\neq e$. (For I with multiplication $*$ this idempotent is $1\in I$). We then give each internal edge a length in M, each root a length in $u*M$ and each twig a length in $M*u$ (hence the trivial trees have lengths in $u*M*u$). The relations (3.1) are the same in the M-version with 0 in (3.1 c) replaced by e. When we compose we give the new internal edge obtained by grafting a twig of length t_1 to a root of length t_2 the length t_1*t_2. Of course, most of our results do not hold for a general M; we have to impose more restrictions.

Proposition 3.3: W : $\mathfrak{Theories} \longrightarrow \mathfrak{Theories}$ is a functor.

Proof: Immediate. ∎

Let W^o be the composite of the free and the forgetful functor $\mathfrak{Theories} \longrightarrow \mathfrak{Eq\ spaces} \longrightarrow \mathfrak{Theories}$, then $W^o\Theta$ is obtained from $FU\Theta$ by imposing the relations (3.1 a,b) forgetting the lengths. Hence we can include $W^o\Theta$ in $W\Theta$ as the subcategory represented by trees whose internal edges all have length 1. We obtain

Proposition 3.4: The inclusion functors $i\Theta : W^o\Theta \subset W\Theta$ define a natural transfomation $i : W^o \longrightarrow W$.

The back adjunctions $\varepsilon\Theta : W^{o}\Theta \longrightarrow \Theta$ extend to a natural transformation $\varepsilon : W \longrightarrow Id_{\mathfrak{Theories}}$ since it is compatible with the relations (3.1).

<u>Definition 3.5</u>: The natural functor $\varepsilon=\varepsilon\Theta : W\Theta \longrightarrow \Theta$ is called the <u>augmentation</u> of Θ. The composite maps $\Theta(\underline{i},k) \xrightarrow[\eta]{} W^{o}\Theta(\underline{i},k) \subset W\Theta(\underline{i},k)$, where η is the front adjunction, are called the <u>standard</u> <u>section</u> of Θ.

<u>Proposition 3.6</u>: If we use (3.1 a*) instead of (3.1 a), the augmentation functor $\varepsilon : W\Theta \longrightarrow \Theta$ is topologically a homotopy equivalence. In fact, there is a fibrewise strong deformation retraction of $W\Theta(\underline{i},k)$ into the standard section, i.e. a strong deformation retraction H_t such that $\varepsilon \circ H_t = \varepsilon$ for all $t \in I$.

<u>Proof</u>: H_t replaces each edge length u by tu, where t runs from 1 to 0. ∎

Relation (3.1 b) shows that for K-coloured split theories Θ over Θ_{cm} or Θ_m with spine $\mathfrak{B}$ we need only consider trees whose vertex labels lie in $\mathfrak{B}$. We deduce that $W\Theta$ is again a split theory over Θ_{cm} respectively Θ_m. If $\mathfrak{B}$ is a PROP the canonical spine $W\mathfrak{B}$ of $W\Theta$ consists of all copses with vertex labels in $\mathfrak{B}$ such that all elements of $\{1,..,n\}$ occur as twig labels in trees with n twigs, subject to the relations (3.1 a, c) and relation (3.1 b), but only for permutations. If $\mathfrak{B}$ is a PRO, the spine $W\mathfrak{B}$ consists of all copses with vertex labels in $\mathfrak{B}$ such that each tree with n twigs has the twig labels 1,2,...,n in clockwise order. Consequently, twig labels may be omitted. The relations are (3.1 a, c), relation (3.1 b) becomes redundant.

So <u>if we refer to relation</u> (3.1 b) <u>in connection with</u> PROPs, <u>we from now on assume that the set operations are permutations</u>. <u>In connection with</u> PROs <u>we will omit it</u>.

2. A FILTRATION OF $W\mathfrak{B}$ FOR PROs AND PROPs

We restrict our attention to the more general case of a PROP. The necessary modifications for PROs are made easily. Just neglect all group actions which will be defined for PROPs in the following.

(3.7) In view of (3.1 c) <u>we assume from now on that for any K-coloured PROP or PRO $\mathfrak{B}$, to which the functor W is applied, each pair</u> $(\mathfrak{B}(k,k),\{id_k\}), k\in K$, is a NDR (cf. Appendix II).

In order to be able to treat the partly homogeneous case simultaneously we take a $(K\times L)$-coloured PROP $\mathfrak{B}$. We consider both $W\mathfrak{B}$ and $HW\mathfrak{B} = H_L W\mathfrak{B}$. Recall that the partly homogeneous case is more general because it includes the inhomogeneous one (take $L=\{*\}$) and also the completely homogeneous one (take $K=\{*\}$). Nevertheless, we start considering the case $W\mathfrak{B}$, because it is easier. The generalization to $HW\mathfrak{B}$ then follows without too much difficulties.

We define the r-skeleton subcategory $W^r\mathfrak{B}$ of $W\mathfrak{B}$ as generated under composition by copses whose trees have at most r internal edges, consider the space M_λ of all trees of a given shape λ in $W\mathfrak{B}$, i.e. the trees of M_λ have the same underlying graphs and the same edge colours (see 2.6). Recall that M_λ has the form

$$M_\lambda = I^r \times \prod_j \mathfrak{B}(\underline{i}_j, k_j)^{m(j)} \times S_n$$

if λ has r internal edges, n twigs and m(j) vertices with labels in $\mathfrak{B}(\underline{i}_j, k_j)$, $k_j \in K\times L$, because a tree of a given shape is specified by its edge lengths, its twig labels which are a permutation of $\{1,..,n\}$, and its vertex labels.

(3.8) An element of M_λ represents a morphism of $W^{r-1}\mathfrak{B}$ iff one of the following conditions hold

(i) Some vertex label is an identity (for then (3.1 a) applies)

(ii) Some internal edge has length 0 (for then (3.1 c) applies)

(iii) Some internal edge has length 1 (for then the tree decomposes)

Let $N_\lambda \subset M_\lambda$ be the subspace of all points satisfying one of these conditions. It remains to account for relation (3.1 b). Let Λ be the set of all tree shapes which can be obtained from λ by an iterated application of (3.1 b). We call Λ the <u>shape orbit</u> of λ. We have a natural group G' acting on $M_\Lambda = \bigcup_{\lambda\in\Lambda} M_\lambda$, which acts on the summand M_λ as follows: The group S_r permutes the coordinates of I^r, the group S_n acts on the set S_n of twig labellings by composition on the right, the group $S_{m(j)}$ permutes the factors of $\mathfrak{B}(\underline{i}_j,k_j)^{m(j)}$, and $(S_q)^{m(j)}$ also acts on $\mathfrak{B}(\underline{i}_j,k_j)^{m(j)}$ if $\underline{i}_j : [q] \longrightarrow K\times L$, by the action of S_q on $\mathfrak{B}(\underline{i}_j,k_j)$ by composition on the right. Let G be the subgroup of G' generated by all elements g which map M_λ into itself and for which the trees g(A) and A are related by a single application of (3.1 b). We call G the <u>symmetry group</u> of the shape λ. The space N_λ is an invariant subspace of M_λ, and the map $N_\lambda \longrightarrow W^{r-1}\mathfrak{B}(\underline{i},k)$ sending trees to their corresponding morphisms factors through a map

$$v_\lambda : N_\lambda/G \longrightarrow W^{r-1}\mathfrak{B}(\underline{i},k)$$

<u>Lemma 3.9</u>: (a) $W^r\mathfrak{B}(\underline{i},k)$ is obtained from $W^{r-1}\mathfrak{B}(\underline{i},k)$ by adjoining spaces M_λ/G relative to N_λ/G with attaching map v_λ, one for each shape orbit of shapes with r internal edges.

(b) $W\mathfrak{B}(\underline{i},k)$ is the colimit (= direct limit) of the $W^r\mathfrak{B}(\underline{i},k)$

(c) If each $\mathfrak{B}(\underline{i},k)$ is Hausdorff, so are $W^r\mathfrak{B}(\underline{i},k)$ and $W\mathfrak{B}(\underline{i},k)$

(d) If $\mathfrak{B}$ is a CW-PROP, so are $W^r\mathfrak{B}$ and $W\mathfrak{B}$

<u>Proof</u>: (a) Since the identities of $\mathfrak{B}$ are closed (3.7), each N_λ is closed in M_λ, and hence $W^{r-1}\mathfrak{B}(\underline{i},k)$ closed in $W^r\mathfrak{B}(\underline{i},k)$. Hence $U \subset W^r\mathfrak{B}(\underline{i},k)$ is closed iff it is closed in $W^{r-1}\mathfrak{B}(\underline{i},k) \cup_{v_\lambda} M_\lambda/G$.

(b) By the argument of (a), $W^r\mathfrak{B}(\underline{i},k)$ is closed in $W\mathfrak{B}(\underline{i},k)$. Given

$U \subset W\mathfrak{B}(\underline{i},k)$ such that $U \cap W^r\mathfrak{B}(\underline{i},k)$ is closed for all r and let V be the set of representing trees of U, then $V \cap M_\lambda$ is closed for all λ. Hence V and therefore U is closed.

(c) It follows from (A 2.3) and (A 2.4) that (M_λ, N_λ) is a G-NDR. Consequently $(M_\lambda/G, N_\lambda/G)$ is a NDR. The result now follows from (A 4.1).

(d) Use that $(M_\lambda/G, N_\lambda/G)$ is a CW-pair and v_λ is skeletal. ∎

The most direct way to construct PROP-functors from $W\mathfrak{B}$ to a PROP $\mathfrak{C}$ is to construct a PROP-Functor from the PROP of copses with edge lengths $\mathfrak{T}$ (see §1) to $\mathfrak{C}$ which factors through the relations (3.1) modified for PROPs. Let M_λ be as above. Composition with permutations π on the right is given by replacing a twig labelling ξ by $\pi^{-1} \circ \xi$. (Recall that composition on the right with π corresponds to composition with the set operation $(\pi^{-1})^*$). This right action of S_n on M_λ commutes with the G-action, so that v_λ actually is a S_n-equivariant map. A PROP-functor $\mathfrak{T} \longrightarrow \mathfrak{C}$ has to be equivariant with respect to the symmetric groups. It induces a PROP-functor $W\mathfrak{B} \longrightarrow \mathfrak{C}$ provided it factors through (3.1 a, c) and the G-actions. To avoid considering the G-action and the S_n-action on M_λ independently we combine the two. We decompose M_λ as

$$M_\lambda = \bigcup_{\xi \in S_n} I^r \times \prod_j \mathfrak{B}(\underline{i}_j, k_j)^{m(j)} \times \xi$$

and we denote the summand associated with ξ by $P_{\lambda,\xi}$. An element $g \in G$ maps $P_{\lambda,\xi}$ onto some $P_{\lambda,\eta}$, and η is of the form $\eta = \xi \cdot \theta(g^{-1})$, where $\theta(g^{-1})$ is a permutation which only depends on g. The correspondence $g \longmapsto \theta(g)$ yields a homomorphism

$$\theta : G \longrightarrow S_n$$

Put $P_\lambda = P_{\lambda,id}$ and define a G-action on P_λ by taking the composite

$$P_\lambda \times G \longrightarrow M_\lambda \longrightarrow P_\lambda$$

whose first map is the G-action on M_λ and whose second map is induced by the homeomorphisms $P_{\lambda,\xi} \longrightarrow P_\lambda$ which forget the S_n-coordinate.

Hence $A \in P_\lambda$ and $g(A) \circ \theta(g) \in P_{\lambda,\theta(g^{-1})}$ are related by iterated applications of (3.1 b).

Let $S_{\underline{i}}$ denote the subgroup of isomorphisms of $\mathfrak{S}_{K\times L}(\underline{i},\underline{i})$. Then $S_{\underline{i}}$ acts on the right of $\mathfrak{B}(\underline{i},k)$ by composition. We note for future use

Lemma 3.10: If each $\mathfrak{B}(\underline{i},k)$ is a free $S_{\underline{i}}$-space, then G acts freely on P_λ. ∎

If $\underline{i}$ is the source of the trees of P_λ and k their target, we can define a G-action on $\mathfrak{C}(\underline{i},k)$ for any PROP $\mathfrak{C}$ by

$$g(a) = a \circ \theta(g^{-1})$$

If we put $Q_{\lambda,\xi} = N_\lambda \cap P_{\lambda,\xi}$ and $Q_\lambda = N_\lambda \cap P_\lambda$, we have a G-equivariant "characteristic" map

$$u_\lambda : (P_\lambda, Q_\lambda) \longrightarrow (W^r\mathfrak{B}(\underline{i},k), W^{r-1}\mathfrak{B}(\underline{i},k))$$

sending trees to their corresponding morphisms. It should be stressed that the G-action on P_λ is not the restriction of the G-action on M_λ. Composing the image of u_λ with all elements of S_n from the right, we account for all morphisms represented by the elements of M_Λ, and as λ runs through a complete set of representatives of shape orbits, we account for all morphisms of $W\mathfrak{B}$.

The treatment of the L-homogeneous case differs only slightly. Let us call k_1 the K-colour and k_2 the L-colour of $k = (k_1,k_2) \in K\times L$. We only consider tree shapes λ for which the twigs have all the same L-colour, because exactly such trees represent morphisms in $HW\mathfrak{B}$. Let M_λ be as before. The elements of M_λ satisfying (3.8) (i) or (ii) represent morphisms in $HW^{r-1}\mathfrak{B}$, but not necessarily those satisfying (3.8) (iii), because the tree might not decompose into representatives of morphisms in $HW\mathfrak{B}$. It does decompose "correctly" if there is a collection of edges of lengths 1 which separates the tree into a copse and a tree whose twigs have the same L-colour. To deal with this phenomenon we refine our filtration. Let $HW^{r,q}\mathfrak{B}$ be the subcategory of

$HW\mathfrak{B}$ generated under composition by copses whose trees represent elements of $HW^{r-1}\mathfrak{B}$ or have exactly r internal edges of which at least r-q have length 1. Note that $HW^{r,o}\mathfrak{B} = HW^{r-1}\mathfrak{B}$ and $HW^{r,r}\mathfrak{B} = HW^{r}\mathfrak{B}$. Consequently, let $M_{\lambda,q}$ be the subspace of those trees of M_λ which have at least r-q edges of lengths 1, $P_{\lambda,\xi,q} = P_{\lambda,\xi} \cap M_{\lambda,q}$ and $P_{\lambda,q} = P_{\lambda,id,q}$.

(3.8*) An element of $M_{\lambda,q}$, $q \geq 0$, represents an element of $HW^{r,q-1}\mathfrak{B}$ (by convention $HW^{r,-1}\mathfrak{B} = HW^{r-1}\mathfrak{B}$ and $HW^{-1}\mathfrak{B}$ contains exactly the permutations) iff one of the following conditions holds:

(i) Some vertex label is an identity

(ii) Some internal edge has length 0

(iii) There is a collection of edges of lengths 1 which separates the tree into a copse and a tree whose twigs all have the same L-colour

(iv) There are more than r-q internal edges of lengths 1.

The first three cases characterize the elements of $M_{\lambda,q}$ representing morphisms in $HW^{r-1}\mathfrak{B}$. Let $Q_{\lambda,q}$ be the set of all elements of $P_{\lambda,q}$ satisfying one of the conditions (3.8*). The G-action on P_λ restricts to a G-action on the pair $(P_{\lambda,q}, Q_{\lambda,q})$, and we again have G-equivariant characteristic maps

$$u_{\lambda,q} : (P_{\lambda,q}, Q_{\lambda,q}) \longrightarrow (HW^{r,q}\mathfrak{B}(\underline{i},k), HW^{r,q-1}\mathfrak{B}(\underline{i},k))$$

An analogue of Lemma 3.9 for the L-homogeneous case can be proved in the same manner.

Let $\mathfrak{C}$ be a topological category with finite products. Let $X_1, X_2, \ldots, X_n \in \text{ob}\,\mathfrak{C}$. A permutation $\pi \in S_n$ defines a map

$$\pi : X_{\pi 1} \times \ldots \times X_{\pi n} \longrightarrow X_1 \times \ldots \times X_n$$

the obvious shuffle corresponding to the set operation $(\pi^{-1})^*$. For each $k \in K$ we take an object X_k of $\mathfrak{C}$ and define $X_{\underline{i}}$ for $\underline{i} = \{i_1, \ldots, i_n\}$ to be the object $X_{i_1} \times \ldots \times X_{i_n}$. We then have a G-action on $\mathfrak{C}(X_{\underline{i}}, Y)$ by $g(a) = a \cdot \theta(g^{-1})$.

<u>Definition 3.11</u>: Let $\mathfrak{C}$ and $\mathfrak{C}'$ be topological categories with finite products, let $\mathfrak{B}$ be a partly homogeneous PROP and $\mathfrak{B}'$ a partly homogeneous PRO. A functor $F : \mathfrak{C} \longrightarrow \mathfrak{C}'$ is called <u>multiplicative</u> if it is continuous and preserves products. A functor $G : \mathfrak{B} \longrightarrow \mathfrak{C}$ is called multiplicative if it is continuous, maps $\oplus$ to the product bifunctor $\times$, and preserves permutations. A functor $H : \mathfrak{B}' \longrightarrow \mathfrak{C}'$ is called multiplicative if it is continuous and carries $\oplus$ to $\times$. The last two cases are equivalent to saying that G is a $\mathfrak{B}$-space and H a $\mathfrak{B}'$-space in $\mathfrak{C}$.

<u>Lemma 3.12</u>: Let $\mathfrak{C}$ be a topological category with finite products and $\mathfrak{B}$ be a $(K\times L)$-coloured PROP as above.

(a) Given a multiplicative functor $F : HW^{r,q-1}\mathfrak{B} \longrightarrow \mathfrak{C}$ and a collection of G-equivariant maps $f_\lambda : P_{\lambda,q} \longrightarrow \mathfrak{C}(F\underline{i}, F(k))$ extending $F \circ (u_{\lambda,q}|Q_{\lambda,q})$, one for each shape orbit of trees with r internal edges, then there is a unique multiplicative functor $F' : HW^{r,q}\mathfrak{B} \longrightarrow \mathfrak{C}$ that extends F and satisfies $F' \circ u_{\lambda,q} = f_\lambda$ for all λ.

(b) Suppose given for each r and each $q>0$ a multiplicative functor $F_{r,q} : HW^{r,q}\mathfrak{B} \longrightarrow \mathfrak{C}$ such that $F_{r,q-1} = F_{r,q}|HW^{r,q-1}\mathfrak{B}$ (here put $F_{r,o} = F_{r-1,r-1}$). Then there exists a unique multiplicative functor $F : HW\mathfrak{B} \longrightarrow \mathfrak{C}$ such that $F|HW^{r,q}\mathfrak{B} = F_{r,q}$ for all r and q.

(c) Both (a) and (b) also hold if we replace $\mathfrak{C}$ by a PROP and the word multiplicative functor by PROP-functor.

<u>Proof</u>:(a) Since F' has to be multiplicative, a representative $A \in P_{\lambda,\xi,q}$ has to be mapped to $f_\lambda(A \circ \xi^{-1})$. This determines a map of the space of all representing trees of $HW^{r,q}\mathfrak{B}$. Since the f_λ are equivariant, extend $F \circ (u_{\lambda,q}|Q_{\lambda,q})$, and each decomposable morphism of $P_{\lambda,q}$ lies in $Q_{\lambda,q}$ this map factors through a functor $HW^{r,q}\mathfrak{B} \longrightarrow \mathfrak{C}$, the required functor F'.

(b) is an immediate consequence of the homogeneous version of (3.7 b).

The proof of (c) uses the same arguments. ∎

Remark: A similar result, with no group actions, holds for PROs.

Definition 3.13: A family of functors $H(t) : \mathfrak{C} \longrightarrow \mathfrak{D}$, $t \in I$, of topological categories is called a homotopy of functors if $H(t)(e)$ is independent of t for all $e \in \operatorname{ob} \mathfrak{C}$ and the functions $\mathfrak{C}(e,e') \times I \longrightarrow \mathfrak{D}(H(0)(e),H(0)(e'))$ given by $(f,t) \longmapsto H(t)(f)$ are continuous.

Let us state a first application of (3.12). Given a subcategory $\mathfrak{D}$ of $HW\mathfrak{B}$, we denote the space of all elements of $P_{\lambda,q}$ which represent a morphism in $\mathfrak{D}$ by $D_{\lambda,q}$. We call $\mathfrak{D}$ an admissible subcategory if each $D_{\lambda,q}$ is closed in $P_{\lambda,q}$ and each pair $(P_{\lambda,q}, Q_{\lambda,q} \cup D_{\lambda,q})$ is a G-NDR and if $a \circ b$ or $a \oplus b$ are in $\mathfrak{D}$ then so are a and b.

Proposition 3.14: Let $\mathfrak{D}$ be an admissible subcategory of $HW\mathfrak{B}$. Suppose given a multiplicative functor F from $HW\mathfrak{B}$ to a topological category $\mathfrak{C}$ with finite products and a homotopy of multiplicative functors $H(t) : \mathfrak{D} \longrightarrow \mathfrak{C}$ such that $H(0) = F|\mathfrak{D}$. Then there exists a homotopy of multiplicative functors $F(t) : HW\mathfrak{B} \longrightarrow \mathfrak{C}$ extending $H(t)$ and F. The same holds if we substitute $\mathfrak{C}$ by a PROP and use PROP-functors.

Proof: Let $F^{-1,q}(t)$ be any homotopy of multiplicative functors from $HW^{-1,q}\mathfrak{B}$ to $\mathfrak{C}$ extending $H(t)|HW^{-1,q}\mathfrak{B} \cap \mathfrak{D}$. Inductively suppose we have defined a homotopy of multiplicative functors $F^{r,q-1}(t) : HW^{r,q-1}\mathfrak{B} \to \mathfrak{C}$ extending the restriction of $H(t)$ to $HW^{r,q-1}\mathfrak{B} \cap \mathfrak{D}$. Using (3.12) we only have to define a homotopy of G-equivariant maps $f_\lambda(t) : P_{\lambda,q} \longrightarrow \mathfrak{C}(F^{r,q-1}(t)(\underline{i}), F^{r,q-1}(t)(k))$ extending $F \circ u_{\lambda,q}$ for $t=0$ and $F^{r,q-1}(t) \circ (u_{\lambda,q}|D_{\lambda,q} \cup Q_{\lambda,q})$. This is possible because $\mathfrak{D}$ is admissible. ∎

3. LIFTING THEOREMS

Let $\mathfrak{B}$ be a (K×L)-coloured PROP as in the previous section. We first show that the augmentation functor $\varepsilon : HW\mathfrak{B} \longrightarrow H\mathfrak{B}$ is a homotopy equivalence. Since $HW\mathfrak{B} \subset W\mathfrak{B}$ and $H\mathfrak{B} \subset \mathfrak{B}$ are full subcategories, this follows from

Proposition 3.15: For each object $\underline{i} : [n] \longrightarrow K\times L$ of $\mathfrak{B}$ and each $k \in K\times L$ the map $\varepsilon : W\mathfrak{B}(\underline{i},k) \longrightarrow \mathfrak{B}(\underline{i},k)$ is a S_n-equivariant homotopy equivalence with the standard section as homotopy inverse. If the identities of $\mathfrak{B}$ are isolated, the S_n-equivariant deformation H_t of $W\mathfrak{B}(\underline{i},k)$ into the standard section can be chosen to be fibrewise.

If we use relation (3.1 a*), the statement follows from the proof of (3.6), but not so if we use (3.1 a). The following proof works for both cases.

Proof: We filter $W\mathfrak{B}(\underline{i},k)$ by the subspaces F_r of morphisms represented by trees with at most r internal edges. An element of M_λ represents a morphism in F_{r-1} iff (3.8)(i) or (ii) holds. Let R_λ be the subspace of M_λ of those elements. We know from (A 2.4) that R_λ is a G-equivariant SDR of M_λ. Hence R_λ/G is a SDR of M_λ/G. The S_n-action on M_λ given by $\pi : P_{\lambda,\xi} \longrightarrow P_{\lambda,\pi^{-1}\cdot\xi}$ makes R_λ/G into an S_n-equivariant SDR of M_λ/G. Since F_r is obtained from F_{r-1} by attaching M_λ/G by the obvious S_n-equivariant map $R_\lambda/G \longrightarrow F_{k-1}$, the space F_{r-1} is an S_n-equivariant SDR of F_r. It follows from (A 4.5) that the standard section F_o is a S_n-equivariant SDR of $W\mathfrak{B}(\underline{i},k)$.

If $\mathfrak{B}$ has isolated identities, we may restrict our attention to the space of those trees which do not have an identity as vertex label, because this space is open and closed in the space of all trees. We then can take the deformation H_t of the proof of (3.6). ∎

Definition 3.16: Let $\mathfrak{C}$ and $\mathfrak{D}$ be topological categories. A continuous functor $F : \mathfrak{C} \longrightarrow \mathfrak{D}$ is called a homotopy equivalence if it is bijective on objects and each $F : \mathfrak{C}(X,Y) \longrightarrow \mathfrak{D}(FX,FY)$ is a homotopy equivalence. If $\mathfrak{C}$ and $\mathfrak{D}$ are PROPs or topological categories with finite products and F a PROP-functor or multiplicative, we call F an equivariant equivalence if it is bijective on objects and each $F : \mathfrak{C}(X,Y) \longrightarrow \mathfrak{D}(FX,FY)$ is an equivariant homotopy equivalence. We call F a fibred homotopy equivalence (equivariant fibred equivalence) if each $F : \mathfrak{C}(X,Y) \longrightarrow \mathfrak{D}(FX,FY)$ has a (equivariant) section and there is a (equivariant) strong deformation retraction $H_t : \mathfrak{C}(X,Y) \longrightarrow \mathfrak{C}(X,Y)$ into the section such that $F \circ H_t = F$ for all $t \in I$.

We use the following theorem to replace naturally occurring PROPs by the artificial bar construction PROP $W\mathfrak{B}$.

Theorem 3.17 (Lifting Theorem): Given a diagram consisting of a $(K\times L)$-coloured PROP $\mathfrak{B}$, L-homogeneous $(K\times L)$-coloured PROPs $\mathfrak{C}$ and $\mathfrak{D}$, an admis-

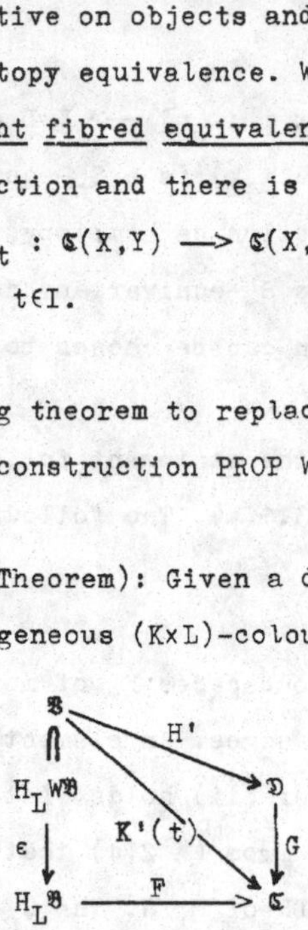

sible subcategory $\mathfrak{B}$ of $H_L W\mathfrak{B}$, PROP-functors F and G, a continuous functor H' and a homotopy of functors $K'(t) : \mathfrak{B} \longrightarrow \mathfrak{C}$ from $F \circ (\varepsilon | \mathfrak{B})$ to $G \circ H'$, both preserving objects, $\oplus$, and permutations. We assume

(i) $G : \mathfrak{D} \longrightarrow \mathfrak{C}$ is an equivariant equivalence for all $\underline{i}$ and k

OR (ii) $G : \mathfrak{D} \longrightarrow \mathfrak{C}$ is a homotopy equivalence and each $\mathfrak{B}(\underline{i},k)$ is a numerable principal $S_{\underline{i}}$-space (see Appendix III) for all $\underline{i}$ and k.

Then:

(A) There exists a PROP-functor $H : H_L W\mathfrak{B} \longrightarrow \mathfrak{D}$ and a homotopy of PROP-functors $K(t) : H_L W\mathfrak{B} \longrightarrow \mathfrak{C}$ from $F \circ \varepsilon$ to $G \circ H$ extending H' and $K'(t)$.

(B) Given two PROP-functors $H_0, H_1 : H_L W\mathfrak{B} \longrightarrow \mathfrak{D}$ and a homotopy of functors $L'(t) : \mathfrak{B} \longrightarrow \mathfrak{D}$ from $H_0|\mathfrak{B}$ to $H_1|\mathfrak{B}$ preserving $\oplus$ and permutations. Further given homotopies of PROP-functors $K_0(t), K_1(t) : H_L W\mathfrak{B} \longrightarrow \mathfrak{C}$ from $F \circ \varepsilon$ to $G \circ H_0$ respectively from $F \circ \varepsilon$ to $G \circ H_1$ and a homotopy of homotopies $K'(t_1,t_2) : \mathfrak{B} \longrightarrow \mathfrak{C}$, $(t_1,t_2) \in I^2$, preserving $\oplus$ and permutations, such that $K'(0,t_2)=K_0(t_2)|\mathfrak{B}, K'(1,t_2)=K_1(t_2)|\mathfrak{B}$, $K'(t_1,0) = F \circ (\varepsilon|\mathfrak{B})$, and $K'(t_1,1) = G \circ L'(t_1)$. Then there exists a homotopy of PROP-functors $L(t) : H_L W\mathfrak{B} \longrightarrow \mathfrak{D}$ from H_0 to H_1 and a homotopy of homotopies of PROP-functors $K(t_1,t_2) : H_L W\mathfrak{B} \longrightarrow \mathfrak{C}$ extending L' and K' and such that $K(t_1,0)=F \circ \varepsilon$ and $K(t_1,1)=L(t_1)$. In particular, H of part (A) is unique up to a homotopy of functors.

Proof: We construct H and $K(t)$ by induction on the skeleton subcategories of $HW\mathfrak{B}$ using Lemma 3.12. Suppose we have defined H and $K(t)$ on $HW^{r,q-1}\mathfrak{B}$. To extend, we need G-equivariant maps $h_{\lambda,q} : P_{\lambda,q} \longrightarrow \mathfrak{D}(\underline{i},k)$ and G-equivariant homotopies $k_\lambda(t) : P_{\lambda,q} \longrightarrow \mathfrak{C}(\underline{i},k)$, already given on $Q_{\lambda,q} \cup V_{\lambda,q}$ and satisfying $k_\lambda(0)=F \circ \varepsilon \circ u_{\lambda,q}$ and $k_\lambda(1)=G \circ h_\lambda$, one for each shape λ of a complete set of shape orbits of trees in $HW^r\mathfrak{B}$. These maps are provided by Theorem A 3.5. To be able to apply the second part of this theorem, we have to verify that $P_{\lambda,q}$ is a numerable principle G-space. Let $\mathfrak{U}(\underline{i},k) = \{U_r\}$ be a $S_{\underline{i}}$-numerable open covering of $\mathfrak{B}(\underline{i},k)$ with numeration $\lambda_{U_r} : \mathfrak{B}(\underline{i},k) \longrightarrow I$. Recall that $P_{\lambda,q} = I^r_q \times \prod_l \mathfrak{B}(\underline{i}_l,k_l) \times \{id\}$, where $I^r_q \subset I^r$ is the subspace of all points with at least r-q coordinates of value 1. The G-numerable cover $\mathfrak{W}$ of $P_{\lambda,q}$ consists of the open sets $W = I^r_q \times \prod_l U_l \times \{id\}$ with $U_l \in \mathfrak{U}(\underline{i}_l,k_l)$ and has the numeration

$$\lambda_W : I_q^r \times \prod_l \mathfrak{B}(\underline{i}_l,k_l) \times \{id\} \longrightarrow I$$
$$(\underline{t},b_1,\ldots,b_n,id) \longmapsto \lambda_{U_1}(b_1)\cdot\ldots\cdot\lambda_{U_n}(b_n)$$

If $gW \cap W = \emptyset$ for $g \in G$, $W \in \mathfrak{M}$, and $\lambda_{gW}(gx) = \lambda_W(x)$, $x \in W$, then $P_{\lambda,q}$ is a numerable principal G-space by (A 3.2). An element $g \in G$ permutes some of the coordinates of I_q^r and some of the factors $\prod_l \mathfrak{B}(\underline{i}_l,k_l)$. Moreover, there is at least one factor $\mathfrak{B}(\underline{i},k)$ which is kept fixed in itself under g but changed by a permutation $\pi \in S_{\underline{i}}$. Since $\mathfrak{B}(\underline{i},k)$ is a numerable principal $S_{\underline{i}}$-space $(U \cdot \pi^*) \cap U = \emptyset$ for $U \in \mathfrak{U}(\underline{i},k)$. Hence $gW \cap W = \emptyset$. Since λ_W is defined factorwise, we obviously have $\lambda_{gW}(gx) = \lambda_W(x)$.

The proof of part (B) is analogous. Just replace the pair $(P_{\lambda,q}, Q_{\lambda,q} \cup V_{\lambda,q})$ by the product $(P_{\lambda,q}, Q_{\lambda,q} \cup V_{\lambda,q}) \times (I,\partial I)$ and observe that $h_\lambda : P_{\lambda,q} \times I \longrightarrow \mathfrak{D}(\underline{i},k)$ and the homotopy $k_\lambda(t) : P_{\lambda,q} \times I \longrightarrow \mathfrak{C}(\underline{i},k)$ are already given on $P_{\lambda,q} \times \partial I \cup (Q_{\lambda,q} \cup V_{\lambda,q}) \times I$ and that $k_\lambda(0)(x,t)$ has to be $F \cdot \varepsilon \cdot u_{\lambda,q}(x)$ for all $t \in I$. ∎

<u>Remark 3.18</u>: The theorem is still true, by the same proof, if we replace $\mathfrak{C}$ and $\mathfrak{D}$ by topological categories with finite products having the same objects and all PROP-functors by multiplicative functors. Then, in addition, we have to assume that $G : \mathfrak{D} \longrightarrow \mathfrak{C}$ preserves objects.

<u>Remark 3.19</u>: By the same proof we can actually show a slight generalization of the lifting theorem, which has some practical value. If we only assume that $G : \mathfrak{D}(\underline{i},k) \longrightarrow \mathfrak{C}(\underline{i},k)$, $\underline{i} : [n] \longrightarrow K \times L$, is an equivariant homotopy equivalence for $n \leq r$, or an ordinary homotopy equivalence and each $\mathfrak{B}(\underline{i},k)$ is a numerable principal $S_{\underline{i}}$-space for $n \leq r$, while all the other assumptions are kept, we can "extend" H' and K'(t) over the PROP-subcategory $Q_L^r W\mathfrak{B}$ of $H_L W\mathfrak{B}$ generated by the morphisms of $H_L W\mathfrak{B}(\underline{i},k)$, $\underline{i} : [n] \longrightarrow K \times L$, with $n \leq r$.

The results (3.17), (3.18) without group actions hold for PROs.

One would obviously like to have $G \circ H = F \circ \varepsilon$. This is in general not possible, but under additional assumptions on F and G we can achieve this.

<u>Theorem 3.20</u>: Given a commutative diagram with a $(K \times L)$-coloured PROP $\mathfrak{B}$, L-homogeneous $(K \times L)$-coloured PROPs $\mathfrak{C}$ and $\mathfrak{D}$, an admissible subcate-

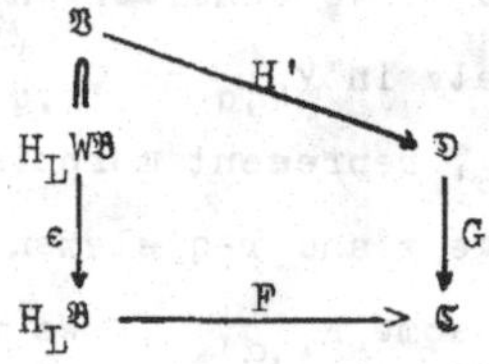

gory $\mathfrak{B}$ of $HW\mathfrak{B}$, PROP-functors F and G, and a continuous functor H' preserving objects, $\oplus$, and permutations. We assume

(i) G is an equivariant fibred equivalence

(ii) each $id_k \in \mathfrak{B}(k,k)$, $k \in K \times L$, has a closed neighbourhood X_k such that $(X_k, \{id_k\} \cup \text{fr } X_k)$ is a NDR and $F(X_k) = \{id_k\} \subset \mathfrak{C}(k,k)$. (fr= frontier in $\mathfrak{B}(k,k)$)

<u>Then</u>:

(A) There exists a PROP-functor $H : H_L W\mathfrak{B} \longrightarrow \mathfrak{D}$ extending H' such that $G \circ H = F \circ \varepsilon$

(B) Given two PROP-functors $H_0, H_1 : H_L W\mathfrak{B} \longrightarrow \mathfrak{D}$ and a homotopy of PROP-functors $K'(t) : \mathfrak{B} \longrightarrow \mathfrak{D}$ from $H_0|\mathfrak{B}$ to $H_1|\mathfrak{B}$ such that $G \circ H_0 = F \circ \varepsilon = G \circ H_1$ and $G \circ K'(t) = F \circ (\varepsilon|\mathfrak{B})$, there exists an extension $K(t): H_L W\mathfrak{B} \longrightarrow \mathfrak{D}$ of K'(t) from H_0 to H_1 such that $G \circ K(t) = F \circ \varepsilon$. In particular, H of part (A) is unique up to a homotopy of PROP-functors.

<u>Proof</u>: For the proof of the theorem a filtration different from the skeleton filtration seems to be more suited: Let $Y_k = X_k - (\{id_k\} \cup \text{fr } X_k)$. Let $F_{p,q}$ be the subcategory of $H_L W\mathfrak{B}$ generated under composition by

copses whose trees represent elements in $\mathfrak{B}$ or have r internal edges of which at least r-q have length 1 and t vertices with labels in the Y_k's, $r + t \le p$. Since $F_{p,q}$ is a closed sub-PROP, it is easy to check that $HW\mathfrak{B}$ is the direct limit of the $F_{p,q}$. We define F_{-1} to be the sub-PROP generated by $\mathfrak{B}$ and the identities of $HW\mathfrak{B}$. Let λ be a tree shape, α a collection of t vertices of λ whose labels lie in the $\mathfrak{B}(k,k)$'s, and β a collection of r-q internal edges. Let $R_{\lambda,\alpha,\beta}$ be the subspace of all points of $P_{\lambda,q}$ for which each vertex of α has a label in some X_k and each edge of β has length 1. We consider only those spaces $R_{\lambda,\alpha,\beta}$ which do not lie completely in $V_{\lambda,q}$ or $Q_{\lambda,q}$ and observe that the elements of $R_{\lambda,\alpha,\beta} \cap R_{\lambda,\alpha',\beta'}$ represent morphisms of some lower filtration if α' and β' also have t and r-q elements. An element of the group G may map $R_{\lambda,\alpha,\beta}$ onto some $R_{\lambda,\alpha',\beta'}$. We take one space in each orbit of spaces under G. Let G' be the subgroup of G whose elements map the collections α and β into themselves. The space $R_{\lambda,\alpha,\beta}$ is of the form $I^q \times X \times Z$, where I^q specifies the lengths of the internal edges not in β, X is an α-indexed product of spaces X_k, and Z is the space of the remaining vertex labels of λ. Then $A \in R_{\lambda,\alpha,\beta}$ represents an element of lower filtration iff $A \in R'_{\lambda,\alpha,\beta} = \partial I^q \times X \times Z \cup I^q \times Y \times Z$, where Y is the (closed) subspace of those points of X with at least one coordinate in some $\{id_k\} \cup \operatorname{fr} X_k$. Note that, the action of G' on $R_{\lambda,\alpha,\beta}$ permutes the coordinates of I^q and X but does not change them, in contrary to the coordinates of Z. The functor H to be constructed is defined already on F_{-1}. Similarly to the proof of (3.17) we inductively have to construct G'-equivariant maps $f = f_{\lambda,\alpha,\beta} : R_{\lambda,\alpha,\beta} \longrightarrow \mathfrak{D}(\underline{i},k)$ already given on $R'_{\lambda,\alpha,\beta}$ such that $G \cdot f = F \cdot \varepsilon \cdot (u_{\lambda,q} | R_{\lambda,\alpha,\beta})$. For part (B) we have to define G'-equivariant maps $h : R_{\lambda,\alpha,\beta} \times I \longrightarrow \mathfrak{D}(\underline{i},k)$ already given on $R_{\lambda,\alpha,\beta} \times \partial I \cup R'_{\lambda,\alpha,\beta} \times I$ such that $G \cdot h(x,t) = F \cdot \varepsilon \cdot (u_{\lambda,q} | R_{\lambda,\alpha,\beta})(x)$. Both maps are provided by our next lemma. ∎

<u>Lemma 3.21</u>: Suppose (X,A) is a G-NDR, B,Y,Z are G-spaces, G operates

on I^n by permuting factors and

$$\begin{array}{ccccccc} A \times B \times I^n \cup X \times B \times \partial I^n & \subset & X \times B \times I^n & \xrightarrow{q} & X \times B & \xrightarrow{q'} & B \\ \downarrow f & & & & \downarrow g & \swarrow g' & \\ Y & & & \xrightarrow{p} & Z & & \end{array}$$

is a commutative diagram of G-equivariant maps with the diagonal action on the products and q,q' projections. Suppose there is an equivariant section s of p and an equivariant homotopy $H : id_Y \simeq s \circ p$ such that $p \circ H_t = p$ for all $t \in I$. Then there exists an extension $h : X \times B \times I^n \longrightarrow Y$ of f such that $p \circ h = g \circ q$.

Proof: Define an equivariant map $F' : (A \times B \times I^n \cup X \times B \times \partial I^n) \times I \longrightarrow Y$ by $F'(x,b,u,t) = H_t(f(x,b,u))$. Then $F'(x,b,u,1) = s \circ p \circ f(x,b,u)$ is independent of u. Hence F' factors through an equivariant map $F : A \times B \times CI^n \cup X \times B \times C\partial I^n \longrightarrow Y$ where C denotes the unreduced cone functor. Let $i : I^n \subset CI^n$ denote the standard inclusion. Define a map $k : I^n \times I \longrightarrow CI^n$ as follows: Each point of $I^n \times I$ lies in a unique line segment from $z = (\frac{1}{2}, \frac{1}{2}, \ldots, \frac{1}{2}, 1)$ to a point $(x,t) \in I^n \times 0 \cup \partial I^n \times I$. Map z to the cone point and (x,t) to $i(x)$ and the rest of line segment linearly (here we identify CI^n with the join of $I^n \times 0$ and $(\frac{1}{2}, \ldots, \frac{1}{2}, 1)$ in $\mathbb{R}^{n+1}$). Let

$$K = F \circ (id \times k) : A \times B \times I^n \times I \cup X \times B \times \partial I^n \times I \cup X \times B \times I^n \times 1 \longrightarrow Y.$$

Since k is symmetric in the coordinates of I^n and F is equivariant, so is K. Since $p \circ F'(x,b,u,t) = p \circ f(x,b,u) = g(x,b)$ and $k(\partial I^n \times I \cup I^n \times 1) = C\partial I^n$, we have $p \circ K(x,b,u,t) = g(x,b)$. Furthermore $K(x,b,u,0) = F(x,b,u,0) = f(x,b,u)$.

Since (X,A) and (I^n, ∂, I^n) are equivariant NDRs, there exists an equivariant retraction $r' : X \times I^n \times I \longrightarrow (A \times I^n \cup X \times \partial I^n) \times I \cup X \times I^n \times 1$, which we extend to an equivariant retraction

$$r : X \times B \times I^n \times I \longrightarrow A \times B \times I^n \times I \cup X \times B \times \partial I^n \times I \cup X \times B \times I^n \times 1$$

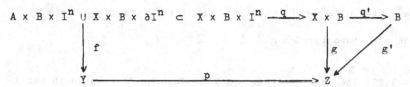

by taking the identity on the factor B. Define $h : X \times B \times I^n \longrightarrow Y$ by $h(x,b,u) = K \circ r(x,b,u,0)$. Let $r'(x,u,0) = (x',u',t)$. Then h extends f, is equivariant and

$$p \circ h(x,b,u) = p \circ K(x',b,u',t) = g(x',b) = g(x,b) = g \circ q(x,b,u)$$

because g factors through g'. ∎

Remark 3.22: The condition (iii) on F in Theorem 3.20 holds in particular if $\mathfrak{B}$ has isolated identities.

Condition (3.20 (iii)) is actually no serious hindrance if we allow to change $\mathfrak{B}$ a bit. Let $\mathfrak{B}'$ be the following PROP: $\mathfrak{B}'(\underline{i},k) = \mathfrak{B}(\underline{i},k)$ if $\underline{i} \neq k$ and $\mathfrak{B}'(k,k) = \mathfrak{B}(k,k) \cup I/\sim$ where $\mathfrak{B}(k,k) \ni id_k \sim 1 \in I$. Let $\circ_{\mathfrak{B}}$ and $\oplus_{\mathfrak{B}}$ denote composition and $\oplus$ in $\mathfrak{B}$. Composition on the right with permutations is the same as in $\mathfrak{B}$. Further, if $b \notin I \subset \mathfrak{B}'(k,k)$ for any k, and a_i are morphisms into basic objects of $\mathfrak{B}'$, we define $b \circ (a_1 \oplus \dots \oplus a_n) = b \circ_{\mathfrak{B}} (a_1' \oplus_{\mathfrak{B}} \dots \oplus_{\mathfrak{B}} a_n')$ with $a_i' = id_k$, if $a_i \in I \subset \mathfrak{B}'(k,k)$, and $a_i' = a_i$ otherwise. If $b = t \in I \subset \mathfrak{B}'(k,k)$, we define $b \circ a = t_* u$ (see(3.1)) if $a = u \in I \subset \mathfrak{B}(k,k)$ and $b \circ a = a$ otherwise. This determines the PROP $\mathfrak{B}'$ completely. There is a PROP functor $\varepsilon' : \mathfrak{B}' \longrightarrow \mathfrak{B}$ given by $\varepsilon'(a) = id_k$ if $a \in I \subset \mathfrak{B}'(k,k)$ and $\varepsilon'(a) = a$ otherwise. The correspondence $\iota'(a) = a$ defines an equivariant non-functorial section $\iota' : \mathfrak{B} \longrightarrow \mathfrak{B}'$, and by shrinking the attached whiskers we obtain an equivariant fibrewise deformation of $\mathfrak{B}'(\underline{i},k)$ into the section.

Given a diagram of PROPs as in (3.20) with the difference that $\mathfrak{B}$ is an admissible subcategory of $H_L W\mathfrak{B}'$ and the condition (iii) is dropped, there exists a PROP-functor $H : HW\mathfrak{B}' \longrightarrow \mathfrak{D}$ extending H' such that $G \circ H = F \circ \bar{\varepsilon}$, where $\bar{\varepsilon}$ is the composite $\varepsilon' \circ \varepsilon(\mathfrak{B}') : HW\mathfrak{B}' \longrightarrow H\mathfrak{B}' \rightarrow H\mathfrak{B}$. The analogue to (3.20 B) also holds. The reason for this is that the composite functor $F \circ \varepsilon'$ satisfies requirement (iii) and (3.20) can be applied with $\mathfrak{B}$ replaced by $\mathfrak{B}'$ and F by $F \circ \varepsilon'$. We should remark that these considerations remain true even if $(\mathfrak{B}(k,k), \{id_k\})$ is not a NDR.

$H_L W\mathfrak{B}'$ also supplies an example that (3.20) is not true if we drop condition (iii). Consequently we cannot expect to obtain commutativity in Theorem 3.17. Suppose we have used relation (3.1 a*) for the definition of $W\mathfrak{B}'$. Then $\varepsilon(\mathfrak{B}') : HW\mathfrak{B}' \longrightarrow H\mathfrak{B}'$ is equivariantly fibre homotopically trivial by (3.6). If condition (iii) of Theorem 3.20 could be dropped, we would have a commutative diagram of PROP-functors

$$\begin{array}{ccc}
H_L W\mathfrak{B} & \xrightarrow{\quad H \quad} & H_L W\mathfrak{B}' \\
 & & \downarrow \varepsilon(\mathfrak{B}') \\
\varepsilon(\mathfrak{B}) \downarrow & & H_L \mathfrak{B}' \\
 & & \downarrow \varepsilon' \\
H_L \mathfrak{B} & \xrightarrow{\quad Id \quad} & H_L \mathfrak{B}
\end{array}$$

by (3.20 A). This is in general impossible by following consideration: Let $\iota(\mathfrak{B}), \iota(\mathfrak{B}')$ denote the standard sections. Then $\varepsilon(\mathfrak{B}') \circ H \circ \iota(\mathfrak{B})$ defines a section of ε' which preserves identities. Since $\varepsilon' | \mathfrak{B}'(k,k)$ is the identity outside the attached whisker and since $0 \in I \subset \mathfrak{B}'(k,k)$ is the new identity in $\mathfrak{B}'$, this section can only be continuous if the identities of $\mathfrak{B}$ are isolated (in which case, of course, (3.20 (iii)) holds).

A more pictorial description of a $W\mathfrak{B}$-action on an object $X \in \mathfrak{Top}_K$ is sometimes useful. Rather than give maps from $W\mathfrak{B}(\underline{i},k)$ to $\mathfrak{Top}(X_{\underline{i}},X_k)$ we consider the maps $W\mathfrak{B}(\underline{i},k) \times X_{\underline{i}} \longrightarrow X_k$ using the full adjointness in the category of k-spaces.

<u>Definition 3.23</u>: A <u>cherry tree</u> on $X \in \mathfrak{Top}_K$ is a representing tree of a morphism in $W\mathfrak{B}$ with a point of X_k instead of a twig label assigned to each twig of colour K. We call this point a <u>cherry</u>.

The set of all cherry trees has an obvious topology: Let $T(\underline{i},k)$, $\underline{i} : [n] \longrightarrow K$, denote the space of all representing trees of $W\mathfrak{B}(\underline{i},k)$. Then the space of all cherry trees is the disjoint union of all spaces

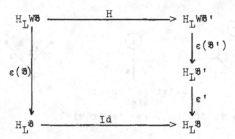

$$T\mathfrak{B}X_k = \bigcup_{\underline{i}\in\mathfrak{B}} T(\underline{i},k) \times X_{\underline{i}}/\sim \quad , \; k\in K$$

with $(A \bullet \sigma^*, x) = (A, \sigma^*(x))$ where $A \in T(\underline{i},k)$, $\sigma \in S_{\underline{i}}$, and $x \in X_{\underline{i}}$.

Examples:

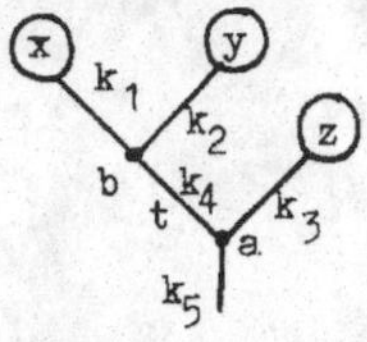

a trivial cherry tree

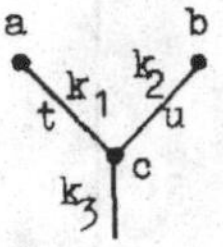

no cherries because stumps

The proof of the following lemma is trivial.

Lemma 3.24: Let $\mathfrak{B}$ be a K-coloured PROP and $X \in \mathfrak{Top}_K$. Then X admits a $W\mathfrak{B}$-action iff there are continuous functions $F_k : T\mathfrak{B}X_k \longrightarrow X_k$ factoring through the following relations

(a) = (3.1 a) for cherry trees

(b) = (3.1 b) for cherry trees and permutations only, the cherries are permuted with the twigs.

(c) = (3.1 c)

(d) = $F_k\Big($  $\Big) = x, \quad x\in X_k$

(e) if the tree A with root colour k has an edge of length 1 coloured l so that $A = A_1 \circ A_2$, then $F_k(A;x_1,\ldots,x_n) = F_k(A_1;x_1,\ldots,x_p,y,x_{q+1},\ldots,x_n)$, where $y = F_l(A_2;x_{p+1},\ldots,x_q)$, and $(x_{p+1},\ldots,x_q)$ and $(x_1,\ldots,x_n)$ are the cherries of A_2 and A in clockwise order.

The relations (a),(b),(c) substitute the relations (3.1) for trees, relation (d) implies that identities act as identities, and relation (e) ensures that composite operations are preserved. ∎

We close this chapter with an application of the lifting theorem.

Proposition 3.25: (a) The loop space ΩY can be made a $W\mathfrak{A}$-space, naturally in Y, i.e. a map $f : Y \longrightarrow Z$ gives rise to a $W\mathfrak{A}$-homomorphism $\Omega f : \Omega Y \longrightarrow \Omega Z$.

(b) An A_n-space, $2\le n\le\infty$, is a $Q^n W\mathfrak{A}$-space (see (3.19)). Note that $Q^\infty W\mathfrak{A} = W\mathfrak{A}$.

(c) An A_n-map, $2\le n\le\infty$, between monoids X and Y is a $Q^n_{\mathfrak{L}_1} W(\mathfrak{A}\otimes\mathfrak{L}_1)$-space, which extends the $W\mathfrak{A}$-action induced on X and Y by $\varepsilon(\mathfrak{A}) : W\mathfrak{A} \longrightarrow \mathfrak{A}$ (recall, a monoid is an $\mathfrak{A}$-space).

These results are an immediate consequence of the results of chapter II,§6 and 7, the lifting theorem and Remark 3.19.

Remark 3.26: The usual loop space $\Omega Y = \mathfrak{Top}((I,\partial I),(Y,*))$ is not a monoid, but the functor Ω preserves product. J.C. Moore modified the definition of a loop space in order to obtain a monoid structure. Moore's loop space $\Omega_M Y$ is the space of all pairs $(\omega,a)\in Y^{\mathbb{R}}\times\mathbb{R}$ with $a\ge 0$ and $\omega : \mathbb{R}\longrightarrow Y$ a map satisfying $\omega(t) = *$, the base point of Y, for $t\le 0$ or $t\ge a$. (As usual, $Y^{\mathbb{R}} = \mathfrak{Top}(\mathbb{R},Y)$ with the function space topology). $\Omega_M Y$ is a monoid under the multiplication defined by $(\omega_1,a_1)\cdot(\omega_2,a_2) = (\omega,a_1+a_2)$, where

$$\omega(t) = \begin{cases} \omega_1(t) & \text{if } 0\le t\le a_1 \\ \omega_2(t-a_1) & \text{if } a_1\le t\le a_2 \end{cases}$$

The usual loop space ΩY is a deformation retract of $\Omega_M Y$. A deformation $H_s : \Omega_M Y \longrightarrow \Omega_M Y$ is given by

$$H_s(\omega,a) = \begin{cases} (\omega,a+s-sa) & a\le 1 \\ (\omega_s,a+s-sa) & a\ge 1 \end{cases}$$

with $\omega_s(t) = \omega(at/(a+s-as))$. The functor Ω_M has in contrary to Ω the disadvantage that it does not preserve products. It is easy to see that <u>no loop space functor</u> L, i.e. <u>a functor</u> L <u>such that</u> $LY \simeq \Omega Y$ <u>for all</u> $Y\in \text{ob}\,\mathfrak{Top}$, <u>can preserve products and be monoid-valued</u>: For otherwise LLY would admit an action of $\Theta_m\otimes\Theta_m \cong \Theta_{cm}$, and a result of Dold and Thom [17; Satz 7.1] asserts that any path connected commutative monoid has the weak homotopy type of a product of Eilenberg-MacLane spaces, which is obviously not the case in general for $\Omega^2 Y$.

We should remark that we are not able to prove an analogue of the lifting theorem for arbitrary theories, because the set operations induced by epimorphisms mess up the skeleton filtration. This is the main reason why we restrict our attention to PROPs and PROs.

IV Chapter

HOMOTOPY HOMOMORPHISMS

1. MAPS BETWEEN W$\mathfrak{B}$-SPACES

Let $\mathfrak{B}$ be a K-coloured PROP. Homomorphisms as maps between W$\mathfrak{B}$-spaces will not do, because if we change the collection of underlying maps $f_k : X_k \longrightarrow Y_k$ of a homomorphism by a homotopy to maps g_k, the g_k do not define a homomorphism in general. We have already seen that a W$\mathfrak{B}$-structure is a $\mathfrak{B}$-structure up to homotopy and all coherence conditions, because the relations of $\mathfrak{B}$ hold in W$\mathfrak{B}$ up to homotopy and the morphism spaces of W$\mathfrak{B}$ have the same homotopy type as those of $\mathfrak{B}$. Similarly we can substitute a homomorphism by a homomorphism up to homotopy and all coherence conditions. Since a $\mathfrak{B} \otimes \mathfrak{L}_1$ action defines a homomorphism of $\mathfrak{B}$-spaces, the construction W suggests to take a W($\mathfrak{B} \otimes \mathfrak{L}_1$)-action on (X,Y) extending the given W$\mathfrak{B}$-actions on X and Y as maps between W$\mathfrak{B}$-spaces. Before we give a rigorous definition, we have to make some notational conventions:

We denote an element b in the standard section of W$\mathfrak{B}$ by its counter image in $\mathfrak{B}$. But we note that the standard section is <u>not</u> a functor and that $\mathfrak{B}$ is <u>not</u> a subcategory of W$\mathfrak{B}$.

Recall that $\mathfrak{L}_n$ is the category with objects 0,1,...,n and exactly one morphism from p to q if p≤q and none otherwise. If c is the morphism from 0 to 1 in $\mathfrak{L}_1$, we denote the morphism $id_k \otimes c \in W(\mathfrak{B} \otimes \mathfrak{L}_1)((k,0),(k,1))$ by j_k. If $\mathfrak{B}$ is monochrome we may drop the index k. The inclusion functors $W\mathfrak{B} \subset W(\mathfrak{B} \otimes \mathfrak{L}_1)$ replacing vertex labels b by $b \otimes id_0$ respectively

$b \otimes id_1$ will be denoted by d^1 respectively d^o.

A $W\mathfrak{B}$-space will from now on be signified by a pair (X,α), where $\alpha : W\mathfrak{B} \longrightarrow \mathfrak{Top}$ is a $W\mathfrak{B}$-space with $X \in \mathfrak{Top}_K$ as underlying space.

Definition 4.1: Let (X,α) and (Y,β) be $W\mathfrak{B}$-spaces. A homotopy homomorphism, for short a $\mathfrak{B}$-map, from (X,α) to (Y,β) is a $W(\mathfrak{B} \otimes \mathfrak{L}_1)$-space $\rho : W(\mathfrak{B} \otimes \mathfrak{L}_1) \longrightarrow \mathfrak{Top}$ such that $\rho \circ d^o = \beta$ and $\rho \circ d^1 = \alpha$. The morphism $\rho(j) = \{\rho(j_k) | k \in K\}$ in $\mathfrak{Top}_K$ is called the underlying map or carrier of ρ. We write a $\mathfrak{B}$-map as pair (f,ρ) where f is the underlying map of ρ.

The definition of a $\mathfrak{B}$-map can be modified in several ways. Our definition allows operations $X \times Y \longrightarrow Y$ and one could argue that we are not really interested in mixed products nor in factorizations of morphisms through them. In chapter II, §7, we have already indicated that partly homogeneous categories are the adequate tool for this modification. Let $HW(\mathfrak{B} \otimes \mathfrak{L}_1) = H_{\mathfrak{L}_1} W(\mathfrak{B} \otimes \mathfrak{L}_1)$.

Definition 4.2: Let (X,α) and (Y,β) be $W\mathfrak{B}$-spaces. A homogeneous homotopy homomorphism, for short a $h\mathfrak{B}$-map, from (X,α) to (Y,β) is a pair (f,ρ) where $\rho : HW(\mathfrak{B} \otimes \mathfrak{L}_1) \longrightarrow \mathfrak{Top}$ is a $HW(\mathfrak{B} \otimes \mathfrak{L}_1)$-space and $f = \{\rho(j_k) | k \in K\} \in \text{mor}\, \mathfrak{Top}_K$ such that $\rho \circ d^o = \beta$ and $\rho \circ d^1 = \alpha$.

Compared to $W(\mathfrak{B} \otimes \mathfrak{L}_1)$, the category $HW(\mathfrak{B} \otimes \mathfrak{L}_1)$ has several drawbacks as we have already mentioned in (II, §7) and (III, §2). Nevertheless, the lifting theorem proves that they are manageable. The main objection one could have is that $HW(\mathfrak{B} \otimes \mathfrak{L}_1)$ is artificial. For example, let 2° denote the object $[2] \longrightarrow \{*\} \times \{0,1\}$ whose image is $(*,0)$ and 1^1 the object $[1] \longrightarrow \{*\} \times \{0,1\}$ whose image is $(*,1)$, then disregarding 0-ary operations the space $HW(\mathfrak{A} \otimes \mathfrak{L}_1)(2^o, 1^1)$ is

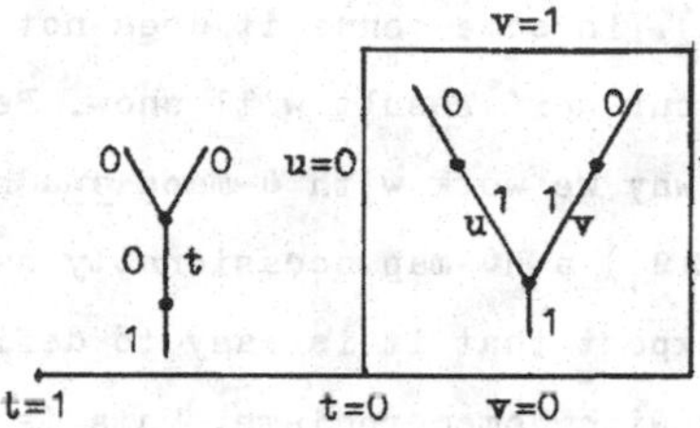

The vertex $t = 1$ represents the operation $(x,y) \longmapsto f(xy)$ and the vertex $u = v = 1$ the operation $(x,y) \longmapsto f(x)f(y)$. One would expect a copy of I divided in the middle instead. We obtain a copy I from our model by restricting the square to its diagonal. This leads to a third definition of maps. We define a subcategory $LW(\mathfrak{B} \otimes \mathfrak{L}_1)$ of $HW(\mathfrak{B} \otimes \mathfrak{L}_1)$ such that $b \in LW(\mathfrak{B} \otimes \mathfrak{L}_1)$ decomposes in $LW(\mathfrak{B} \otimes \mathfrak{L}_1)$ iff it decomposes in $W(\mathfrak{B} \otimes \mathfrak{L}_1)$- so it does not have one of the drawbacks of $HW(\mathfrak{B} \otimes \mathfrak{L}_1)$ - and moreover the morphism spaces are the intuitively correct ones. Although $LW(\mathfrak{B} \otimes \mathfrak{L}_1)$ is manageable it is too complicated to work with.

<u>Definition 4.3</u>: (inductive) We call a tree in $HW(\mathfrak{B} \otimes \mathfrak{L}_1)$ <u>level</u> if it has no or one vertex, or if the following holds: When we stretch the tree by dividing all its edge lengths by the length of its longest edge, so as to produce some edges of length 1 and a decomposition of the tree in $W(\mathfrak{B} \otimes \mathfrak{L}_1)$, we require this decomposition to be a decomposition into level trees. Let $LW(\mathfrak{B} \otimes \mathfrak{L}_1)$ be the subcategory of $HW(\mathfrak{B} \otimes \mathfrak{L}_1)$ generated by level trees.

Note in particular that the decomposition has to be a decomposition into trees of $HW(\mathfrak{B} \otimes \mathfrak{L}_1)$. It follows that the t-edge and the diagonal of the square are the only level trees in the space of the previous example. We now can define a third type of map between $W\mathfrak{B}$-spaces using $LW(\mathfrak{B} \otimes \mathfrak{L}_1)$ instead of $HW(\mathfrak{B} \otimes \mathfrak{L}_1)$.

Since $LW(\mathfrak{B} \otimes \mathfrak{L}_1)$ is too complicated, we will only consider maps as defined in (4.1) and (4.2). In some sense it does not matter which sort of maps we take, as our next result will show. Before we state it, let us give a reason why we work with $\mathfrak{B}$-maps <u>and</u> $h\mathfrak{B}$-maps. In spite of the drawbacks of $HW(\mathfrak{B}\otimes\mathfrak{L}_1)$ a $h\mathfrak{B}$-map occasionally has advantages over a $\mathfrak{B}$-map. One would expect that it is easy to define the composite of a $\mathfrak{B}$-map or $h\mathfrak{B}$-map with homomorphisms. This is not quite true for $\mathfrak{B}$-maps.

<u>Definition 4.4:</u> Let $f:(X,\alpha) \longrightarrow (Y,\beta)$ and $h:(Z,\gamma) \longrightarrow (W,\delta)$ be homomorphisms of $W\mathfrak{B}$-spaces. Let $(g,\rho):(Y,\beta) \longrightarrow (Z,\gamma)$ be a $\mathfrak{B}$-map, and $(p,\pi):(Y,\beta) \longrightarrow (Z,\gamma)$ a $h\mathfrak{B}$-map. Define composites $(g\circ f,\overline{\rho})=(g,\rho)\circ f$, $(p\circ f,\overline{\pi}) = (p,\pi)\circ f$, and $(h\circ p,\pi') = h\circ(p,\pi)$ as follows: Let $a \in W(\mathfrak{B}\otimes\mathfrak{L}_1)(\underline{i},k)$ and $b \in H_{\mathfrak{L}_1} W(\mathfrak{B}\otimes\mathfrak{L}_1)(\underline{j},k)$. Then $\overline{\rho}(a), \overline{\pi}(b)$, and $\pi'(b)$ are given by α,β,γ, or δ unless k has $\mathfrak{L}_1$-colour 1, and at least one $i_r \in \underline{i} = \{i_1,\ldots,i_n\}$ or each $j_r \in \underline{j} = \{j_1,\ldots,j_n\}$ has $\mathfrak{L}_1$-colour 0. In this case we define

$$\overline{\rho}(a) = \rho(a)\circ(f_1'\times\ldots\times f_n')$$

$$\overline{\pi}(b) = \pi(b)\circ f^n$$

$$\pi'(b) = h\circ\pi(b)$$

where $f_r' = id_Z$ respectively f if i_r has $\mathfrak{L}_1$-colour 1 or 0. We call the so-defined composites the <u>canonical composites of a $\mathfrak{B}$-map or $h\mathfrak{B}$-map with a homomorphism</u>.

<u>Remark 4.5</u>: We cannot in general define a composite $h\circ(g,\rho) = (h\circ g,\rho')$, for let $\rho(a): Y\times Z \longrightarrow Z$ be the action of a particular $a \in W(\mathfrak{B}\otimes\mathfrak{L}_1)$ under ρ. Then $\rho'(a)$ is a map $Y\times W \longrightarrow W$ which we cannot obtain from h and $\rho(a)$. All we get is that the required map $\rho'(a)$ has to make the square

$$\begin{array}{ccc} X\times Y & \xrightarrow{\ \rho(a)\ } & Y \\ \downarrow{\scriptstyle id\times h} & & \downarrow{\scriptstyle h} \\ X\times Z & \xrightarrow[\ \rho'(a)\ ]{} & Z \end{array}$$

commute. So we say that $(h \cdot g, \rho')$ is a <u>canonical</u> <u>composite</u> $h \cdot (g,\rho)$ if the following holds: Let $a \in W(\mathfrak{B} \otimes \mathfrak{L}_1)(\underline{i}, l)$ where l has $\mathfrak{L}_1$-colour 1, let $y_r = z_r \in X_k$ if $\underline{i}(r) = (k,0) \in K \times \text{ob } \mathfrak{L}_1$, and $y_r \in Y_k$ and $z_r = h(y_r)$ if $\underline{i}(r) = (k,1)$. Then

$$\rho'(a)(z_1,\ldots,z_n) = h(\rho(a)(y_1,\ldots,y_n))$$

<u>Proposition 4.6</u>: Given a map of $W\mathfrak{B}$-spaces $f : (X,\alpha) \longrightarrow (Y,\beta)$. Then

(a) f admits a $\mathfrak{B}$-map structure iff it admits a $h\mathfrak{B}$-map structure

(b) f admits a $h\mathfrak{B}$-map structure iff it admits a level-tree map structure, at least if the category $LW(\mathfrak{B} \otimes \mathfrak{L}_1)$ is constructed by using relation (3.1 a*).

<u>Proof</u>: $LW(\mathfrak{B} \otimes \mathfrak{L}_1) \subset HW(\mathfrak{B} \otimes \mathfrak{L}_1) \subset W(\mathfrak{B} \otimes \mathfrak{L}_1)$. Hence if f admits a $\mathfrak{B}$-map structure it admits a $h\mathfrak{B}$-map structure, which implies that it admits a level-tree map structure.
Conversely given a functor $\rho : LW(\mathfrak{B} \otimes \mathfrak{L}_1) \longrightarrow \mathfrak{Top}$ such that $\rho \cdot d^1 = \alpha$ and $\rho \cdot d^0 = \beta$. The restriction of the deformation H_t of (3.6) stays inside $LW(\mathfrak{B} \otimes \mathfrak{L}_1)$. Therefore we can apply the lifting theorem with $G = \varepsilon : LW(\mathfrak{B} \otimes \mathfrak{L}_1) \longrightarrow H(\mathfrak{B} \otimes \mathfrak{L}_1)$, $\mathfrak{B} = LW(\mathfrak{B} \otimes \mathfrak{L}_1)$ and H' the identity. Hence there exists a retraction functor $R : HW(\mathfrak{B} \otimes \mathfrak{L}_1) \longrightarrow LW(\mathfrak{B} \otimes \mathfrak{L}_1)$ and therefore an extension of ρ to $HW(\mathfrak{B} \otimes \mathfrak{L}_1)$.
Now suppose $(f,\rho) : (X,\alpha) \longrightarrow (Y,\beta)$ is a $h\mathfrak{B}$-map. Let $\mathfrak{D}$ be the subcategory of $W(\mathfrak{B} \otimes \mathfrak{L}_1)$ generated under composition and $\oplus$ by the morphisms of $HW(\mathfrak{B} \otimes \mathfrak{L}_1)$. Then ρ extends to an action $\overline{\rho} : \mathfrak{D} \longrightarrow \mathfrak{Top}$. Let $\varepsilon : \mathfrak{D} \longrightarrow \mathfrak{B} \otimes \mathfrak{L}_1$ be the restriction of the augmentation $W(\mathfrak{B} \otimes \mathfrak{L}_1) \to \mathfrak{B} \otimes \mathfrak{L}_1$. We show that each morphism space $\mathfrak{D}(\underline{i},k)$, $\underline{i} = \{i_1,\ldots,i_n\}$ can be deformed equivariantly into a suitable section of $\varepsilon | \mathfrak{D}(\underline{i},k)$. If all $i_r \in \underline{i}$ and k have the same $\mathfrak{L}_1$-colour the deformation is given by Prop. 3.15. Suppose that k has $\mathfrak{L}_1$-colour 1 and at least one i_r the $\mathfrak{L}_1$-colour 0. Then each representing tree of an element of $\mathfrak{D}(\underline{i},k)$ can be decomposed in $\mathfrak{D}$ into $A \cdot (C_1 \oplus \ldots \oplus C_n) \cdot \xi^*$, where A is a tree whose

twigs all have $\mathfrak{L}_1$-colour 1 (A may be the identity), the twigs of each tree C_i all have the same $\mathfrak{L}_1$-colour 0 or 1, and ξ is a permutation. The deformation is defined in steps. We first shrink all edges of $\mathfrak{L}_1$-colour 0 using the deformation of Prop. 3.15. We end in the space T of all morphisms of $\mathfrak{D}(\underline{i},k)$ representable by trees whose <u>internal</u> edges all have $\mathfrak{L}_1$-colour 1. The next step replaces each twig of colour $(k,0)$, $k \in K$, by

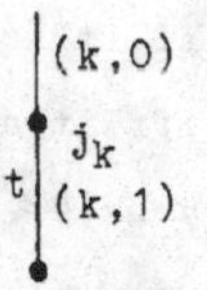

where t runs from 0 to 1. Finally we shrink all internal edges which are not the outgoing edge of a vertex with label j_k, **using the** deformation of (3.15). All three deformations stay inside $\mathfrak{D}(\underline{i},k)$. The composite deformation deforms $\mathfrak{D}(\underline{i},k)$ equivariantly into a subspace which is mapped homeomorphically onto $\mathfrak{B} \otimes \mathfrak{L}_1(\underline{i},k)$ by ε. We now apply the inhomogeneous lifting theorem with $\mathfrak{B}$ generated by $d^1 W\mathfrak{B} \cup d^0 W\mathfrak{B} \cup \{j_k | k \in K\}$ and H' the inclusion. (We cannot take $\mathfrak{B} = \mathfrak{D}$ because an indecomposable $a \in \operatorname{mor} \mathfrak{D}$ could be decomposable in $W(\mathfrak{B} \otimes \mathfrak{L}_1)$). If $H : W(\mathfrak{B} \otimes \mathfrak{L}_1) \longrightarrow \mathfrak{D}$ is an extension of H', then $\bar{\rho} \circ H$ defines a $\mathfrak{B}$-map structure on f. ∎

2. COMPOSITION AND THE HOMOTOPY CATEGORY

Unfortunately we cannot take $\mathfrak{B}$-maps or h$\mathfrak{B}$-maps as morphisms in a category for lack of a definition of the composite of two morphisms, unless we are in the situation of Def. 4.4. The phenomenon is seen at its simplest in the case of h$\mathfrak{M}$-maps $f : X \longrightarrow Y$ and $g : Y \longrightarrow Z$ between monoids. There we are given homotopies $H : f \circ \lambda_2 \simeq \lambda_2 \circ (f \oplus f)$ and $K : g \circ \lambda_2 \simeq \lambda_2 \circ (g \oplus g)$. We deduce that $h \circ \lambda_2 \simeq \lambda_2 \circ (h \oplus h)$, where $h = g \circ f$, but not by any homotopy that is going to make composition associative.

Instead of a category we can define a simplicial class having $W\mathfrak{B}$-spaces are vertices and $\mathfrak{B}$-maps respectively $h\mathfrak{B}$-maps as 1-simplexes.

Observe that a functor $\mathfrak{L}_n \longrightarrow \mathfrak{L}_m$ is uniquely determined by an order-preserving map of the sets of objects $\{0,1,\ldots,n\} \longrightarrow \{0,1,\ldots,m\}$. Let $\delta_n^i : \mathfrak{L}_{n-1} \longrightarrow \mathfrak{L}_n$ and $\sigma_n^i : \mathfrak{L}_{n+1} \longrightarrow \mathfrak{L}_n$, $i = 0,1,\ldots,n$, be the functors corresponding to the maps

$$\{0,1,\ldots,n-1\} \ni j \longmapsto \begin{cases} j & j<i \\ j+1 & j\geq i \end{cases} \in \{0,1,\ldots,n\}$$

respectively

$$\{0,1,\ldots,n+1\} \ni j \longmapsto \begin{cases} j & j\leq i \\ j-1 & j>1 \end{cases} \in \{0,\ldots,n\}$$

Definition 4.7: We define simplicial classes $\mathfrak{K}_{\mathfrak{B}}$ and $\mathfrak{K}_{h\mathfrak{B}}$ by taking as n-simplexes all $W(\mathfrak{B}\otimes\mathfrak{L}_n)$-spaces $W(\mathfrak{B}\otimes\mathfrak{L}_n) \longrightarrow \mathfrak{Top}$ respectively all $H_{\mathfrak{L}_n}W(\mathfrak{B}\otimes\mathfrak{L}_n)$-spaces $H_{\mathfrak{L}_n}W(\mathfrak{B}\otimes\mathfrak{L}_n) \longrightarrow \mathfrak{Top}$. The face and degeneracy operations d^i and s^i in $\mathfrak{K}_{\mathfrak{B}}$ and $\mathfrak{K}_{h\mathfrak{B}}$ are given by $d_n^i(\alpha) = \alpha \circ W(\mathrm{Id}\otimes\delta_n^i)$ and $s_n^i(\alpha) = \alpha \circ W(\mathrm{Id}\otimes\sigma_n^i)$.

Recall that d^i and s^i have to satisfy following identities which follow from the dual formulae for δ_n^i and σ_n^i:

$$d^id^j = d^{j-1}d^i \qquad i<j$$

$$s^is^j = s^{j+1}s^i \qquad i\leq j$$

$$d^is^j = \begin{cases} s^{j-1}d^i & i<j \\ \mathrm{id} & i,i-1=j \\ s^jd^{i-1} & i>j+1 \end{cases}$$

In order to simplify the notation we give an alternative description of the representing trees of $W(\mathfrak{B}\otimes\mathfrak{L}_n)$. In general, a tree is given by its underlying graph and its twig and vertex labels, because the vertex labels determine the edge colours (we identify the trivial trees with the trees having exactly one vertex and an identity as vertex label). In our case, each vertex label has the form $a \otimes b$ and $b \in \mathfrak{L}_n$ is uniquely determined by its source and target. So if we specify the $\mathfrak{L}_n$-colours of the incoming and outgoing edges of this vertex and use only the $\mathfrak{B}$-part a of $a \otimes b$ as vertex label we can recover the original vertex label and hence the original tree. We use this new description, which works for all categories $\mathfrak{B} \otimes \mathfrak{C}$, where $\mathfrak{C}$ is a discrete topological category with at most one morphism between any two objects. Given a $W(\mathfrak{B} \otimes \mathfrak{C})$-space ρ we also frequently colour tree edges by the underlying space $\rho((k,c))$ instead of (k,c).

Definition 4.8: A simplicial class $\mathfrak{R}$ satisfies the restricted Kan condition if given (n-1)-simplexes $x_0, x_1, \ldots, x_{r-1}, x_{r+1}, \ldots, x_n$, where $0<r<n$, such that $d^{j-1}x_i = d^i x_j$ for $0 \le i < j \le n$, $i,j \neq r$, then there exists an n-simplex x such that $d^i x = x_i$ for $i \neq r$. In other words, $\mathfrak{R}$ satisfies the usual Kan extension condition, except that the omitted face in the data is not allowed to be the first or the last.

Our next result implies that the simplicial classes $\mathfrak{R}_{\mathfrak{B}}$ and $\mathfrak{R}_{n\mathfrak{B}}$ are good substitutes for categories.

Theorem 4.9: The simplicial classes $\mathfrak{R}_{\mathfrak{B}}$ and $\mathfrak{R}_{n\mathfrak{B}}$ satisfy the restricted Kan condition.

Proof: Since the argument is the same in both cases with exception that we use $H_{\mathfrak{L}_n} W(\mathfrak{B} \otimes \mathfrak{L}_n)$ for $\mathfrak{R}_{n\mathfrak{B}}$ instead of $W(\mathfrak{B} \otimes \mathfrak{L}_n)$, we only prove the statement for $\mathfrak{R}_{\mathfrak{B}}$. Suppose we are given (n-1)-simplexes $\rho_0, \ldots, \rho_{r-1}, \rho_{r+1}, \ldots, \rho_n : W(\mathfrak{B} \otimes \mathfrak{L}_{n-1}) \longrightarrow \mathfrak{Top}$ for $r \neq 0,n$ such that $d^{j-1}\rho_i = d^i \rho_j$ for $0 \le i < j \le n$. Let $\mathfrak{C}$ be the subcategory of $W(\mathfrak{B} \otimes \mathfrak{L}_n)$ generated under composition and $\oplus$ by the "faces" $d^i W(\mathfrak{B} \otimes \mathfrak{L}_n)$, $i \neq k$, the images of $W(\mathfrak{B}\otimes\mathfrak{L}_{n-1})$ under $W(Id\otimes\delta^i_n)$. So $d^iW(\mathfrak{B}\otimes\mathfrak{L}_n)$ is the subcategory of trees containing no edge coloured i. Define a multiplicative functor $\overline{\rho} : \mathfrak{C} \longrightarrow \mathfrak{Top}$ by $\overline{\rho}|d^iW(\mathfrak{B}\otimes\mathfrak{L}_n) = \rho_i$. (This is possible since $W(Id\otimes\delta^i_n)$ is an inclusion). We show that there is a multiplicative retraction functor $W(\mathfrak{B}\otimes\mathfrak{L}_n) \longrightarrow \mathfrak{C}$.

Consider the pairs of spaces $(W(\mathfrak{B}\otimes\mathfrak{L}_n)(\underline{i},k),\mathfrak{C}(\underline{i},k))$. We deform $W(\mathfrak{B}\otimes\mathfrak{L}_n)(\underline{i},k)$ equivariantly into $\mathfrak{C}(\underline{i},k)$ in steps. We first shrink all internal edges coloured 0 by the deformation of Prop. 3.15. We next replace each twig coloured 0 by

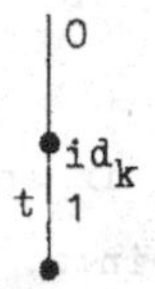

(k is the $\mathfrak{B}$-colour of the twig determined by the vertex label at its bottom) where t runs from 0 to 1 (compare the proof of 4.6). At the end of this deformation the tree can be decomposed into a tree with no edge of colour 0 and a copse with no edge of colour n and hence represents a morphism of $\mathfrak{C}$. Therefore the composite homotopy deforms $W(\mathfrak{B}\otimes\mathfrak{L}_n)(\underline{i},k)$ into $\mathfrak{C}(\underline{i},k)$, keeping $\mathfrak{C}(\underline{i},k)$ inside $\mathfrak{C}(\underline{i},k)$. So the inclusion functor $i : \mathfrak{C} \subset W(\mathfrak{B}\otimes\mathfrak{L}_n)$ is an equivariant equivalence and hence, by Prop. 3.15, the composite $\epsilon \cdot i : \mathfrak{C} \longrightarrow \mathfrak{B}\otimes\mathfrak{L}_n$, too. We now apply the lifting theorem (3.17) with $\mathfrak{B} = \mathfrak{C}$ and H' the identity to obtain a retraction functor $R : W(\mathfrak{B}\otimes\mathfrak{L}_n) \longrightarrow \mathfrak{C}$. The n-simplex $\rho=\overline{\rho}\cdot R$ satisfies $d^i(\rho) = \rho_i$, $i \neq r$, as desired. ∎

Theorem 4.9 provides us with all we need. Given $\mathfrak{B}$-maps or $h\mathfrak{B}$-maps $(f,\rho) : (X,\alpha) \longrightarrow (Y,\beta)$ and $(g,\varkappa) : (Y,\beta) \longrightarrow (Z,\gamma)$ there is a 2-simplex $\sigma : W(\mathfrak{B}\otimes\mathfrak{L}_2) \longrightarrow \mathfrak{Top}$ respectively $\sigma : H_{\mathfrak{L}_2}W(\mathfrak{B}\otimes\mathfrak{L}_2) \longrightarrow \mathfrak{Top}$ such that $d^0(\sigma) = \varkappa$ and $d^2(\sigma) = \rho$. The third edge $d^1(\sigma) : W(\mathfrak{B}\otimes\mathfrak{L}_1) \longrightarrow \mathfrak{Top}$ respectively $H_{\mathfrak{L}_1}W(\mathfrak{B}\otimes\mathfrak{L}_1) \longrightarrow \mathfrak{Top}$, which is some $\mathfrak{B}$-map or $h\mathfrak{B}$-map $(h,\pi) : (X,\alpha) \longrightarrow (Y,\beta)$, is called <u>a composite of</u> f <u>and</u> g. Of course, this composite need not be unique.

Definition 4.10: Given any simplicial class $\mathfrak{K}$ satisfying the restricted Kan condition, we call two edges f and g in $\mathfrak{K}$ homotopic and write $f \simeq g$, if there is a 2-simplex σ with $d^2(\sigma) = f$, $d^1(\sigma) = g$ and $d^0(\sigma)$ is degenerate. This implies in particular that $d^i(f) = d^i(g)$, i=0,1.

From now on we assume that any simplicial class we consider satisfies the condition that any collection of edges with given end points forms a set.

Lemma 4.11: The notion of homotopy is an equivalence relation on the set of all edges with given end points.

Proposition 4.12: Let $\mathfrak{K}$ be a simplicial class satisfying the restricted Kan condition. Then there is a category, the fundamental category of $\mathfrak{K}$, which has the vertices of $\mathfrak{K}$ as objects and the homotopy classes of edges f with $d^0 f = y$ and $d^1 f = x$ as morphisms from x to y.

The proofs of these results are fairly standard (compare the theory of the fundamental groupoid of a Kan complex). We include them for the sake of completeness.

Proof of 4.11:
(a) Given an edge f, then there are 2-simplexes σ and τ such that $d^0(\sigma)$ and

d=degenerate

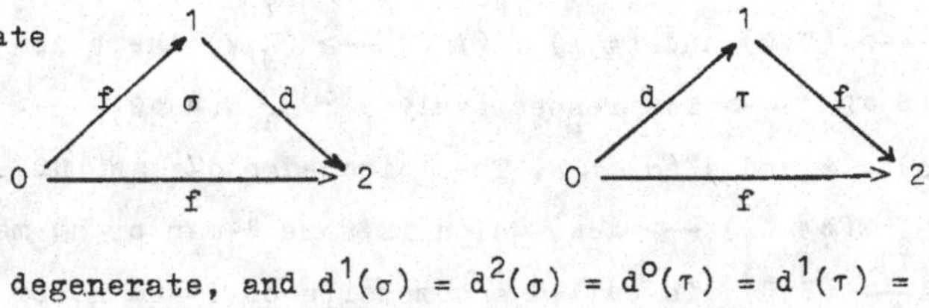

$d^2(\tau)$ are degenerate, and $d^1(\sigma) = d^2(\sigma) = d^0(\tau) = d^1(\tau) = f$, namely

$\sigma = s^1(f)$ and $\tau = s^0(f)$. Hence $f \simeq f$. (For the diagrams note that the face d^i is opposite the vertex i).

(b) Suppose $f \simeq g$. By assumption and part (a), the faces d^0, d^2, and d^3 of the following 3-simplex are given.

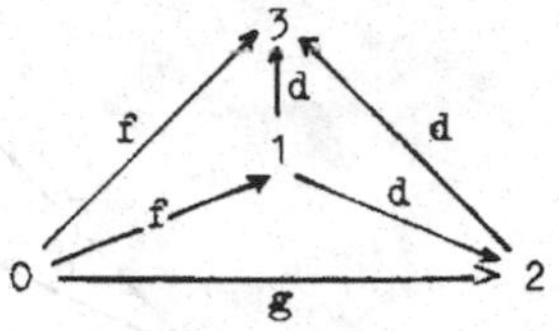

Hence we can fill in the 3-simplex by the extension condition, and the face d^1 provides a homotopy $g \simeq f$.

(c) Suppose $f \simeq g \simeq h$. By assumption and part (a), the faces d^0, d^1, d^3 of the following 3-simplex are given

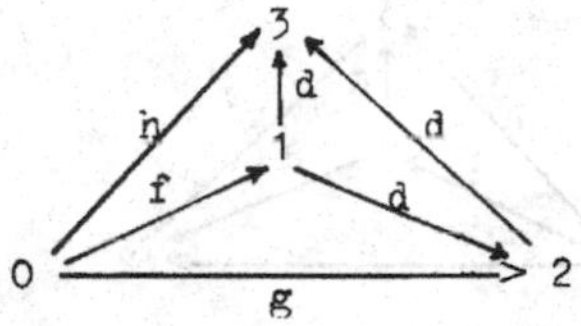

Hence we can fill in the 3-simplex by the extension condition, and the face d^2 provides a homotopy $f \simeq h$. ∎

<u>Proof of 4.12</u>: We define composition in the same way as for $\mathfrak{B}$-maps. Given edges $f : x \longrightarrow y$ and $g : y \longrightarrow z$, there exists a 2-simplex σ with $d^0\sigma = g$ and $d^2\sigma = f$. We call $d^1\sigma = h : x \longrightarrow z$ <u>a</u> <u>composite</u> of f and g.

(a) If h and k are composites of f and g, then $h \simeq k$: The faces d^0, d^2, and d^3 of the following 3-simplex are given, the last two by the assumption that h and k are composites of f and g.

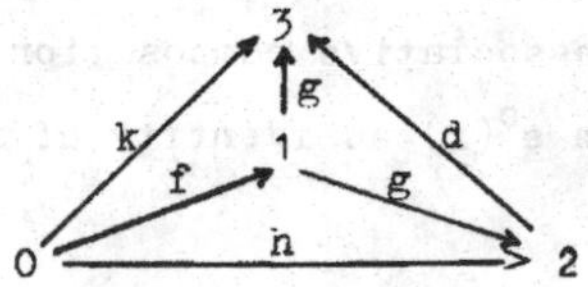

Hence we can fill in the 3-simplex, and d^1 defines a homotopy $h \simeq k$.

(b) $f, g : x \longrightarrow y$ are homotopic iff there exists a 2-simplex σ with $d^1(\sigma) = g$, $d^0(\sigma) = f$, and $d^2(\sigma)$ generate: One way the faces d^0, d^1, d^3 of the first simplex, the other way the faces d^0, d^2, d^3 of the second simplex are given.

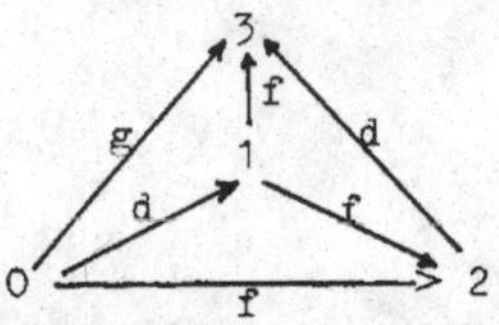

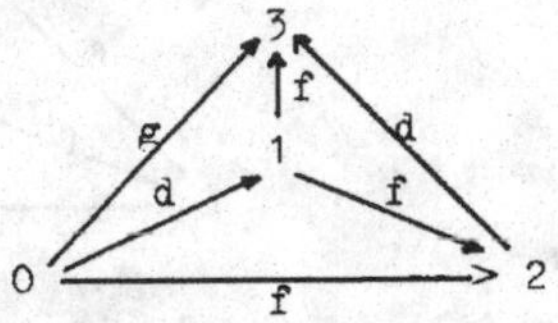

We fill in, and the faces d^2 respectively d^1 give the required result.

(c) Given $f, f' : x \longrightarrow y$ and $g, g' : y \longrightarrow z$ such that $f \simeq f'$ and $g \simeq g'$. Then $g \circ f \simeq g' \circ f'$: Consider

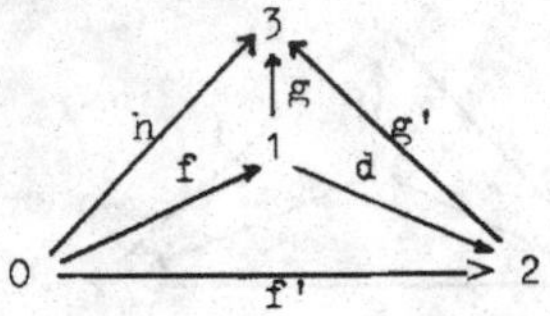

The faces d^0 and d^3 are given by assumption and part (b). If h is a composite of f and g, then d^2 is given, we can fill in and find, that h is a composite of f' and g'.

(d) $h \circ (g \circ f) \simeq (h \circ g) \circ f$ if defined: The faces d^0, d^1, and d^3 of the following simplex are given.

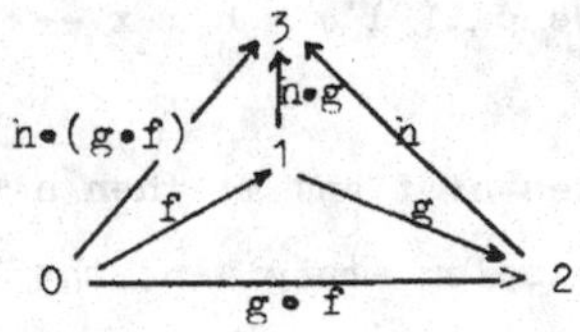

We fill in and find that $h \circ (g \circ f)$ serves as composite of f with $h \circ g$.

So we have defined an associative composition of the homotopy classes of edges of $\mathfrak{K}$ with $s^0(x)$ as identity of x. ∎

We denote the fundamental categories of $\mathfrak{K}_{\mathfrak{B}}$ and $\mathfrak{K}_{h\mathfrak{B}}$ by $\mathfrak{Map}_{\mathfrak{B}}$ respectively $\mathfrak{Map}_{h\mathfrak{B}}$. Their objects are $W\mathfrak{B}$-spaces and their morphisms homotopy classes, in the simplicial sense, of $\mathfrak{B}$-maps respectively $h\mathfrak{B}$-maps. There is a more obvious definition of homotopy, which in the case of A_∞-maps can be found in the literature (e.g. [21]). We could call two $\mathfrak{B}$-maps [$h\mathfrak{B}$-maps] $(f,\rho),(g,\varkappa) : (X,\alpha) \longrightarrow (Y,\beta)$ homotopic if there is a homotopy through $\mathfrak{B}$-maps [$h\mathfrak{B}$-maps] from ρ to $\varkappa$. The next result shows that the two notions coincide.

<u>Lemma 4.13</u>: Two $\mathfrak{B}$-maps [$h\mathfrak{B}$-maps] $(f,\rho),(g,\varkappa) : (X,\alpha) \longrightarrow (Y,\beta)$ are homotopic in the simplicial sense iff there is a homotopy through $\mathfrak{B}$-maps [$h\mathfrak{B}$-maps] $(h_t,H_t) : (X,\alpha) \longrightarrow (Y,\beta)$ with $H_o = \rho$ and $H_1 = \varkappa$.

<u>Proof</u>: Again we only prove the $\mathfrak{B}$-map case because the proof for $h\mathfrak{B}$ is completely analogous.

Suppose there is a homotopy of multiplicative functors H_t as stated. Let $\mathfrak{D}$ be the subcategory of $W(\mathfrak{B}\otimes\mathfrak{L}_2)$ generated by the faces $d^1W(\mathfrak{B}\otimes\mathfrak{L}_2)$. We identify $d^iW(\mathfrak{B}\otimes\mathfrak{L}_2)$ with $W(\mathfrak{B}\otimes\mathfrak{L}_1)$ using $W(\mathrm{Id}\otimes\sigma^j)$, where $j = 0$ if $i = 0$ and $j = 1$ if $i = 1,2$. Define a homotopy of multiplicative functors $K_t : \mathfrak{D} \longrightarrow \mathfrak{Top}$ by

$$K_t|d^oW(\mathfrak{B}\otimes\mathfrak{L}_2) = \rho : W(\mathfrak{B}\otimes\mathfrak{L}_1) \longrightarrow \mathfrak{Top}$$

$$K_t|d^1W(\mathfrak{B}\otimes\mathfrak{L}_2) = H_t : W(\mathfrak{B}\otimes\mathfrak{L}_1) \longrightarrow \mathfrak{Top}$$

$$K_t|d^2W(\mathfrak{B}\otimes\mathfrak{L}_2) = s^o(\alpha) : W(\mathfrak{B}\otimes\mathfrak{L}_1) \longrightarrow \mathfrak{Top}$$

The functor $F = s^o(\rho) : W(\mathfrak{B}\otimes\mathfrak{L}_2) \longrightarrow \mathfrak{Top}$ extends K_o. By (3.14), there exists a homotopy of multiplicative functors $F_t : W(\mathfrak{B}\otimes\mathfrak{L}_2) \longrightarrow \mathfrak{Top}$ extending K_t and F. By part (b) of the proof of (4.12), the functor $F_1 : W(\mathfrak{B}\otimes\mathfrak{L}_2) \longrightarrow \mathfrak{Top}$ provides the required simplicial homotopy.

Conversely, suppose $(f,\rho) \simeq (g,\varkappa)$. Then there is an action $\sigma : W(\mathfrak{B}\otimes\mathfrak{L}_2) \longrightarrow \mathfrak{Top}$ such that $d^o(\sigma) = \rho$, $d^1(\sigma) = \varkappa$, and $d^2(\sigma) = s^o(\alpha)$. Let $\mathfrak{C}$ be the quotient category of $W(\mathfrak{B}\otimes\mathfrak{L}_2)$ by following additional

relation on the trees:
A tree A whose root has $\mathfrak{L}_2$-colour 1 is related to $A' \circ (a_1 \oplus \ldots \oplus a_n)$, where A' is obtained from A by changing all edge colours to 1, $a_i = id_{(k,1)}$ if the i-th twig of A has colour (k,1), and

$$a_i = j_k =$$

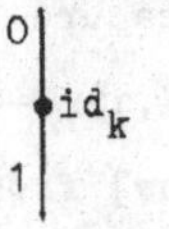

if the i-th twig of A has colour (k,0).

Let $\mathfrak{D}$ be the full sub-PROP of $\mathfrak{C}$ consisting of all objects $\underline{i} = \{i_1,\ldots,i_r\}$ such that the $\mathfrak{L}_2$-colour of each i_q is 0 or 2. Since $d^2(\sigma) = s^0(\alpha)$, the action σ factors through $\mathfrak{C}$ and hence induces an action $\tau : \mathfrak{D} \longrightarrow \mathfrak{Top}$. The functor $s^0 : W(\mathfrak{B} \otimes \mathfrak{L}_2) \longrightarrow W(\mathfrak{B} \otimes \mathfrak{L}_1)$ also factors through $\mathfrak{C}$ and induces a functor $\pi : \mathfrak{D} \longrightarrow W(\mathfrak{B} \otimes \mathfrak{L}_1)$. Define two functors H_o, $H_1 : W(\mathfrak{B} \otimes \mathfrak{L}_1) \longrightarrow \mathfrak{D}$. The functor H_o is induced by $d^1 : W(\mathfrak{B} \otimes \mathfrak{L}_1) \longrightarrow W(\mathfrak{B} \otimes \mathfrak{L}_2)$. The functor H_1 maps trees whose edges have all colour 0 or all colour 1 by d^1 too. Now let A be a tree with root colour 1 and at least one twig of colour 0. Then H_1 maps A to $A' \circ (a_1 \oplus \ldots \oplus a_n)$ where A' is obtained from A by changing edge colours 0 to 1 and 1 to 2 and where

$$a_i = \begin{cases} id_{(k,2)} & \text{if the i-th twig of A has colour } (k,1) \\ j_k & \text{if the " " " " " " } (k,0) \end{cases}$$

Because of the additional relation H_1 is a functor. Note that $\tau \circ H_o = \sigma \circ d^1 = \varkappa$ and $\tau \circ H_1 = \sigma \circ d^0 = \rho$. If $\overline{\varepsilon} = \varepsilon \circ \pi : \mathfrak{D} \longrightarrow W(\mathfrak{B} \otimes \mathfrak{L}_1) \longrightarrow \mathfrak{B} \otimes \mathfrak{L}_1$, then $\overline{\varepsilon} \circ H_o = \overline{\varepsilon} \circ H_1 = \varepsilon$, because $\pi \circ H_o = id = \pi \circ H_1$. Provided that $\overline{\varepsilon}$ is an equivariant equivalence, we can apply part (B) of the lifting theorem (3.17) with $\mathfrak{B} = d^0 W(\mathfrak{B} \otimes \mathfrak{L}_1) \cup d^1 W(\mathfrak{B} \otimes \mathfrak{L}_1)$ and $L'(t) = H_o | \mathfrak{B}$ to obtain a homotopy of PROP-functors $H_t : W(\mathfrak{B} \otimes \mathfrak{L}_1) \longrightarrow \mathfrak{D}$ from H_o to H_1 with $H_t | \mathfrak{B} = H_o(\mathfrak{B})$. Then $\tau \circ H_t$ is the required homotopy through $\mathfrak{B}$-maps from (f,ρ) to $(g,\varkappa)$.

To show that $\overline{\varepsilon}$ is an equivariant equivalence, we deform each space

$\mathfrak{D}(\underline{i},k)$ equivariantly into a suitable section of $\overline{\varepsilon}$. If k has the $\mathfrak{L}_2$-colour 0, then each $i_r \in \underline{i} = \{i_1,\ldots,i_n\}$ has $\mathfrak{L}_2$-colour 0 and $\mathfrak{D}(\underline{i},k) \cong W(\mathfrak{B} \otimes \mathfrak{L}_1)(\underline{i},k)$ with $\overline{\varepsilon} = \varepsilon$. Hence (3.15) provides the deformation. Suppose k has $\mathfrak{L}_2$-colour 2, then we first shrink all internal edges of colour 2 using the deformation of (3.15). We then replace each incoming edge of colour l, l = 0,1

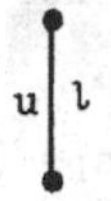

u = length

of the root vertex by

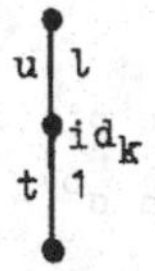

where t runs from 0 to 1. Because of the additional relation we end in the space of all trees whose internal edges all have colour 1. Now shrink all internal edges of these trees using the deformation of (3.15). ∎

3. HOMOTOPY INVARIANCE AND HOMOTOPY EQUIVALENCES

We first show that admitting the structure of a $\mathfrak{B}$-map or $h\mathfrak{B}$-map is an invariance of the homotopy class.

<u>Proposition 4.14:</u> Let $(f,\rho) : (X,\alpha) \longrightarrow (Y,\beta)$ be a $\mathfrak{B}$-map [$h\mathfrak{B}$-map] and $g : X \longrightarrow Y$ a morphism of $\mathfrak{T}op_K$ homotopic to f, i.e. for each $k \in K$ there is a homotopy $f_k \simeq g_k$. Then g admits a $\mathfrak{B}$-map [$h\mathfrak{B}$-map] structure $(g,\varkappa) : (X,\alpha) \longrightarrow (Y,\beta)$ such that $(f,\rho) \simeq (g,\varkappa)$.

Proof: Let $\mathfrak{D} \subset W(\mathfrak{B} \otimes \mathfrak{Q}_1)$ be the subcategory generated under composition and $\oplus$ by $d^0 W(\mathfrak{B} \otimes \mathfrak{Q}_1)$, $d^1 W(\mathfrak{B} \otimes \mathfrak{Q}_1)$ and $\{j_k | k \in K\}$. Define a multiplicative functor $H(t) : \mathfrak{D} \longrightarrow \mathfrak{Top}$ by $H(t) | d^0 W(\mathfrak{B} \otimes \mathfrak{Q}_1) = \beta$, $H(t) | d^1 W(\mathfrak{B} \otimes \mathfrak{Q}_1) = \alpha$, and $H(t)(j_k) = h_k(t)$ where $h_k(t)$ is any homotopy $f_k \simeq g_k$. By Prop. 3.14, there exists an extension $F(t)$ of $H(t)$ such that $F(0) = \rho$. Each $F(t)$ defines a $\mathfrak{B}$-map from (X,α) to (Y,β), and $\varkappa = F(1)$ has g as carrier. Hence $(f,\rho) \simeq (g,\varkappa)$ by (4.13). The proof for $h\mathfrak{B}$-maps is the same. ∎

Corollary 4.15: Given $\mathfrak{B}$-maps or $h\mathfrak{B}$-maps $(f,\rho) : (X,\alpha) \longrightarrow (Y,\beta)$ and $(g,\varkappa) : (Y,\beta) \longrightarrow (Z,\gamma)$. Then there is a composite $(h,\eta) : (X,\alpha) \longrightarrow (Z,\gamma)$ of (f,ρ) and $(g,\varkappa)$ such that $h = g \circ f$.

Proof: Let (h',η') be any composite of (f,ρ) and (g,β). Then there is an action $\sigma : W(\mathfrak{B} \otimes \mathfrak{Q}_2) \longrightarrow \mathfrak{Top}$ with $d^0(\sigma) = \beta$, $d^1(\sigma) = \eta'$, and $d^2(\sigma) = \rho$. Hence

$$\sigma \begin{bmatrix} & 0 \\ & \bullet \mathrm{id}_k \\ t & 1 \\ & \bullet \mathrm{id}_k \\ & 2 \end{bmatrix} \qquad t \in I$$

is a homotopy from $g_k \cdot f_k$ to h'_k. Now apply (4.14). ∎

We next investigate the question how much of a $W\mathfrak{B}$-structure on $X \in \mathfrak{Top}_K$ survives if we change X by a homotopy equivalence.

Let $\mathfrak{J}$ be the category $\mathfrak{J}\mathfrak{s}_1$ of two objects 0 and 1 such that each morphism space contains exactly one element. So an $\mathfrak{J}$-space is a homeomorphism $X_0 \cong X_1$.

Lemma 4.16: Suppose $p : X \longrightarrow Y$ is a homotopy equivalence in $\mathfrak{Top}_K$. Then p carries a $W(K \otimes \mathfrak{J})$-structure (or a $W\mathfrak{J}$-structure in $\mathfrak{Top}_K$ instead of $\mathfrak{Top}$), where we consider K as the category with K as set of objects

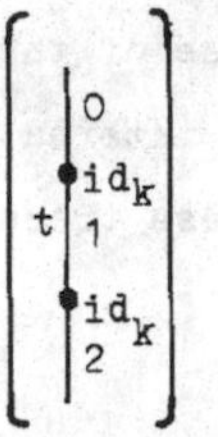

and only identity morphisms.

<u>Proof</u>: Since the identities of $K \otimes \mathfrak{J}$ are isolated, we need only consider trees which are simplified by relation (3.1 a). So the trees in question are vertical linear trees with edges coloured alternately X_k and Y_k (for $(k,0)$ and $(k,1)$). Let $\mathfrak{C}_r$ be the subcategory of $W(K \otimes \mathfrak{J})$ generated by all trees with root colour X_k, some k, and at most $r-1$ internal edges, or with root colour Y_k, some k, and at most r internal edges. The homotopy equivalence p defines an action $\mathfrak{C}_0 \longrightarrow \mathfrak{T}op$.

Suppose we are given an action of $\mathfrak{C}_{2n}$. To extend over $\mathfrak{C}_{2n+1}$ we need the actions of the trees

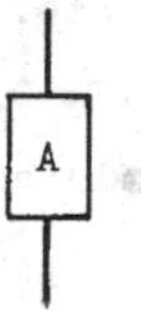

$: Y_k \longrightarrow X_k$

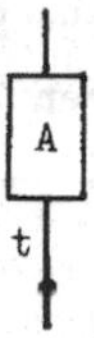

$: Y_k \longrightarrow Y_k$

where A stands for a tree with twig colour Y_k, root colour X_k, and $2n$ internal edges, which we represent by its edge lengths as a point in I^{2n}. Hence we require maps

$$f_k : I^{2n} \times Y_k \longrightarrow X_k \qquad h_k(t) : I^{2n} \times Y_k \longrightarrow Y_k$$

which are already given on $\partial I^{2n} \times Y_k$, because $x \in I^{2n}$ represents a morphism in $\mathfrak{C}_{2n}$ iff $x \in \partial I^{2n}$. Now $h_k(0)$ is known in terms of $\mathfrak{C}_{2n}$ because of relation (3.1 c), and we require $h_k(1) = p_k \circ f_k$. The maps f_k and $h_k(t)$ are provided by (A 3.5).

Similarly extend from $\mathfrak{C}_{2n+1}$ to $\mathfrak{C}_{2n+2}$. ∎

This categorical description of a homotopy equivalence turns out to be useful for the study of homotopy equivalent $W\mathfrak{B}$-spaces. Let

$$d^0, d^1 : W\mathfrak{B} \longrightarrow W(\mathfrak{B} \otimes \mathfrak{J}) \qquad u, v : W(\mathfrak{B} \otimes \mathfrak{L}_1) \longrightarrow W(\mathfrak{B} \otimes \mathfrak{J})$$

and the corresponding partly homogeneous versions be the functors induced by $\bar{d}^0, \bar{d}^1 : \mathfrak{L}_0 \longrightarrow \mathfrak{J}$, where $\bar{d}^0(0) = 1$, $\bar{d}^1(0) = 0$, and $\bar{u}, \bar{v} : \mathfrak{L}_1 \to \mathfrak{J}$, where $\bar{u}(i) = i$, $i = 0,1$ and $\bar{v}(0) = 1$, $\bar{v}(1) = 0$.

<u>Lemma 4.17</u>: Given an action $\rho : W(\mathfrak{B} \otimes \mathfrak{J}) \longrightarrow \mathfrak{Top}$ $[\rho : H_{\mathfrak{J}}W(\mathfrak{B} \otimes \mathfrak{J}) \longrightarrow \mathfrak{Top}]$ then $\rho \circ u$ and $\rho \circ v : W(\mathfrak{B} \otimes \mathfrak{L}_1) \longrightarrow \mathfrak{Top}$ $[H_{\mathfrak{L}_1}W(\mathfrak{B} \otimes \mathfrak{L}_1) \longrightarrow \mathfrak{Top}]$ are $\mathfrak{B}$-maps [$h\mathfrak{B}$-maps], which are homotopy inverse to each other, i.e. $\rho \circ u$ represents an isomorphism in the category $\mathfrak{Map}_{\mathfrak{B}}$ [$\mathfrak{Map}_{h\mathfrak{B}}$] whose inverse is represented by $\rho \circ v$.

<u>Proof</u>: Again we only prove the statement for $\mathfrak{B}$-maps. We have to define actions $\mu, \nu : W(\mathfrak{B} \otimes \mathfrak{L}_2) \longrightarrow \mathfrak{Top}$ such that $d^0(\mu) = \rho \circ v$, $d^2(\mu) = \rho \circ u$, $d^0(\nu) = \rho \circ u$, $d^2(\nu) = \rho \circ v$ and $d^1(\mu)$ and $d^1(\nu)$ are degenerate.

Let $k, l : \mathfrak{L}_2 \longrightarrow \mathfrak{J}$ be given by $k(i) = 0$, $l(i) = 1$ for $i = 0,2$ and $k(1) = 1$, $l(1) = 0$. Then $\mu = \rho \circ W(\mathrm{Id} \otimes k)$ and $\nu = \rho \circ W(\mathrm{Id} \otimes l)$ are actions as required. ∎

We now pass to the main results of this section.

<u>Theorem 4.18</u>: Let $\mathfrak{D}$ be a sub-PROP of $\mathfrak{B}$ such that each $(\mathfrak{B}(\underline{i},k), \mathfrak{D}(\underline{i},k))$ is an $S_{\underline{i}}$-NDR, and let $p : X \longrightarrow Y$ be a homotopy equivalence in $\mathfrak{Top}_K$. Suppose (Y,β) is a $W\mathfrak{B}$-space and p admits a $W(\mathfrak{D} \otimes \mathfrak{J})$-structure ρ' such that $\rho' \circ d^0 = \beta | W\mathfrak{D}$. Then there exists an extension $\rho : W(\mathfrak{B} \otimes \mathfrak{J}) \longrightarrow \mathfrak{Top}$ of ρ' such that $\rho \circ d^0 = \beta$. The same holds for the partly homogeneous version.

<u>Theorem 4.19</u>: Let $\mathfrak{D}$ be a sub-PROP of $\mathfrak{B}$ such that $(\mathfrak{B}(\underline{i},k), \mathfrak{D}(\underline{i},k))$ is an $S_{\underline{i}}$-NDR. Suppose $(p,\pi) : (X,\alpha) \longrightarrow (Y,\beta)$ is a $\mathfrak{B}$-map whose underlying map p is a homotopy equivalence in $\mathfrak{Top}_K$ and suppose p admits a $W(\mathfrak{D} \otimes \mathfrak{J})$-structure ρ' such that $\rho' \circ (u | W(\mathfrak{D} \otimes \mathfrak{L}_1)) = \pi | W(\mathfrak{D} \otimes \mathfrak{L}_1)$. Then there is an extension $\rho : W(\mathfrak{B} \otimes \mathfrak{J}) \longrightarrow \mathfrak{Top}_K$ of ρ' such that $\rho \circ u = \pi$. The same holds for the partly homogeneous version.

Before we prove the theorems let us deduce some important consequences.

<u>Corollary 4.20</u>: Let (X,α') be a $W\mathfrak{D}$-space and (Y,β) a $W\mathfrak{B}$-space such that (X,α') and $(Y,\beta\circ i)$ are homotopy equivalent as $W\mathfrak{D}$-spaces ($i : W\mathfrak{D} \subset W\mathfrak{B}$ is the inclusion functor). Then the $W\mathfrak{D}$-action on X extends to a $W\mathfrak{B}$-action α and the $\mathfrak{D}$-homotopy equivalence $(X,\alpha') \longrightarrow (Y,\beta\circ i)$ to a homotopy equivalence $(X,\alpha) \longrightarrow (Y,\beta)$ of $W\mathfrak{B}$-spaces.

<u>Proof</u>: By assumption there is a $\mathfrak{D}$-map $(p,\pi') : (X,\alpha) \longrightarrow (Y,\beta\circ i)$ whose underlying map is a homotopy equivalence. By (4.16), p admits a $W(K\otimes\mathfrak{J})$-structure. Now apply (4.19) with $\mathfrak{D} = K$ to extend π' to an action $\rho' : W(\mathfrak{D}\otimes\mathfrak{J}) \longrightarrow \mathfrak{Top}$, which in turn can be extended to an action $\rho : W(\mathfrak{B}\otimes\mathfrak{J}) \longrightarrow \mathfrak{Top}$ such that $\rho\circ d^0 = \beta$, by (4.18). Then $\alpha = \rho\circ d^1$ is the required $W\mathfrak{B}$-structure on X. ∎

<u>Corollary 4.21</u>: Let $(p,\pi) : (X,\alpha) \longrightarrow (Y,\beta)$ be a $\mathfrak{B}$-map and p a homotopy equivalence. Let q be any homotopy inverse of p carrying a $\mathfrak{D}$-map structure $(q,\varkappa') : (Y,\beta\circ i) \longrightarrow (X,\alpha\circ i)$, which is homotopy inverse to $(p,\pi\circ(i\otimes id))$. Then $(q,\varkappa')$ can be extended to a $\mathfrak{B}$-map $(q,\varkappa):(Y,\beta)\to(X,\alpha)$ which is homotopy inverse to (p,π). The same holds for $h\mathfrak{B}$-maps.

<u>Proof</u>: The action π can be extended to an action $\rho : W(\mathfrak{B}\otimes\mathfrak{J}) \longrightarrow \mathfrak{Top}$ by (4.16) and (4.19). Hence there exists a homotopy inverse $(q',\varkappa'')$ of (p,π) namely $\varkappa'' = \rho\circ v$. In particular, $\rho\circ i : W(\mathfrak{D}\otimes\mathfrak{J}) \longrightarrow \mathfrak{Top}$ provides a homotopy inverse (q',ξ) of $(p,\pi\circ i)$, which has to be homotopic to $(q\ ,\varkappa')$. By (4.13), there is a homotopy through $\mathfrak{D}$-maps $(q_t,\xi_t) : (Y,\beta\circ i) \longrightarrow (X,\alpha\circ i)$ with $(q_o,\xi_o) = (q',\xi)$ and $(q_1,\xi_1)=(q,\varkappa')$. Let $\mathfrak{B} \subset W(\mathfrak{B}\otimes\mathfrak{L}_1)$ be the subcategory generated under $\oplus$ and composition by $d^iW\mathfrak{B}$, $i = 0,1$, and $W(\mathfrak{D}\otimes\mathfrak{L}_1)$. Then (q_t,ξ_t) and the constant homotopies on α and β define a homotopy $H(t) : \mathfrak{B} \longrightarrow \mathfrak{Top}$ of multiplicative

functors such that $\varkappa''$ extends $H(0)$. By (3.14), there is a $\mathcal{B}$-map
$(q,\varkappa) : (Y,\beta) \longrightarrow (X,\alpha)$ extending $(q,\varkappa')$ and homotopic to $(q',\varkappa'')$. ∎

Remark: Corollary 4.21 includes a result of Fuchs [21, Satz 4.1] (see
also (1.17) of these notes). Moreover, we provide the first complete
proof of this result available in the literature.

To prove the theorems, we seek to give p a $W(\mathcal{B}\otimes\mathcal{J})$-structure, where
we are in effect given the action of a sub-PROP $\mathfrak{C}$ of $W(\mathcal{B}\otimes\mathcal{J})$. We extend by applying the lifting theorem (3.17) with $\mathcal{B} = \mathfrak{C}$ to obtain a
retraction functor $W(\mathcal{B}\otimes\mathcal{J}) \longrightarrow \mathfrak{C}$, for which we need only show that
the restriction of the augmentation $\varepsilon : W(\mathcal{B}\otimes\mathcal{J}) \longrightarrow \mathcal{B}\otimes\mathcal{J}$ to $\mathfrak{C}$ is an
equivariant equivalence. We only prove the theorems for the inhomo-
geneous version, because they are similar for the homogeneous one.
To make the argument more transparent we denote the elements of $\mathfrak{C}$
often by their images under the given actions. For example, the colours
$(k,0)$ and $(k,1)$ are identified with X_k and Y_k and simply denoted by
X and Y, and the trees

$$p_k = \overset{0}{\underset{1}{\bullet}}\,id_k \qquad \text{and} \qquad q_k = \overset{1}{\underset{0}{\bullet}}\,id_k$$

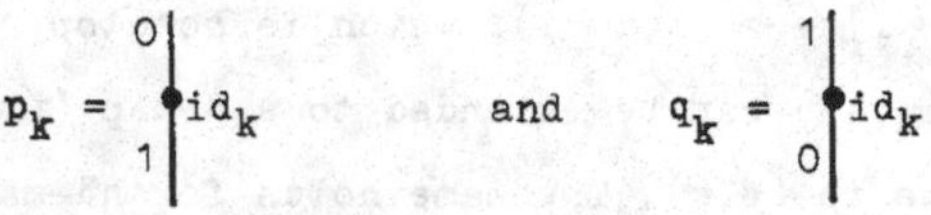

with $p_k : X_k \longrightarrow Y_k$ are often denoted by p and q. We again label the
vertices by elements in $\mathcal{B}$. In view of (4.16) we can assume that $\mathcal{D} \subset \mathcal{B}$
contains all identities. To make life easier we prove

Lemma 4.22: Let $\mathfrak{C}$ be a sub-PROP of $W(\mathcal{B}\otimes\mathcal{J})$ containing $W(K\otimes\mathcal{J})$. Then
$\varepsilon : \mathfrak{C}(\underline{i},k) \longrightarrow (\mathcal{B}\otimes\mathcal{J})(\underline{i},k)$ is an equivariant homotopy equivalence
provided it is one for $\underline{i} = \{i_1,\ldots,i_n\}$ and k such that k and each i_r
has $\mathcal{J}$-colour 1.

Proof: Substitute each X-coloured twig of a tree in an arbitrary

morphism space $\mathfrak{C}(\underline{i},k)$ by

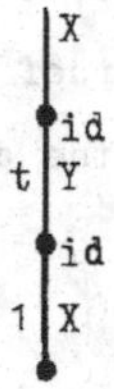

where t runs from 0 to 1. Similarly substitute a X-coloured root by

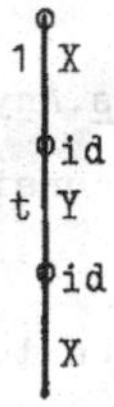

This deforms $\mathfrak{C}(\underline{i},k)$ into the subspace of trees of the form

$$v \bullet A \bullet (u_1 \oplus \ldots \oplus u_n)$$

where $v = q$ if k has colour X and $v = id$ otherwise, $u_r = p$ if i_r has colour X and $u_r = id$ otherwise, and A is a tree whose root and twigs all have colour Y. By assumption, ε is an equivariant homotopy equivalence on the space of all trees A. ∎

<u>Proof of 4.18</u>: Take $\mathfrak{C}$ to be the subcategory of $W(\mathfrak{B} \otimes \mathfrak{J})$ generated by $W(\mathfrak{D} \otimes \mathfrak{J})$ and $d^o(W\mathfrak{B})$. Then $\mathfrak{C}$ satisfies the requirements for $\mathfrak{B}$ in Theorem 3.17, and ρ' and β define an action $\mathfrak{C} \longrightarrow \mathfrak{T}op$. To show that $\varepsilon : \mathfrak{C} \to \mathfrak{B}\otimes\mathfrak{J}$ is an equivariant equivalence, we take a space $\mathfrak{C}(\underline{i},k)$ whose trees have only Y-coloured root and twigs and contract all internal X-edges using the deformation of (3.15). This deforms $\mathfrak{C}(\underline{i},k)$ into the subspace of all trees with no X-edge, i.e. into $W\mathfrak{B}(\underline{i},k)$, which is what we need. ∎

<u>Proof of 4.19:</u> This case is considerably more complicated. Take $\mathfrak{C}$ to

be the subcategory of $W(\mathfrak{B}\otimes\mathfrak{J})$ generated by $W(\mathfrak{D}\otimes\mathfrak{J})$ and $u(W(\mathfrak{B}\otimes\mathfrak{L}_1))$. Then $\mathfrak{C}$ satisfies the requirements for $\mathfrak{B}$ in Theorem 3.17, ρ' and π define an action $\mathfrak{C} \longrightarrow \mathfrak{Top}$. For the proof that $\epsilon : \mathfrak{C} \longrightarrow \mathfrak{B}\otimes\mathfrak{J}$ is an equivariant equivalence, we modify the description of $W(\mathfrak{B}\otimes\mathfrak{J})$.

The vertices

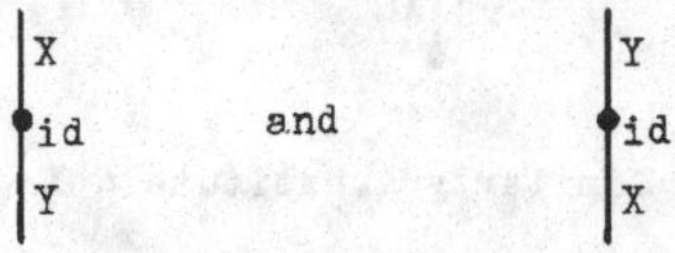

are called p-vertices and q-vertices. Any other vertex not labelled by an identity is called e-vertex. We consider trees that have only two kinds of vertices:

(i) p-vertices and q-vertices (which imply changes of the $\mathfrak{J}$-colour)

(ii) vertices in which all incoming edges have the same $\mathfrak{J}$-colour as the outgoing edge.

The relations among these trees are the same as (3.1) with the exception that edges of length 0 with an e-vertex on one end and a p- or q-vertex on the other cannot be shrunk. Instead, we have the notion of "pushing a p-vertex (or similarly a q-vertex) up through an e-vertex": Given an edge of length 0 with an e-vertex at the top and a p-vertex at the bottom. We replace the p-vertex below the e-vertex by k p-vertices just above the e-vertex, separated from it by edges of length 0, one for each incoming edge to the e-vertex.

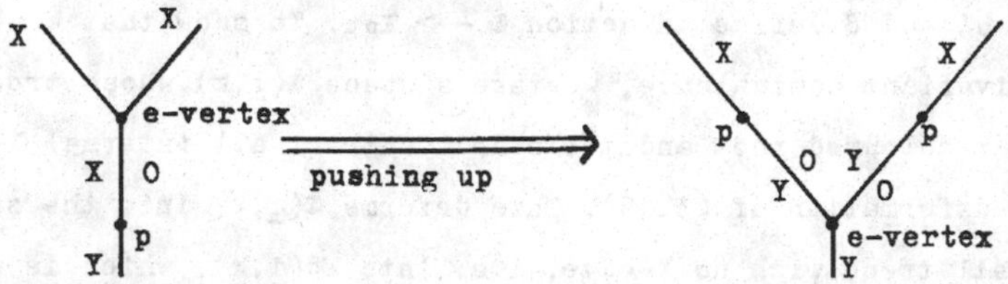

Then $W(\mathfrak{B}\otimes\mathfrak{J})$ is the quotient of the space of these representing trees modulo the relations (3.1) with the above modification of (3.1 c).

This can be seen as follows: Any representing tree can in a canonical way be brought into a tree of this form by introducing redundant edges of lengths 0 with extra p-vertices and q-vertices. If an incoming edge of length t of an e-vertex has the wrong colour, X say, we substitute it by

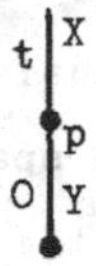

Similarly, use q, if the wrong colour is Y. Starting at the root vertex and working upwards we can change each representing tree inside its equivalence class to one of the required form.

The set of all e-vertices divides into two classes: b-<u>vertices</u> are those with labels $a \in \mathfrak{B} - \mathfrak{D}$, and d-<u>vertices</u> with $a \in \mathfrak{D}$, $a \neq \mathrm{id}$. In terms of the alternative description of $W(\mathfrak{B}\otimes\mathfrak{J})$, the trees that lie in $\mathfrak{C}$ are precisely those that satisfy the <u>separation condition</u>: In the directed edge path between any q-vertex and any b-vertex, there is an edge of length 1.

We filter $W(\mathfrak{B}\otimes\mathfrak{J})$. In view of (4.22), we only consider trees whose twigs and root have colour Y. Define the <u>height</u> of any vertex of a tree to be the number of vertices between it and the root, and the e-<u>height</u> as the number of these that are e-vertices. For $j = 0,1,2,\ldots$ let m_j be the number of p-vertices and q-vertices with e-height exactly j, and let n be the number of e-vertices in a tree. Then order the tree shapes lexicographically according to the sequences $(n, m_0, m_1, m_2, \ldots)$. This ordering is not as infinite as it looks, because $m_j = 0$ for $j>n$. We obtain an induced filtration F_ν of $W(\mathfrak{B}\otimes\mathfrak{J})$, which has the advantage that it strictly reduces by the application of any relation (3.1 a,c) in the modified form. To make the definition compatible with the topology we have to allow identities as "degenerate" e-vertices. If $\mu<\nu$ in the above ordering, it is easy to see, that the inclusion

of the representing trees of F_μ into the space of representing trees of F_ν is a closed equivariant cofibration. Hence it suffices to show
(a) Let F be the space of trees of a given shape in $\mathfrak{C}$ with order $(n,m_o,m_1,\ldots)$, some $m_j \neq 0$, and $F' \subset F$ the subspace of trees of lower filtration. Then F' is an equivariant SDR of F.
(b) Let R be the space of all trees of order $(n,0,0,\ldots)$, $n = 1,2,\ldots$, then $\epsilon|R$ is an equivariant homotopy equivalence.
Statement (b) holds, because R is the space of representing trees of $W\mathfrak{B}$. The idea for the proof of (a) is the following: Since roots and twigs have colour Y, there is at least one q-vertex between a p-vertex and a twig above it. Move all p-vertices as high as possible, to cancel out the q-vertices. This reduces $\mathfrak{C}(\underline{i},k)$ to $W\mathfrak{B}(\underline{i},k)$, which is what we need.

Consider the space of trees of a given shape with Y-root and no X-twigs and suppose the shape has ordering $(n,m_o,m_1,\ldots)$ with some $m_j \neq 0$. Let k be the minimal height of a p-vertex and let $P_1,\ldots,P_{r+s}$ denote the p-vertices of height k. Their e-height is also k. Hence $m_j = 0$ for $j<k$, and $m_k \geq r+s$. We first assume that $k>0$. Let u_i be the length of the incoming edge of P_i, and v_i the length of the outgoing edge. To take care of the separation condition we induct on the number of P_i such that u_i or v_i is 1. To be precise, we refine the filtration: Let F be the space of all trees we are considering at present. Let F_r be the subspace of those trees such that at least r of $P_1,\ldots,P_{r+s}$ (index them $P_1,\ldots,P_r$) have an incoming or outgoing edge of length 1. Then (F_{r-1},F_r) is an equivariant NDR. Let $J = I\times 1 \cup 1\times I \subset I^2$. Then F_r has the form $J^r\times(I^2)^s\times H$, where J^r is the space of lengths $(u_1,v_1,\ldots,u_r,v_r)$, $(I^2)^s$ is the space of lengths $(u_{r+1},v_{r+1},\ldots,u_{r+s},v_{r+s})$ and H is a large product of copies $\mathfrak{B}(\underline{j},l)$ and I taking care of all other parameters. To obtain the elements of $F_r \cap \mathfrak{C}$, we restrict the H-coordinate to lie in a certain subspace H' (depending on r) of H. What we want is an equivariant strong deformation retraction of $F_r \cap \mathfrak{C}$ into

$Q = F_{r+1} \cap \mathfrak{C} \cup$ {all elements of $F_r \cap \mathfrak{C}$ related to a tree of lower ordering}. A tree lies in this space if its H-coordinate lies in a certain subspace H" of H', if $(u_i,v_i) \in J$ for some $i>r$, or if any $u_i = 0$ (but not if some $v_i = 0$!, because we then cannot reduce inside our modified space of representatives). Hence Q is of the form $Q = J^r \times (I^2)^s \times H'' \cup D(J^r \times (I^2)^s) \times H'$, where $D(J^r \times (I^2)^s) \subset J^r \times (I^2)^s$ is the subspace of all points $(u_1,v_1,\ldots,u_{r+s},v_{r+s})$ with some $u_i = 0$ or some $u_j = 1$ or $v_j = 1$ for $j>r$. We require an equivariant strong deformation retraction

$$r : J^r \times (I^2)^s \times H' \longrightarrow J^r \times (I^2)^s \times H'' \cup D(J^r \times (I^2)^s) \times H'$$

Now H' and H" are unions of subproducts of H, where we substitute certain factors $\mathfrak{B}(\underline{j},l)$ by $\mathfrak{D}(\underline{j},l)$ and I by 0 or 1, and H" is the subspace of H' obtained by substituting a factor $\mathfrak{B}(l,l)$ or $\mathfrak{D}(l,l)$ by an identity or a suitable factor I by 0. Since $(\mathfrak{D}(k,k),\{id_k\})$ is a NDR, (H',H") is an equivariant NDR. Since $\{(0,1)\}$ is a SDR of J and since $J \cup 0 \times I$ is a SDR of I^2, the space $D(J^r \times (I^2)^s)$ is an equivariant SDR of $J^r \times (I^2)^s$. Hence r exists. This settles the case $k>0$. If k is large enough, e.g. $k>n$, we have no p-vertices left and hence no q-vertices. Therefore we are in the situation (b) as desired.

If $k=0$, the root vertex is a p-vertex and the space of all trees of this type in $\mathfrak{C}$ is of the form $I \times H'$, where I is the space of lengths u of the incoming edge to the root vertex, and H' takes care of all other paramaters. A tree in $I \times H'$ is related to a tree of lower ordering if its H'-coordinate lies in some subspace H" of H' as above or if $u=0$. Since (H',H") is an equivariant NDR, there is an equivariant strong deformation retraction

$$I \times H' \longrightarrow 0 \times H' \cup I \times H''$$

as required. This completes the proof of (4.19). ∎

4. RELATING $\mathfrak{B}$-MAPS AND $h\mathfrak{B}$-MAPS TO $W\mathfrak{B}$-HOMOMORPHISMS

Since $\mathfrak{B}$-maps and $h\mathfrak{B}$-maps are difficult to work with, it is desirable to substitute them by homomorphisms.

Theorem 4.23: For any $W\mathfrak{B}$-space (X,α) **there exists a $W\mathfrak{B}$-space $U(X,\alpha)=(UX,\alpha^*)$, a $\mathfrak{B}$-map $(q_\alpha,\xi_\alpha) : (X,\alpha) \longrightarrow U(X,\alpha)$,** and a $h\mathfrak{B}$-map $(q_\alpha,\eta_\alpha) : (X,\alpha) \longrightarrow U(X,\alpha)$ such that

(a) The map q embeds X as SDR into UX.

(b) Any $\mathfrak{B}$-map $(f,\rho) : (X,\alpha) \longrightarrow (Y,\beta)$ is a canonical composite in the sense of (4.5) of (q_α,ξ_α) and a unique $W\mathfrak{B}$-homomorphism $h : U(X,\alpha) \longrightarrow (Y,\beta)$.

(c) Any $h\mathfrak{B}$-map $(f,\rho) : (X,\alpha) \longrightarrow (Y,\beta)$ is the canonical composite $h \circ (q_\alpha,\eta_\alpha)$ of a unique $W\mathfrak{B}$-homomorphism $h : U(X,\alpha) \longrightarrow (Y,\beta)$ and (q,η).

(d) If we change (f,ρ) inside its homotopy class, then the induced homomorphism h stays inside its homotopy class.

Although the definition of a homotopy of homomorphisms is obvious let us state it.

Definition 4.24: Let $\mathfrak{B}$ be a PROP. Two homomorphisms $h_0, h_1 : X \longrightarrow Y$ of $\mathfrak{B}$-spaces are called homotopic if there is a homotopy of homomorphisms $h_t : X \longrightarrow Y$ from h_0 to h_1.

Remark: Since the composite of a $\mathfrak{B}$-map and a $W\mathfrak{B}$-homomorphism is in general only defined up to homotopy, part (b) is not particularly useful.

Proof of the theorem:

For $\underline{i} = \{i_1,\ldots,i_n\} \in \text{ob } \mathfrak{B}$, let $\underline{i}^\varepsilon = \{(i_1,\varepsilon),\ldots,(i_n,\varepsilon)\} \in \text{ob } (\mathfrak{B} \otimes \mathfrak{L}_1)$, $\varepsilon = 0,1$. Define

$$(4.25) \qquad (UX)_k = \bigcup_{\underline{i} \in \text{ob}\mathfrak{B}} HW(\mathfrak{B} \otimes \mathfrak{L}_1)(\underline{i}^0, k^1) \times X_{\underline{i}} / \sim$$

with $(c \circ a;\ x_1,\ldots,x_n) \sim (c;\ a(x_1,\ldots,x_n))$, $c \in HW(\mathfrak{B} \otimes \mathfrak{L}_1)(\underline{j}^o, k^1)$, $a \in HW(\mathfrak{B} \otimes \mathfrak{L}_1)(\underline{i}^o, \underline{j}^o)$. Here we identify a with its image under $\alpha : W\mathfrak{B} \longrightarrow \mathfrak{Top}$, the $W\mathfrak{B}$-structure on X. We give $UX \in \mathfrak{Top}_K$ a $W\mathfrak{B}$-structure α^* by

$$b[(c_1;\ \underline{x}_1),\ldots,(c_r;\ \underline{x}_r)] = [b \circ (c_1 \oplus \ldots \oplus c_r);\ \underline{x}_1,\ldots,\underline{x}_r]$$

where $b \in W\mathfrak{B}$ is identified with its image under $d^o : W\mathfrak{B} \longrightarrow HW(\mathfrak{B} \otimes \mathfrak{L}_1)$ and $\underline{x}_i$ stands for $x_{i1},\ldots,x_{in_i}$.

To define the $\mathfrak{B}$-map $(q_\alpha, \xi_\alpha) : (X,\alpha) \longrightarrow (UX,\alpha^*)$ consider $HW(\mathfrak{B} \otimes \mathfrak{L}_1)$ as subcategory of $W(\mathfrak{B} \otimes \mathfrak{L}_1)$. The functors α and α^* determine ξ_α on $d^oW\mathfrak{B}$ and $d^1W\mathfrak{B}$. On elements $d \in W(\mathfrak{B} \otimes \mathfrak{L}_1)(\underline{r}, l)$ where $\underline{r} = \{r_1,\ldots,r_n\}$ contains at least one element with $\mathfrak{L}_1$-colour 0 and l has $\mathfrak{L}_1$-colour 1, the action ξ_α is given by

$$(4.26) \qquad d(y_1,\ldots,y_n) = (d \circ (a_1 \oplus \ldots \oplus a_n);\ z_1,\ldots,z_n)$$

where

$$(a_i, z_i) = \begin{cases} (id_k, y_i) & \text{if } r_i = (k,0) \in K \times ob\mathfrak{L}_1 \\ (c_i, \underline{x}_i) & \text{if } r_i = (k,1) \in K \times ob\mathfrak{L}_1 \text{ and } y_i = (c_i; \underline{x}_i) \in (UX)_k \end{cases}$$

The underlying map $q_\alpha = \{q_k : X_k \longrightarrow (UX)_k\}$ is given by $q_k(x) = (j_k; x)$. Recall that j_k **is represented by** the tree

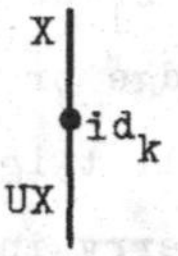

<u>Proof of (b)</u>: For a $\mathfrak{B}$-map $(f,\rho) : (X,\alpha) \longrightarrow (Y,\beta)$ define a $W\mathfrak{B}$-homomorphism $h : U(X,\alpha) \longrightarrow (Y,\beta)$ by

$$(4.27) \qquad h(c;\ \underline{x}) = \rho(c)(\underline{x})$$

The homomorphism $h : UX \longrightarrow Y$ necessarily has to satisfy (4.27), because it expresses the condition that (f,ρ) be a canonical composite $h \circ (q_\alpha, \xi_\alpha)$ for elements $c \in HW(\mathfrak{B} \otimes \mathfrak{L}_1)$. It remains to check that condition (4.5) is satisfied for $a \in W(\mathfrak{B} \otimes \mathfrak{L}_1)(\underline{r}, l)$ with $\underline{r}$ and l as above. Let $y_i = w_i \in X_k$ if $r_i = (k,0)$ and $y_i = (c_i, \underline{x}_i)$, $w_i = h(c_i; \underline{x}_i)$ if

$r_i = (k,1)$. Then

$$\rho(a)(w_1,\ldots,w_n) = h(\xi(a)(y_1,\ldots,y_n))$$

because

$$\begin{aligned} h(\xi(a)(y_1,\ldots,y_n)) &= h(a\circ(a_1 \oplus \ldots \oplus a_n);z_1,\ldots,z_n) \text{ with } a_i, z_i \text{ as in (4.26)} \\ &= \rho(a\circ(a_1 \oplus \ldots \oplus a_n))(z_1,\ldots,z_n) \\ &= \rho(a)(\rho(a_1)(z_1),\ldots,\rho(a_n)(z_n)) \\ &= \rho(a)(w_1,\ldots,w_n) \end{aligned}$$

Proof of (c): The $h\mathfrak{B}$-map $(q_\alpha, \eta_\alpha) : (X,\alpha) \longrightarrow U(X,\alpha)$ is given by $\xi_\alpha | HW(\mathfrak{B} \otimes \mathfrak{L}_1)$. For any $h\mathfrak{B}$-map $(f,\rho) : (X,\alpha) \longrightarrow (Y,\beta)$ define the $W\mathfrak{B}$-homomorphism h as in (4.27).

Proof of (d): Let $(f_o,\rho_o) \simeq (f_1,\rho_1)$. Then there is a homotopy through $\mathfrak{B}$-maps $\rho_t : (X,\alpha) \longrightarrow (Y,\beta)$. Define the homotopy through homomorphisms by $h_t(c;\, \underline{x}) = \rho_t(c)(\underline{x})$. Of course, this works also for $h\mathfrak{B}$-maps.

Proof of (a): As usually when working with deformations we use the tree language. So we first express UX in terms of trees. As in the previous section we call an edge of $\mathfrak{L}_1$-colour 0 an X-edge or X-coloured and an edge of $\mathfrak{L}_1$-colour 1 a Y-edge or Y-coloured. Then UX is the space of all trees with a Y-root, all twigs are X-twigs and to each twig of K-colour k is assigned a cherry in X_k subject to the relations (compare (3.24))

(4.28) (a) = (3.1 a)

(b) = (3.1 b) for permutations. The cherries are permuted along with the twigs.

(c) = (3.1 c)

(d) if the tree A has an X-edge of length 1 so that $A = A_1 \circ A_2$, then

$$(A;\, x_1,\ldots,x_n) \sim (A_1;\, x_1,\ldots,x_p,y,x_{q+1},\ldots,x_n)$$

where $y = \alpha(A_2)(x_{p+1},\ldots,x_q)$, and $(x_{p+1},\ldots,x_q)$ and $(x_1,\ldots,x_n)$ are the cherries of A_2 and A in clockwise order.

We filter each space $(UX)_k$ by subspaces F_n of trees having at most n internal edges. In a similar manner as for Lemma 3.9 we can show

(4.29) (a) $(UX)_k$ is the direct limit of the subspaces F_n

(b) (F_{n+1}, F_n) is a NDR

(c) if each $\mathfrak{B}(\underline{i},k)$ is Hausdorff, so is each $(UX)_k$

(d) if $\mathfrak{B}$ is a CW-PROP, then $(UX)_k$ is a CW-complex.

The deformation $UX \longrightarrow X$ is constructed in two steps. Let $MX \subset UX$ be the subspace of all trees having no internal Y-edge.

<u>Step 1</u>: Deform UX into MX. For this it suffices to deform $F_n \cup MX$ into $F_{n-1} \cup MX$. Let $Q \times I^k \times I^l \times X_{\underline{i}}$ be a space of trees A in UX, where Q is the space of vertex labels, I^k and I^l the spaces of lengths of X-edges and Y-edges, and $X_{\underline{i}}$ the space of cherries. We intend to shrink the Y-edges of A. Let $DQ \subset Q$ be the subspace of elements which can be reduced by (4.28 a) and let $LI^l \subset I^l$ be the union of lower faces. Then $A \in Q \times I^k \times I^l \times X_{\underline{i}}$ represents an element of lower filtration iff

$$A \in DQ \times I^k \times I^l \times X_{\underline{i}} \cup Q \times \partial I^k \times I^l \times X_{\underline{i}} \cup Q \times I^k \times LI^l \times X_{\underline{i}}$$

On DQ and on the lower faces of $I^k \times I^l$ we may reduce A by (4.28 a and c) while we may reduce it by (4.28 d) on the upper faces of I^k. So we need an equivariant (because of (4.18 b)) strong deformation retraction

$$Q \times I^k \times I^l \times X_{\underline{i}} \longrightarrow DQ \times I^k \times I^l \times X_{\underline{i}} \cup Q \times \partial I^k \times I^l \times X_{\underline{i}} \cup Q \times I^k \times LI^l \times X_{\underline{i}} \text{ ,}$$

which exists by (A 2.4) for $l > 0$.

<u>Step 2</u>: Next deform MX into X. Note that the subspace X of MX consists of the elements represented by the trees

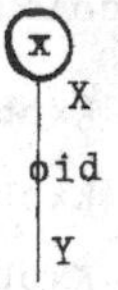

where x is a cherry. Substitute the roots of trees in MX by

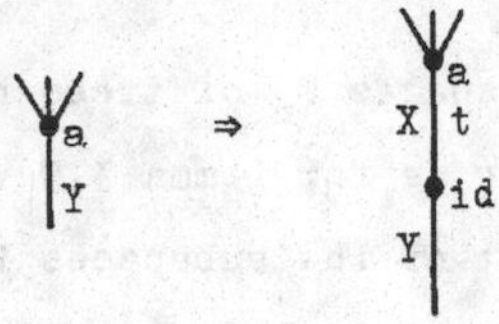

where t runs from 0 to 1. For t = 1, relation (4.28 d) applies and reduces the tree to one in X. Because of (4.28 a), the points of X are kept fixed under the deformation. ∎

Let $\mathfrak{Hom}_{W\mathfrak{B}}$ denote the category of $W\mathfrak{B}$-spaces and homotopy classes of $W\mathfrak{B}$-homomorphisms. We are going to define functors

$$\mathfrak{Map}_{\mathfrak{B}} \xleftarrow{\ J\ } \mathfrak{Hom}_{W\mathfrak{B}} \xrightarrow{\ J'\ } \mathfrak{Map}_{h\mathfrak{B}}$$

Let $f : (X,\alpha) \longrightarrow (Y,\beta)$ be a $W\mathfrak{B}$-homomorphism. Define a $\mathfrak{B}$-map $Jf = (f,f_*) : (X,\alpha) \longrightarrow (Y,\beta)$ by $f_*|d^1W\mathfrak{B} = \alpha$ and on $a \in W(\mathfrak{B}\otimes\mathfrak{L}_1)(\underline{i},k)$, where k has $\mathfrak{L}_1$-colour 1, by $f_*(a) = \beta(s^0(a))\circ(h_1\times\ldots\times h_n)$ with $h_r = id_{Y_l}$ if $\underline{i}(r) = (l,1)$ and $h_r = f_l$ if $\underline{i}(r) = (l,0)$. The definition of the induced $h\mathfrak{B}$-map (f,f_*') is the same. It is immediate from the definitions that J and J' preserve identities. It remains to check that they preserve composition:

<u>Lemma 4.30</u>: Let $f : (X,\alpha) \longrightarrow (Y,\beta)$ and $g : (Y,\beta) \longrightarrow (Z,\gamma)$ be $W\mathfrak{B}$-homomorphisms. Then $(g\circ f,(g\circ f)_*)$ is a composite $\mathfrak{B}$-map $(g,g_*)\circ(f,f_*)$. The analogous result holds for $h\mathfrak{B}$-maps.

<u>Proof</u>: Define $\rho : W(\mathfrak{B}\otimes\mathfrak{L}_2) \longrightarrow \mathfrak{Top}$ by $\rho|d^2W(\mathfrak{B}\otimes\mathfrak{L}_2) = f_*$ and on morphisms $a \in W(\mathfrak{B}\otimes\mathfrak{L}_2)(\underline{i},k)$ where k has $\mathfrak{L}_2$-colour 2 by $\rho(a)=\gamma(s^0s^0(a))\,(h_1\times\ldots\times h_n)$ with

$$h_r = \begin{cases} id_{Z_l} & \text{if } \underline{i}(r) = (l,2)\in K\times ob\,\mathfrak{L}_2 \\ g_l & \text{if } \underline{i}(r) = (l,1)\in K\times ob\,\mathfrak{L}_2 \\ g_l\circ f_l & \text{if } \underline{i}(r) = (l,0)\in K\times ob\,\mathfrak{L}_2 \end{cases}$$

Then $\rho \cdot d^0 = g_*$, $\rho \cdot d^1 = (g \cdot f)_*$ and $\rho \cdot d^2 = f_*$. Hence $(g \cdot f)_* = g_* \cdot f_*$ as $\mathfrak{B}$-map. The proof for $h\mathfrak{B}$-maps is the same. ∎

As a direct consequence of Theorem 4.23 we obtain

<u>Proposition 4.31</u>: The functors J and J' have fully faithful left adjoints

$$U : \mathfrak{Map}_{\mathfrak{B}} \longrightarrow \mathfrak{Hom}_{W\mathfrak{B}} \qquad U' : \mathfrak{Map}_{h\mathfrak{B}} \longrightarrow \mathfrak{Hom}_{W\mathfrak{B}}$$

i.e. there is a natural bijection $\mathfrak{Map}_{\mathfrak{B}}((X,\alpha),(Y,\beta)) \cong \mathfrak{Hom}_{W\mathfrak{B}}(U(X,\alpha),(Y,\beta))$ and the function $U : \mathfrak{Map}_{\mathfrak{B}}((X,\alpha),(Y,\beta) \longrightarrow \mathfrak{Hom}_{\mathfrak{B}}(U(X,\alpha),U(Y,\beta))$ is bijective; and analogously for U'.

<u>Proof</u>: On objects, U and U' are given by the construction of (4.23) and on morphisms by

$$\begin{array}{ccc}
(X,\alpha) & \xrightarrow{(q_\alpha,\xi_\alpha)} & U(X,\alpha) \\
{\scriptstyle (f,\rho)}\downarrow & & \downarrow{\scriptstyle U(f,\rho)} \\
(Y,\beta) & \xrightarrow{(q_\beta,\xi_\beta)} & U(Y,\beta)
\end{array}$$

We use the universal property of (q_α,ξ_α). Since $U(f,\rho)$ is unique, it follows that U is a functor. By (4.23 b), it is left adjoint to J with

$$(q_\alpha,\xi_\alpha) : (X,\alpha) \longrightarrow (UX,\alpha^*) = JU(X,\alpha)$$

as front adjunction. Since the (q_α,ξ_α) are isomorphisms, U is fully faithful by adjoint functor nonsense. The proof for U' is the same. ∎

As a corollary we obtain

<u>Proposition 4.32</u>: The inclusion functor $i : HW(\mathfrak{B} \otimes \mathfrak{L}_1) \subset W(\mathfrak{B} \otimes \mathfrak{L}_1)$ induces a functor $H : \mathfrak{Map}_{\mathfrak{B}} \longrightarrow \mathfrak{Map}_{h\mathfrak{B}}$, which is an isomorphism of categories.

Proof: Let $j : HW(\mathfrak{B} \otimes \mathfrak{L}_2) \subset W(\mathfrak{B} \otimes \mathfrak{L}_2)$ be the inclusion. Let $\sigma : W(\mathfrak{B} \otimes \mathfrak{L}_2) \longrightarrow \mathfrak{Top}$ define a homotopy between two $\mathfrak{B}$-maps (f,ρ) and (f',ρ') and $\tau : W(\mathfrak{B} \otimes \mathfrak{L}_2) \longrightarrow \mathfrak{Top}$ a composite $(g \circ f,\lambda)$ of two $\mathfrak{B}$-maps $(g,\varkappa)$ and (f,ρ). Then $\sigma \circ j$ defines a homotopy between $(f,\rho \circ i)$ and $(f',\rho' \circ i)$ and $\tau \circ j$ shows that $(g \circ f,\lambda \circ i)$ is a composite of $(g,\varkappa \circ i)$ and $(f,\rho \circ i)$. Hence H is indeed a functor.

Let $(f,\rho) : (X,\alpha) \longrightarrow (Y,\beta)$ be a $\mathfrak{B}$-map. By construction, the induced $W\mathfrak{B}$-homomorphism $h : U(X,\alpha) \longrightarrow (Y,\beta)$ of Theorem 4.23 is the same for the $\mathfrak{B}$-map (f,ρ) and the $h\mathfrak{B}$-map $(f,\rho \circ i)$. By definition, $U'(X,\alpha) = U(X,\alpha) = (UX,\alpha^*)$. The morphism $U[f,\rho]$, where $[f,\rho]$ denotes the homotopy class of (f,ρ), is represented by the $W\mathfrak{B}$-homomorphism h induced by a composite $(q_\alpha \circ f,\lambda)$ of (f,ρ) and (q_α,ξ_α) while $U'[f,\rho \circ i]$ is represented by the $W\mathfrak{B}$-homomorphism h' induced by a composite $(q_\alpha \circ f,\lambda')$ of $(f,\rho \circ i)$ and (q_α,η_α). By construction, $\eta_\alpha = \xi_\alpha \circ i$. Hence $(q_\alpha \circ f,\lambda \circ i)$ is a composite of $(f,\rho \circ i)$ and (q_α,η_α), and h also represents $U'[f,\rho \circ i]$. We therefore have a commutative diagram

$$\begin{array}{ccc} \mathfrak{Map}_{\mathfrak{B}}((X,\alpha),(Y,\beta)) & \cong & \mathfrak{Hom}_{W\mathfrak{B}}(U(X,\alpha),(Y,\beta)) \\ \downarrow H & & \| \\ \mathfrak{Map}_{h\mathfrak{B}}((X,\alpha),(Y,\beta)) & \cong & \mathfrak{Hom}_{W\mathfrak{B}}(U'(X,\alpha),(Y,\beta)), \end{array}$$

which implies the proposition. ∎

5. MAPS FROM $W\mathfrak{B}$-SPACES TO $\mathfrak{B}$-SPACES

Any $\mathfrak{B}$-space Y is canonically a $W\mathfrak{B}$-space by means of $\varepsilon : W\mathfrak{B} \longrightarrow \mathfrak{B}$. Hence we have a good concept of maps from a $W\mathfrak{B}$-space to a $\mathfrak{B}$-space, and this is one type of maps we study in this section. On the other hand, the edge lengths are irrelevant in the given $W\mathfrak{B}$-action on Y. This suggests considering the quotient of $W(\mathfrak{B} \otimes \mathfrak{L}_1)$ in which the action of any tree is independent of the lengths of its Y-edges, which might as well be 0. Since we work with a $W\mathfrak{B}$-space X and a $\mathfrak{B}$-space Y, we feel that in this connection it seems reasonable to avoid mixed products $X \times Y$ although we can treat those, too. So we stick to a sort of homogeneous version. Let us make this precise: In general, we consider sequences of maps

$$X_o \longrightarrow X_1 \longrightarrow \ldots \longrightarrow X_{n-1} \longrightarrow X_n$$

where $X_o,\ldots,X_{n-1}$ are $W\mathfrak{B}$-spaces and X_n is a $\mathfrak{B}$-space. We allow mixed terms of $X_o,\ldots,X_{n-1}$, but products including X_n have only factors X_n.

Let $\mathfrak{C}$ be the full subcategory of $W(\mathfrak{B}\otimes\mathfrak{L}_n)$ consisting of all objects $\underline{i} = \{i_1,\ldots,i_r\}$ where the $\mathfrak{L}_n$-colour of any i_p is less than n or $i_1,\ldots,i_r$ all have $\mathfrak{L}_n$-colour n. Define $W_r(\mathfrak{B}\otimes\mathfrak{L}_n)$ to be the quotient category of $\mathfrak{C}$ under the following additional relation on the trees.

(4.33) a tree A is related to the tree obtained from A by changing the lengths of all X_n-edges to 0 (X_n-edges = edges of $\mathfrak{L}_n$-colour n).

In view of this relation, we need only consider trees which have no X_n-edge with exception of possibly the root. We call such trees target reduced or simply <u>reduced</u>.

The inclusion functors $d^i : W(\mathfrak{B}\otimes\mathfrak{L}_{n-1}) \longrightarrow W(\mathfrak{B}\otimes\mathfrak{L}_n)$, $0 \leq i \leq n$, induce inclusion functors

$$d^i : W_r(\mathfrak{B}\otimes\mathfrak{L}_{n-1}) \longrightarrow W_r(\mathfrak{B}\otimes\mathfrak{L}_n) \qquad 0 \leq i < n$$
$$d^n : W(\mathfrak{B}\otimes\mathfrak{L}_{n-1}) \longrightarrow W_r(\mathfrak{B}\otimes\mathfrak{L}_n)$$

<u>Definition 4.34</u>: Let (X,α) be a $W\mathfrak{B}$-space and (Y,β) a $\mathfrak{B}$-space. A <u>reduced</u> $\mathfrak{B}$-<u>map</u> $(f,\rho) : (X,\alpha) \longrightarrow (Y,\beta)$ consists of an action $\rho : W_r(\mathfrak{B}\otimes\mathfrak{L}_1) \longrightarrow \mathfrak{Top}$ and its underlying map $f : X \longrightarrow Y$ in $\mathfrak{Top}_K$ such that $\rho \circ d^1 = \alpha : W\mathfrak{B} \longrightarrow \mathfrak{Top}$ and $\rho \circ d^o = \beta : \mathfrak{B} \longrightarrow \mathfrak{Top}$.

Observe that $H_{\mathfrak{L}_1} W_r(\mathfrak{B}\otimes\mathfrak{L}_1) = W_r(\mathfrak{B}\otimes\mathfrak{L}_1)$ and $W_r(\mathfrak{B}) = \mathfrak{B}$.

Before we investigate reduced $\mathfrak{B}$-maps let us prove a variant of

Theorem 4.20 for our usual definition of $\mathfrak{B}$-maps.

Theorem 4.35: For any $W\mathfrak{B}$-space (X,α) there exists a $\mathfrak{B}$-space $N(X,\alpha) = (NX,\alpha^*)$ and a $\mathfrak{B}$-map $(p_\alpha,\xi_\alpha) : (X,\alpha) \longrightarrow (NX,\alpha^* \circ \varepsilon)$ and a $h\mathfrak{B}$-map $(p_\alpha,\eta_\alpha) : (X,\alpha) \longrightarrow (NX,\alpha^* \circ \varepsilon)$, where $\varepsilon : W\mathfrak{B} \longrightarrow \mathfrak{B}$ is the augmentation, such that

(a) The map p_α is a homotopy equivalence in $\mathfrak{Top}_K$.

(b) Any $\mathfrak{B}$-map $(f,\rho) : (X,\alpha) \longrightarrow (Y,\beta \circ \varepsilon)$, where (Y,β) is a $\mathfrak{B}$-space, is a canonical composite in the sense of (4.5) of (p_α,ξ_α) and a unique homomorphism of $\mathfrak{B}$-spaces $N(X,\alpha) \longrightarrow (Y,\beta)$

(c) Any $h\mathfrak{B}$-map $(f,\rho) : (X,\alpha) \longrightarrow (Y,\beta \circ \varepsilon)$, where (Y,β) is a $\mathfrak{B}$-space, is the canonical composite of (p_α,η_α) and a unique $\mathfrak{B}$-homomorphism $h : N(X,\alpha) \longrightarrow (Y,\beta)$

(d) If we change (f,ρ) inside its homotopy class, then the induced homomorphism h stays inside its homotopy class.

Proof: The $\mathfrak{B}$-space $NX = \{NX_k\}$ is the quotient of UX under the following relation:

(4.36) $\qquad (c \circ a;\ x_1,\ldots,x_n) \sim (\varepsilon(c) \circ a;\ x_1,\ldots,x_n) \qquad \varepsilon$=augmentation

$c \in HW(\mathfrak{B} \otimes \mathfrak{L}_1)(\underline{j}^1,k^1)$, $a \in HW(\mathfrak{B} \otimes \mathfrak{L}_1)(\underline{i}^o,\underline{j}^1)$ with the notation of (4.25). The $\mathfrak{B}$-structure on NX is given by

$$b[(c_1;\ \underline{x}_1),\ldots,(c_r;\ \underline{x}_r)] = [(\iota b)\circ(c_1 \oplus \ldots \oplus c_r;\ \underline{x}_1,\ldots,\underline{x}_r]$$

where $\iota : \mathfrak{B} \longrightarrow W\mathfrak{B} \longrightarrow HW(\mathfrak{B} \otimes \mathfrak{L}_1)$ is the composite of the standard section and the functor d^o. By (4.36),

$$[\iota(b_1 \circ b_2)\circ(c_1 \oplus \ldots \oplus c_r);\underline{x}_1,\ldots,\underline{x}_r] \sim [\iota(b_1)\circ\iota(b_2)\circ(c_1 \oplus \ldots \oplus c_r);\underline{x}_1,\ldots,\underline{x}_r]$$

so that this is indeed a $\mathfrak{B}$-structure. The $\mathfrak{B}$-map (p_α,ξ_α) and $h\mathfrak{B}$-map (p_α,η_α) are induced by the corresponding maps $(X,\alpha) \longrightarrow (UX,\alpha^*)$. The $\mathfrak{B}$-homomorphism $h : N(X,\alpha) \longrightarrow (Y,\beta)$ is defined as in (4.23) and part (b),(c), and (d) are proved in the same manner as in (4.23), only the argument of part (a) is different. We will prove it at a later stage. ∎

As an immediate consequence of (4.35) and (4.21) we have the following generalization of the theorem of Adams (see 1.11).

<u>Theorem 4.37</u>: A space $X \in \mathfrak{Top}_K$ admits a $W\mathfrak{B}$-structure iff it is of the homotopy type of a $\mathfrak{B}$-space. Precisely, if X is of the homotopy type of a $\mathfrak{B}$-space then it admits a $W\mathfrak{B}$-structure such that the homotopy equivalence carries a $\mathfrak{B}$-map structure and any $W\mathfrak{B}$-space X is homotopy equivalent to a $\mathfrak{B}$-space NX such that the homotopy equivalence carries a $\mathfrak{B}$-map structure. ▮

Using reduced $\mathfrak{B}$-maps we shall construct a $\mathfrak{B}$-space MX for any $W\mathfrak{B}$-space X containing X as SDR, which is more closely related to Adams' construction.

Let $\mathfrak{Hom}_{\mathfrak{B}}$ be the category of $\mathfrak{B}$-spaces and homotopy classes of homomorphisms. Define functors

$$\mathfrak{Map}_{\mathfrak{B}} \xleftarrow{\quad J \quad} \mathfrak{Hom}_{\mathfrak{B}} \xrightarrow{\quad J' \quad} \mathfrak{Map}_{n\mathfrak{B}}$$

on objects by $(X,\alpha) \longrightarrow (X,\alpha \circ \varepsilon)$. A representing homomorphism $\mathfrak{B} \otimes \mathfrak{L}_1 \longrightarrow \mathfrak{Top}$ is mapped to its composition with the augmentation $W(\mathfrak{B} \otimes \mathfrak{L}_1) \longrightarrow \mathfrak{B} \otimes \mathfrak{L}_1 \longrightarrow \mathfrak{Top}$. Since a composite in $\mathfrak{Hom}_{\mathfrak{B}}$ is given by a functor $\mathfrak{B} \otimes \mathfrak{L}_2 \longrightarrow \mathfrak{Top}$ the definition is functorial. Extending the correspondence $X \longrightarrow NX$ to functors $N : \mathfrak{Map}_{\mathfrak{B}} \longrightarrow \mathfrak{Hom}_{\mathfrak{B}}$ and $N' : \mathfrak{Map}_{n\mathfrak{B}} \longrightarrow \mathfrak{Hom}_{\mathfrak{B}}$ in the same way as in the proof of (4.31), we obtain

<u>Proposition 4.38</u>: The functors N and N' are left adjoint to J and J'. Moreover, they are fully faithful. ▮

We now return to reduced $\mathfrak{B}$-maps. To be able to work with them we need variants of the extension result (3.14) and the lifting theorem (3.17). We want to prove them using the analogue of Lemma 3.12 for reduced trees. Define $W_r^p(\mathfrak{B} \otimes \mathfrak{L}_n)$ as the subcategory of $W_r(\mathfrak{B} \otimes \mathfrak{L}_n)$ generated by reduced trees with at most p internal edges. Let P_λ be a space of reduced trees of shape λ with p internal edges as defined in (III, §2). Considered as representative of $W^p(\mathfrak{B}\otimes\mathfrak{L}_n)$, a tree A in P_λ decomposes only if it is related to a tree of lower filtration p, i.e. A must lie in Q_λ. This is different for $W_r^p(\mathfrak{B}\otimes\mathfrak{L}_n)$. If λ has no edge of $\mathfrak{L}_n$-colour n, then $A \in P_\lambda$ represents a decomposable element of $W_r(\mathfrak{B}\otimes\mathfrak{L}_n)$ iff $A \in Q_\lambda$. But if λ has a root of colour n and q incoming edges to the root vertex, then a tree in P_λ decomposes canonically as the composite of some morphism $b \in W_r^o(\mathfrak{B}\otimes\mathfrak{L}_n)$ with twigs and root of colour n and a copse of q trees with roots of colour n and an identity as label of the root vertex:

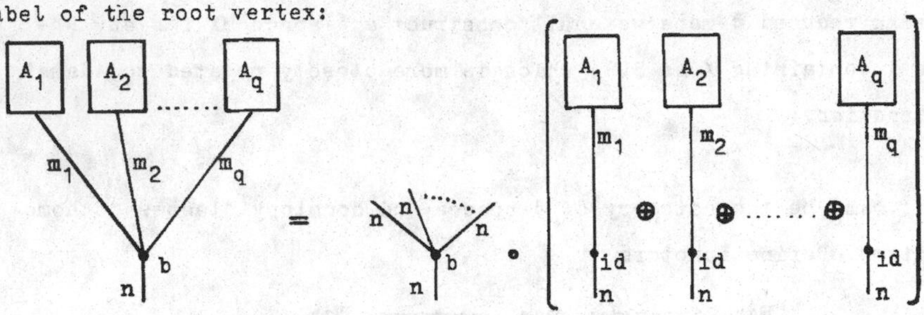

This requires a modification of Lemma 3.12 (a). Let $T_n \subset \mathrm{ob}\,(\mathfrak{B}\otimes\mathfrak{L}_n)$ be the subset of all objects $\underline{i} = \{i_1,\dots,i_q\}$ such that each i_r has $\mathfrak{L}_n$-colour n. Call a shape orbit essential if its shapes belong to reduced trees with no edge of $\mathfrak{L}_n$-colour n unless the space of labels of the root vertex is of the form $\mathfrak{B}(k,k)$, $k\in K$. For such a shape λ define $P'_\lambda = P_\lambda$ if λ has no edge of $\mathfrak{L}_n$-colour n, and $P'_\lambda \subset P_\lambda$ is the subspace of all trees whose root vertex label is id_k if λ has an n-coloured root and the space of root vertex labels is $\mathfrak{B}(k,k)$.

Lemma 3.12*: Let $\mathfrak{C}$ be a topological category with finite products and let $\mathfrak{D} \subset W_r(\mathfrak{B}\otimes\mathfrak{L}_n)$ be the full subcategory of objects in T_n.

(a) Given multiplicative functors $F : W_r^{p-1}(\mathfrak{B}\otimes\mathfrak{L}_n) \longrightarrow \mathfrak{C}$ and $H : \mathfrak{D} \to \mathfrak{C}$ which coincide on $W_r^{p-1}(\mathfrak{B}\otimes\mathfrak{L}_n) \cap \mathfrak{D}$, and a colloection of G-equivariant maps $f_\lambda : P'_\lambda \longrightarrow \mathfrak{C}(F\underline{i}, F(k))$ extending $F\circ(u_\lambda | Q_\lambda \cap P'_\lambda)$, one for each essential shape orbit. Then there is a unique multiplicative functor $F' : W_r^p(\mathfrak{B}\otimes\mathfrak{L}_n) \longrightarrow \mathfrak{C}$ extending F, coinciding with H on $\mathfrak{D} \cap W_r^p(\mathfrak{B}\otimes\mathfrak{L}_n)$, and satisfying $F'\circ(u_\lambda|P'_\lambda) = f_\lambda$ for all λ considered. The same holds for the $\mathfrak{L}_n$-homogeneous case if we replace P'_λ by $P'_{\lambda,q}$.

(b) Suppose given multiplicative functors $H : \mathfrak{D} \longrightarrow \mathfrak{C}$ and $F_p : W_r^p(\mathfrak{B}\otimes\mathfrak{L}_n) \longrightarrow \mathfrak{C}$, one for each $p>0$, such that F_p coincides with H on $W_r^p(\mathfrak{B}\otimes\mathfrak{L}_n) \cap \mathfrak{D}$ and with F_{p-1} on $W_r^{p-1}(\mathfrak{B}\otimes\mathfrak{L}_n)$. Then there exists a unique multiplicative functor $F : W_r(\mathfrak{B}\otimes\mathfrak{L}_n) \longrightarrow \mathfrak{C}$ such that $F|\mathfrak{D} = H$ and $F|W_r^p(\mathfrak{B}\otimes\mathfrak{L}_n) = F_p$. A similar result holds for the $\mathfrak{L}_n$-homogeneous case.

(c) Both (a) and (b) hold if we replace $\mathfrak{C}$ by a PROP and the word "multiplicative functor" by "PROP-functor".

<u>Proof</u>: Let λ be a shape of reduced trees with $\mathfrak{L}_n$-root. A tree A in P_λ decomposes canonically and continuously into $b\circ A'$ as illustrated above. Define $f_\lambda : P_\lambda \longrightarrow \mathfrak{C}$ by $f_\lambda(A) = H(b)\circ f(A')$ where $f(A')$ is given by the f_λ of the assumption. This defines G-equivariant maps $f_\lambda : P_\lambda \longrightarrow \mathfrak{C}(F\underline{i},F(k))$, one for each shape orbit of reduced trees and compatible with relation (4.32). Now proceed as in the proof of Lemma 3.12. ∎

Using Lemma 3.12* instead of (3.12), we obtain the following variants of the extension proposition and the lifting theorem (we state them in the generality needed although more general results hold).

<u>Lemma 4.39</u>: Let $\mathfrak{D} \subset W_r(\mathfrak{B}\otimes\mathfrak{L}_n)$ be a subcategory with the following properties

(a) If $x \in \mathfrak{D}$ is of the form $x = y\circ z$ or $x = y\oplus z$, then y and z are in $\mathfrak{D}$.

(b) The full subcategory of $W_r(\mathfrak{B}\otimes\mathfrak{L}_n)$ of objects in T_n is contained in $\mathfrak{D}$.

(c) $D_\lambda \cap P'_\lambda$ is closed in P'_λ and $(P'_\lambda,(D_\lambda \cup Q_\lambda) \cap P'_\lambda)$ is a G-NDR for all essential λ (we use the notation of (3.14) and (3.12*)).

Suppose given a multiplicative functor F from $W_r(\mathfrak{B}\otimes\mathfrak{L}_n)$ to a topological category $\mathfrak{C}$ with finite products and a homotopy of multiplicative functors $H(t) : \mathfrak{D} \longrightarrow \mathfrak{C}$ such that $H(0) = F|\mathfrak{D}$, then there exists a homotopy of multiplicative functors $F(t) : W_r(\mathfrak{B}\otimes\mathfrak{L}_n) \longrightarrow \mathfrak{C}$ extending F and $H(t)$.
The same holds for the $\mathfrak{L}_n$-homogeneous version if we substitute $D_\lambda \cup Q_\lambda \subset P'_\lambda$ by $D_{\lambda,q} \cup Q_{\lambda,q} \subset P'_{\lambda,q}$. ∎

Lemma 4.40: Given a diagram consisting of a K-coloured PROP $\mathfrak{B}$, a $(K\times \mathrm{ob}\mathfrak{L}_n)$-coloured PROP $\mathfrak{C}$, a sub-PROP $\mathfrak{A}$ of $W_r(\mathfrak{B}\otimes\mathfrak{L}_n)$ generated by some

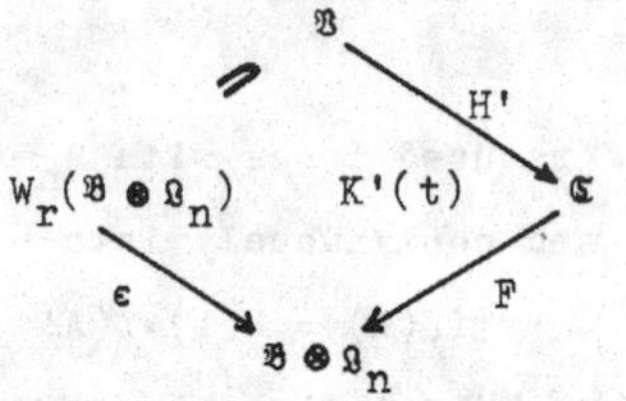

of the faces $d^iW_r(\mathfrak{B}\otimes\mathfrak{L}_n)$, PROP-functors F and H', and a homotopy of PROP-functors $K'(t) : \mathfrak{A} \longrightarrow \mathfrak{B}\otimes\mathfrak{L}_n$ from $\varepsilon|\mathfrak{A}$ to $F\circ H'$. Let $\mathfrak{A}' \subset \mathfrak{A}$ and $\mathfrak{C}' \subset \mathfrak{C}$ be the full subcategories whose objects lie in T_n. We require
(a) F is an equivariant equivalence and on $\mathfrak{C}'$ an isomorphism
(b) $H'|\mathfrak{A}' = F^{-1}\circ(\varepsilon|\mathfrak{A}')$
Then there exists a PROP-functor $H : W_r(\mathfrak{B}\otimes\mathfrak{L}_n) \longrightarrow \mathfrak{C}$ and a homotopy of PROP-functors $K(t) : W_r(\mathfrak{B}\otimes\mathfrak{L}_n) \longrightarrow \mathfrak{C}$ from ε to $F\circ H$ extending H' and $K(t)$. Moreover, any two such extensions H_0 and H_1 of H' are homotopic through a homotopy of PROP-functors $H(t) : W_r(\mathfrak{B}\otimes\mathfrak{L}_n) \longrightarrow \mathfrak{C}$ such that $H(t)|\mathfrak{A} = H'$.

The same holds for the $\mathfrak{L}_n$-homogeneous version.

Proof: To be able to apply Lemma 3.12* we substitute $\mathfrak{A}$ by the sub-PROP $\mathfrak{D}$ of $W_r(\mathfrak{B}\otimes\mathfrak{L}_n)$ generated by $\mathfrak{A}$ and the full subcategory $\mathfrak{E}$ of $W_r(\mathfrak{B}\otimes\mathfrak{L}_n)$

whose objects are in T_n. The functor H' is substituted by $H'' : \mathfrak{D} \to \mathfrak{C}$ given on $\mathfrak{B}$ by H' and on $\mathfrak{C}$ by $F^{-1} \cdot (\varepsilon | \mathfrak{C})$. We now proceed as in the proof of (3.17) using (3.12*) instead of (3.12). ∎

In analogy to (IV; §2) we define

<u>Definition 4.41</u>: Two reduced $\mathfrak{B}$-maps $(f,\rho),(g,\varkappa) : (X,\alpha) \to (Y,\beta)$ from a $W\mathfrak{B}$-space (X,α) to a $\mathfrak{B}$-space (Y,β) are called <u>homotopic</u>, if there is an action $\sigma : W_r(\mathfrak{B} \otimes \mathfrak{L}_2) \to \mathfrak{Top}$ such that $d^0(\sigma) = \rho$, $d^1(\sigma) = \varkappa$, and $d^2(\sigma) = s^0(\alpha)$.

<u>Definition 4.42</u>: Let (X,α) and (Y,β) be $W\mathfrak{B}$-spaces and (Z,γ) be a $\mathfrak{B}$-space. A reduced $\mathfrak{B}$-map $(h,\lambda) : (X,\alpha) \to (Z,\gamma)$ is called a <u>composite</u> of the $\mathfrak{B}$-map [$h\mathfrak{B}$-map] $(f,\rho) : (X,\alpha) \to (Y,\beta)$ with the reduced $\mathfrak{B}$-map $(g,\varkappa) : (Y,\beta) \to (Z,\gamma)$, if there exists an action $\sigma : W_r(\mathfrak{B} \otimes \mathfrak{L}_2) \to \mathfrak{Top}$ [$\sigma : HW_r(\mathfrak{B} \otimes \mathfrak{L}_2) \to \mathfrak{Top}$] such that $d^0(\sigma) = \varkappa$, $d^1(\sigma) = \lambda$, and $d^2(\sigma) = \rho$.

If we apply the proof of Lemma 4.13 to reduced trees, we obtain

<u>Lemma 4.43</u>: There exist actions $\sigma : W_r(\mathfrak{B} \otimes \mathfrak{L}_2) \to \mathfrak{Top}$ [$\sigma : HW_r(\mathfrak{B} \otimes \mathfrak{L}_2) \to \mathfrak{Top}$] with $d^0(\sigma) = \rho$, $d^1(\sigma) = \varkappa, d^2(\sigma) = s^0(\alpha)$ iff there is a homotopy through reduced $\mathfrak{B}$-maps $H(t) : W_r(\mathfrak{B} \otimes \mathfrak{L}_1) \to \mathfrak{Top}$ from (X,α) to (Y,β), where $\beta = d^0 d^0(\sigma)$, such that $H(0) = \rho$ and $H(1) = \varkappa$. ∎

<u>Corollary 4.44</u>: (a) The notion of homotopy of reduced $\mathfrak{B}$-maps is an equivalence relation. ∎

The proof of Theorem 4.9 also carries over to give

<u>Lemma 4.45</u>: Let $\mathfrak{C} \subset W_r(\mathfrak{B} \otimes \mathfrak{L}_n)$ be the sub-PROP generated by the faces $d^i W_r(\mathfrak{B} \otimes \mathfrak{L}_n)$, $i = 0,1,\ldots,k-1,k+1,\ldots,n$ with $k \neq 0,n$. Then there exists

a retraction PROP-functor

$$W_r(\mathfrak{B} \otimes \mathfrak{L}_n) \longrightarrow \mathfrak{C}$$

The same holds for the $\mathfrak{L}_n$-homogeneous version. ∎

As in (IV,2) this implies

<u>Proposition 4.46</u>: Given a $\mathfrak{B}$-map [$h\mathfrak{B}$-map] $(f,\rho) : (X,\alpha) \longrightarrow (Y,\beta)$ and a reduced $\mathfrak{B}$-map $(g,\varkappa) : (Y,\beta) \longrightarrow (Z,\gamma)$. Then there exists a composite reduced $\mathfrak{B}$-map $(h,\lambda) : (X,\alpha) \longrightarrow (Z,\gamma)$ of (f,ρ) and $(g,\varkappa)$, and its homotopy class depends only on the homotopy classes of (f,ρ) and $(g,\varkappa)$. ∎

As a second application of Lemma 4.39, we can carry out the proof of (4.14) for reduced categories and obtain

<u>Proposition 4.47</u>: Let $(f,\rho) : (X,\alpha) \longrightarrow (Y,\beta)$ be a reduced $\mathfrak{B}$-map and $g : X \longrightarrow Y$ a morphism in $\mathfrak{Top}_K$ homotopic to f. Then g carries a reduced $\mathfrak{B}$-map structure $(g,\varkappa) : (X,\alpha) \longrightarrow (Y,\beta)$ such that $(f,\rho) \simeq (g,\varkappa)$. ∎

<u>Corollary 4.48</u>: Given a $\mathfrak{B}$-map [$h\mathfrak{B}$-map] $(f,\rho) : (X,\alpha) \longrightarrow (Y,\beta)$ and a reduced $\mathfrak{B}$-map $(g,\varkappa) : (Y,\beta) \longrightarrow (Z,\gamma)$. Then there exists a composite reduced $\mathfrak{B}$-map $(h,\lambda) : (X,\alpha) \longrightarrow (Z,\gamma)$ of (f,ρ) and $(g,\varkappa)$ such that $h = g \circ f$. ∎

We next prove the analogue of (4.34) for reduced $\mathfrak{B}$-maps. In contrary to Theorem 4.34, part (a) can be shown easily, and we shall see that (4.34 a) is a consequence of this result.

<u>Theorem 4.49</u>: For any $W\mathfrak{B}$-space (X,α) there exists a $\mathfrak{B}$-space $M(X,\alpha) = (MX,\bar{\alpha})$ and a reduced $\mathfrak{B}$-map $(i_\alpha,\nu_\alpha) : (X,\alpha) \longrightarrow M(X,\alpha)$ such that

(a) The map $i_\alpha : X \longrightarrow MX$ embeds X as SDR into MX

(b) Any reduced $\mathfrak{B}$-map $(f,\rho) : (X,\alpha) \longrightarrow (Y,\beta)$ is the canonical compo-

site of (f_α, ν_α) and a unique $\mathfrak{B}$-homomorphism $h : M(X,\alpha) \longrightarrow (Y,\beta)$.
(c) If we change (f,ρ) inside its homotopx class, the induced $\mathfrak{B}$-homomorphism h stays inside its homotopy class.

Proof: Define $MX = \{MX_k\}$ by

(4.50) $$MX_k = \bigcup_{\underline{i}\in\mathfrak{B}} W_r(\mathfrak{B} \otimes \mathfrak{L}_1)(\underline{i}^0, k^1) \times X_{\underline{i}}/\sim$$

with $(c \circ a; x_1, \ldots, x_n) \sim (c; a(x_1, \ldots, x_n))$, $c \in W_r(\mathfrak{B} \otimes \mathfrak{L}_1)(\underline{j}^0, k^1)$, $a \in W_r(\mathfrak{B} \otimes \mathfrak{L}_1)(\underline{i}^0, \underline{j}^0)$. For the definition of the $\mathfrak{B}$-structure on MX note that $d^0 : \mathfrak{B} \longrightarrow W_r(\mathfrak{B} \otimes \mathfrak{L}_1)$ is an inclusion functor. Define

$$b[(c_1; \underline{x}_1), \ldots, (c_n; x_n)] = [b \circ (c_1 \oplus \ldots \oplus c_n); \underline{x}_1, \ldots, \underline{x}_n]$$

$b \in \mathfrak{B}$. The canonical maps $W_r(\mathfrak{B} \otimes \mathfrak{L}_1)(\underline{i}^0, k^1) \times X_{\underline{i}} \longrightarrow MX_k$ define a reduced $\mathfrak{B}$-map $(i_\alpha, \nu_\alpha) : (X,\alpha) \longrightarrow M(X,\alpha)$ whose underlying map i is given by

$x \longmapsto$ 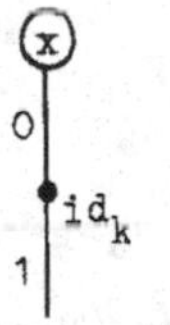 $x \in X_k$

in cherry tree notation.

Given a reduced $\mathfrak{B}$-map $(f,\rho) : (X,\alpha) \longrightarrow (Y,\beta)$, the induced $\mathfrak{B}$-homomorphism $h : M(X,\alpha) \longrightarrow (Y,\beta)$ is given by

$$h(c; \underline{x}) = \rho(c)(\underline{x})$$

It is the unique homomorphism satisfying $(f,\rho) = h \circ (i_\alpha, \nu_\alpha)$ with the canonical composition on the right. In view of Lemma 4.43, a change of (f,ρ) by a homotopy changes ρ by a homotopy of functors and hence h by a homotopy through homomorphisms. It remains to prove (a): As in the proof of (4.23), we express MX in terms of cherry trees. Then MX is the space of all reduced cherry trees, i.e. reduced trees whose roots have $\mathfrak{L}_1$-colour 1, whose twigs have $\mathfrak{L}_1$-colour 0, and there is a cherry in X_k assigned to each twig of K-colour k. On this space we have the relations (4.28), but for reduced trees only:

(a) = 3.1 (a)

(b) = 3.1 (b) for permutations only, and the cherries are permuted along with the twigs

(c) = 3.1 (c)

(d) If a reduced tree A has an internal edge of length 1, i.e. A decomposes into $A_1 \circ A_2$, then $(A; x_1,\dots,x_n) \sim (A_1; x_1,\dots,x_p, y, x_{q+1},\dots,x_n)$ where $(x_1,\dots,x_n)$ and $(x_{p+1},\dots,x_q)$ are the cherries of A and A_2 in clockwise order, and $y = \alpha(A_2)(x_{p+1},\dots,x_q)$.

The deformation retraction MX ⟶ X is given by step 2 of the proof of (4.23 a). ∎

If $\mathfrak{B} = \mathfrak{A}$, the PRO belonging to the theory of monoids, and (X,α) is a $W\mathfrak{A}$-space then $(X,\alpha) \longrightarrow (MX,\bar{\alpha})$ is essentially the construction of Adams mentioned in chapter I.

Let

$$\mathfrak{Map}_{\mathfrak{B}} \xleftarrow{\;J\;} \mathfrak{Hom}_{\mathfrak{B}} \xrightarrow{\;J'\;} \mathfrak{Map}_{h\mathfrak{B}}$$

be the functors previously defined. Using (4.46) and (4.49 b), we can extend the correspondence $(X,\alpha) \longrightarrow M(X,\alpha)$ to functors $M : \mathfrak{Map}_{\mathfrak{B}} \longrightarrow \mathfrak{Hom}_{\mathfrak{B}}$ and $M' : \mathfrak{Map}_{h\mathfrak{B}} \longrightarrow \mathfrak{Hom}_{\mathfrak{B}}$ in the same way as in the proof of (4.31).

We now want to compare the functors N and N' with M and M'. In view of Proposition 4.32 we restrict our attention to N' and M'. Recall that N' is left adjoint to the functor $J' : \mathfrak{Hom}_{\mathfrak{B}} \longrightarrow \mathfrak{Map}_{\mathfrak{B}}$. The front adjunction

$$\eta : Id \longrightarrow J'N'$$

is given by the $\mathfrak{B}$-maps $(p_\alpha, \eta_\alpha) : (X,\alpha) \longrightarrow (NX, \alpha^* \circ \epsilon) = J'N'(X,\alpha)$. For any $h\mathfrak{B}$-map $(f,\rho) : (X,\alpha) \longrightarrow (Y,\beta)$ we have a diagram where r is induced by the universal property of η and the reduced $\mathfrak{B}$-map (i,ν) is considered as $h\mathfrak{B}$-map.

$$\begin{array}{ccccc}
 & & (X,\alpha) & & \\
 & \swarrow^{(i_\alpha,\nu_\alpha)} & \downarrow{\scriptstyle (f,\rho)} & \searrow^{(p_\alpha,\eta_\alpha)} & \\
J'M'(X,\alpha) & & \xleftarrow{\quad J'r_\alpha \quad} & & J'N'(X,\alpha) \\
\downarrow{\scriptstyle J'M'(f,\rho)} & & (Y,\beta) & & \downarrow{\scriptstyle J'N'(f,\rho)} \\
 & \swarrow^{(i_\beta,\nu_\beta)} & & \searrow^{(p_\beta,\eta_\beta)} & \\
J'M'(Y,\beta) & & \xleftarrow{\quad J'r_\beta \quad} & & J'N'(Y,\beta)
\end{array}$$

Since the backward squares commute, so does the front square because of the universal property of η_α. Hence

$$r : N' \longrightarrow M'$$

is a natural transformation. We want to show that it is a natural equivalence. By (4.49 b) there is a $\mathfrak{B}$-homomorphism $s_\alpha : M'J'N'(X,\alpha) \to N'(X,\alpha)$ induced by

$$\begin{array}{ccc}
J'N'(X,\alpha) & \xrightarrow{\quad (i,\nu) \quad} & J'M'J'N'(X,\alpha) \\
 & {\scriptstyle Id}\searrow \qquad \swarrow{\scriptstyle J's_\alpha} & \\
 & J'N'(X,\alpha) &
\end{array}$$

because the identity is apparently a reduced map. The homomorphism

$$s_\alpha \circ M'(p_\alpha,\eta_\alpha) : M'(X,\alpha) \longrightarrow M'J'N'(X,\alpha) \longrightarrow N'(X,\alpha)$$

is an inverse of r_α in $\mathfrak{Hom}_{h\mathfrak{B}}$ because

$$\begin{aligned}
J'(s_\alpha \circ M'(p_\alpha,\eta_\alpha) \circ r_\alpha) \circ (p_\alpha,\eta_\alpha) &= J's_\alpha \circ J'M'(p_\alpha,\eta_\alpha) \circ (i_\alpha,\nu_\alpha) \\
&= J's_\alpha \circ (i,\nu) \circ (p_\alpha,\eta_\alpha) \\
&= (p_\alpha,\eta_\alpha)
\end{aligned}$$

and

$$\begin{aligned}
J'(r_\alpha \circ s_\alpha \circ M'(p_\alpha,\eta_\alpha)) \circ (i_\alpha,\nu_\alpha) &= J'r_\alpha \circ J's_\alpha \circ (i,\nu) \circ (p_\alpha,\eta_\alpha) \\
&= J'r_\alpha \circ (p_\alpha,\eta_\alpha) \\
&= (i_\alpha,\nu_\alpha)
\end{aligned}$$

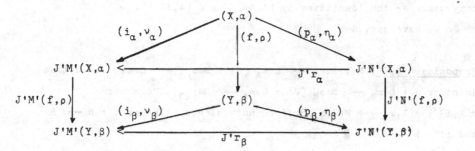

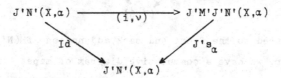

so that $s_\alpha \circ M'(p_\alpha,\eta_\alpha) \circ r_\alpha$ and $r_\alpha \circ s_\alpha \circ M'(p_\alpha,\eta_\alpha)$ are in the same homotopy class as the identities by (4.35 c) and (4.49 b).

So we have proved

<u>Proposition 4.51</u>: $M : \mathfrak{Map}_{\mathfrak{B}} \longrightarrow \mathfrak{Hom}_{\mathfrak{B}}$ $[M' : \mathfrak{Map}_{n\mathfrak{B}} \longrightarrow \mathfrak{Hom}_{\mathfrak{B}}]$ is left adjoint to $J : \mathfrak{Hom}_{\mathfrak{B}} \longrightarrow \mathfrak{Map}_{\mathfrak{B}}$ $[J' : \mathfrak{Hom}_{\mathfrak{B}} \longrightarrow \mathfrak{Map}_{n\mathfrak{B}}]$. Moreover $r(X,\alpha) : N(X,\alpha) \longrightarrow M(X,\alpha)$ defines natural equivalences $r : N \longrightarrow M$ and $r' : N' \longrightarrow M'$. ∎

The front and back adjunction of the adjoint pair (M',J') are given by $(i_\alpha,\nu_\alpha) : (X,\alpha) \longrightarrow J'M'(X,\alpha)$ of (4.49) and the homomorphisms $\mu(Y,\beta) : M'J'(Y,\beta) \longrightarrow (Y,\beta)$ determined by the diagram

$$\begin{array}{ccc} J'(Y,\beta) & \xrightarrow{\nu J'(Y,\beta)} & J'M'J'(Y,\beta) \\ & {}_{Id}\searrow \quad \swarrow_{J'\mu(Y,\beta)} & \\ & J'(Y,\beta) & \end{array} \qquad (Y,\beta) \in \mathfrak{Hom}_{\mathfrak{B}}$$

They are related to the front and back adjunction of (N',J') by r'. In particular, we have a commutative diagram of maps

$$\begin{array}{ccc} X & \xrightarrow{p_\alpha} & NX \\ & {}_{i_\alpha}\searrow \quad \swarrow_{r_\alpha} & \\ & MX & \end{array}$$

for each $W\mathfrak{B}$-space (X,α). Since r_α and i_α are homotopy equivalences, r_α is a homotopy equivalence, which fills the gap left in the proof of (4.35).

Evidently, $r(X,\alpha) : N(X,\alpha) \longrightarrow M(X,\alpha)$ is induced by the projections $\pi : HW(\mathfrak{B} \otimes \mathfrak{L}_1)(\underline{i}^o,k^1) \longrightarrow W_r(\mathfrak{B} \otimes \mathfrak{L}_1)(\underline{i}^o,k^1)$. If $Rmap_{\mathfrak{B}}((X,\alpha),J'(Y,\beta))$ denotes the set of homotopy classes of reduced $\mathfrak{B}$-maps $(X,\alpha) \longrightarrow (Y,\beta)$, we therefore have a commutative diagram

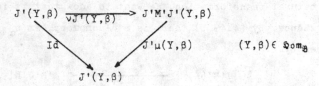

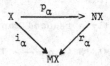

$$\begin{array}{ccc}
\mathfrak{Map}_{h\mathfrak{B}}((X,\alpha),J'(Y,\beta)) & \cong & \mathfrak{Hom}_{\mathfrak{B}}(N'(X,\alpha),(Y,\beta)) \\
\uparrow{\scriptstyle \pi^*} & & \uparrow{\scriptstyle r(X,\alpha)^*} \\
\mathrm{Rmap}_{\mathfrak{B}}((X,\alpha),J'(Y,\beta)) & \cong & \mathfrak{Hom}_{\mathfrak{B}}(M'(X,\alpha),(Y,\beta))
\end{array}$$

and Proposition 4.51 implies

<u>Corollary 4.52</u>: The projection functors $\pi_1 : W(\mathfrak{B}\otimes\mathfrak{L}_1) \longrightarrow W_r(\mathfrak{B}\otimes\mathfrak{L}_1)$ and $\pi_2 : HW(\mathfrak{B}\otimes\mathfrak{L}_1) \longrightarrow W_r(\mathfrak{B}\otimes\mathfrak{L}_1)$ induce bijections

$$\mathfrak{Map}_{\mathfrak{B}}((X,\alpha),J(Y,\beta)) \xleftarrow[\pi_1^*]{} \mathrm{Rmap}_{\mathfrak{B}}((X,\alpha),J'(Y,\beta)) \xrightarrow[\pi_2^*]{} \mathfrak{Map}_{h\mathfrak{B}}((X,\alpha),J'(Y,\beta))$$

In particular, any $\mathfrak{B}$-map or $h\mathfrak{B}$-map into a $\mathfrak{B}$-space is homotopic to a reduced $\mathfrak{B}$-map. ∎

<u>Remark</u>: One might be tempted to dualize Theorem 4.35 and 4.49. That is, given a $W\mathfrak{B}$-space (X,α), one might want to construct a $\mathfrak{B}$-space (VX,α') and a $h\mathfrak{B}$-map $(p,\varkappa) : (VX,\alpha'\circ\varepsilon) \longrightarrow (X,\alpha)$ such that any $h\mathfrak{B}$-map $(f,\rho) : (Y,\beta\circ\varepsilon) \longrightarrow (X,\alpha)$ from a $\mathfrak{B}$-space (Y,β) to (X,α) factors uniquely as

$$\begin{array}{ccc}
 & J'(Y,\beta) & \\
{\scriptstyle J'h}\swarrow & & \searrow{\scriptstyle (f,\rho)} \\
J'(VX,\alpha') & \xrightarrow[(p,\varkappa)]{} & (X,\alpha)
\end{array}$$

where $h : (Y,\beta) \longrightarrow (VX,\alpha')$ is a $\mathfrak{B}$-homomorphism. More precisely, (f,ρ) is the canonical composite of $(p,\varkappa)$ and a $\mathfrak{B}$-homomorphism. The following example shows that this is not possible in general. Let $\mathfrak{B}=\mathfrak{A}$, the PRO belonging to the theory of monoids, let (X,α) be a $W\mathfrak{A}$-space, (U,μ) a monoid and $(p,\varkappa) : (U,\mu\circ\varepsilon) \longrightarrow (X,\alpha)$ an $\mathfrak{A}$-map. Then $(p,\varkappa)$ fails to have the required universal property: Suppose $(f,\rho) : (Y,\beta\circ\varepsilon) \longrightarrow (X,\alpha)$ is an $\mathfrak{A}$-map. Let C be the cyclic monoid on one generator c and $i_y : C \longrightarrow (Y,\beta)$ the homomorphism defined by $i_y(c) = y$ for some $y\in Y$. If $(p,\varkappa)$ were universal, (f,ρ) lifts to a homomorphism $h : (Y,\beta) \longrightarrow (U,\mu)$, and $h\circ i_y$ is the unique homomorphism lifting the canonical composite $(f,\rho)\circ i_y : C \longrightarrow (X,\alpha)$. As y varies, the collection of $h\mathfrak{B}$-maps $(f,\rho)\circ i_y$ determine h and therefore the whole structure of (f,ρ) uniquely, which is absurd, because the adjoints

$$HW(\mathfrak{A}\otimes\mathfrak{L}_1)(n^o,1^1)\times Y^n \longrightarrow X$$

(n denotes the unique object $[n] \longrightarrow *$) of $\rho : HW(\mathfrak{A}\otimes\mathfrak{L}_1)(n^o,1^1) \longrightarrow \mathfrak{Top}(Y^n,X)$ are not determined by the $(f,\rho)\circ i_y$ on the elements $(a; y_1,\ldots,y_n)$, where a is indecomposable and not each y_i of the form $y_i = z^{r_i}$ with z fixed for all i.

6. AN EQUIVALENCE OF CATEGORIES

In this section we show that the categories $\mathfrak{Map}_{\mathfrak{B}}$ and $\mathfrak{Map}_{h\mathfrak{B}}$ are equivalent to a category of fractions of the category $\mathfrak{Mor}_{\mathfrak{B}}$ of $\mathfrak{B}$-spaces and $\mathfrak{B}$-homomorphisms.

Let $\mathfrak{C}$ be an arbitrary category and Σ a class of morphisms in $\mathfrak{C}$. The category of fractions $\mathfrak{C}/\Sigma$ (see [23]) has the same objects as $\mathfrak{C}$. Its morphisms are words in words in following generators

(a) the morphisms of $\mathfrak{C}$

(b) a morphism $\overline{g} : X \longrightarrow Y$ for each morphism $g : Y \longrightarrow X$ in Σ.

The relations are

(i) $[f|g] = [f\circ g]$ if $f,g \in \text{mor}\,\mathfrak{C}$

(ii) $[g|\overline{g}] = id$, $[\overline{g}|g] = id$

(iii) $[id] = id$

There is a canonical functor $P = P_\Sigma : \mathfrak{C} \longrightarrow \mathfrak{C}/\Sigma$ which is the identity on objects and which sends a morphism f to its equivalence class in $\mathfrak{C}/\Sigma$. The functor P has the universal property that given a functor $F : \mathfrak{C} \longrightarrow \mathfrak{D}$ such that $F(g)$ is an isomorphism in $\mathfrak{D}$ for each $g \in \Sigma$, there exists a unique functor $G : \mathfrak{C}/\Sigma \longrightarrow \mathfrak{D}$ such that $G\circ P = F$.

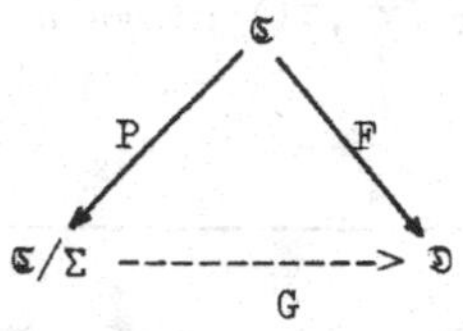

In our case we take $\Sigma \subset \mathfrak{Mor}_{\mathfrak{B}}$ or $T \subset \mathfrak{Hom}_{\mathfrak{B}}$ to be the class of all homomorphisms respectively of all homotopy classes of homomorphisms, whose underlying maps are homotopy equivalences in $\mathfrak{Top}_K$. The following proposition is an immediate consequence of the results of the previous section (see also [23; Prop. 1.3,p.7]).

<u>Proposition 4.53</u>: The functor $P \circ M' : \mathfrak{Map}_{h\mathfrak{B}} \longrightarrow \mathfrak{Hom}_{\mathfrak{B}} \longrightarrow \mathfrak{Hom}_{\mathfrak{B}}/T$ is an equivalence of categories.

<u>Proof</u>: If $f \in T$, then $J'(f)$ is an isomorphism in $\mathfrak{Map}_{h\mathfrak{B}}$ by (4.21). Hence there exists a unique functor $F : \mathfrak{Hom}_{\mathfrak{B}}/T \longrightarrow \mathfrak{Map}_{h\mathfrak{B}}$ such that $F \circ P = J'$. The front adjunction $\nu : Id \longrightarrow J' \circ M' = F \circ (P \circ M')$ is a natural equivalence, and the back adjunction $\mu : M' \circ J' \longrightarrow Id$ induces a natural equivalence $P\mu : (P \circ M') \circ J' \longrightarrow Id$, because each $\mu(Y,\beta)$ is a homotopy equivalence considered as morphism of $\mathfrak{Top}_K$. ∎

If $f \in \mathfrak{Mor}_{\mathfrak{B}}$ is a homotopy equivalence as morphism in $\mathfrak{Top}_K$, then $J' \circ H(f)$ is an isomorphism, where $H : \mathfrak{Mor}_{\mathfrak{B}} \longrightarrow \mathfrak{Hom}_{\mathfrak{B}}$ is the functor sending each $\mathfrak{B}$-homomorphism to its homotopy class. Hence there is a unique functor

$$G : \mathfrak{Mor}_{\mathfrak{B}}/\Sigma \longrightarrow \mathfrak{Map}_{h\mathfrak{B}}$$

such that $G \circ P = J' \circ H$. More interesting, but considerably harder to prove is

<u>Proposition 4.54</u>: The functor $G : \mathfrak{Mor}_{\mathfrak{B}}/\Sigma \longrightarrow \mathfrak{Map}_{h\mathfrak{B}}$ is an equivalence of categories.

Proof: Again the adjoint pair (M',J') plays an essential role. Consider

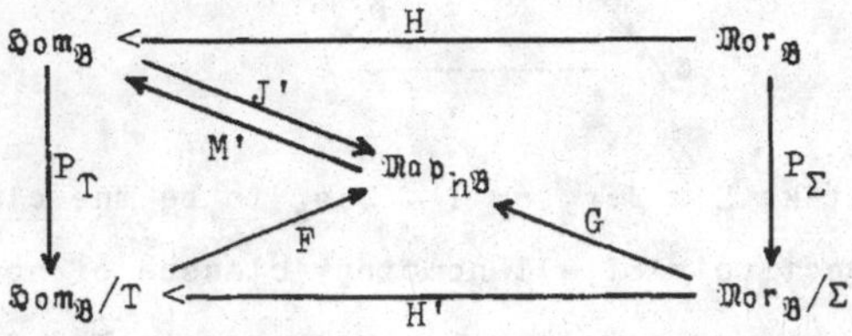

where H' is the functor induced by H. Using the universal property of P_Σ we find that $G = F \circ H'$.

Claim: The functor $P_T \circ M'$ factors through $\mathfrak{Mor}_{\mathfrak{B}}/\Sigma$

Proof: By (4.49), each reduced $\mathfrak{B}$-map $(f,\rho) : (X,\alpha) \longrightarrow (Y,\beta)$ induces a unique $\mathfrak{B}$-homomorphism $h : M(X,\alpha) \longrightarrow (Y,\beta)$, such that (f,ρ) is the canonical composite $h \circ (i_\alpha, \nu_\alpha)$. Suppose $(f',\rho') : (X,\alpha) \longrightarrow (Y,\beta)$ is a reduced $\mathfrak{B}$-map homotopic to (f,ρ) and h' its induced $\mathfrak{B}$-homomorphism. Suppose we knew that $P_\Sigma(h) = P_\Sigma(h')$, we could define a functor

$$R : \mathfrak{Map}_{h\mathfrak{B}} \longrightarrow \mathfrak{Mor}_{\mathfrak{B}}/\Sigma$$

by $R(X,\alpha) = M(X,\alpha)$ on objects, and on $h\mathfrak{B}$-maps $(g,\varkappa) : (X,\alpha) \longrightarrow (Z,\gamma)$ by $R(g,\varkappa) = P_\Sigma(r)$, where r is the $\mathfrak{B}$-homomorphism induced by some composite of $(g,\varkappa)$ and (i_γ, ν_γ). By our supposition, this definition is independent of the choice of the representative $(g,\varkappa)$ and of the composite. Using the universal property of (i_α, ν_α) we find that R is a functor, and evidently $H' \circ R = P \circ M'$.

Let $u_k \in HW_r(\mathfrak{B} \circledast \mathfrak{L}_2)$ be the morphism

$u_k =$ 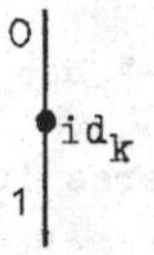

If it is clear from the context, we drop the index k. Let $\mathfrak{C}$ be the quotient category of $HW_r(\mathfrak{B} \circledast \mathfrak{L}_2)$ modulo the relation (compare the proof of (4.13)):

A tree A whose root has $\mathfrak{L}_2$-colour 1 and whose twigs have $\mathfrak{L}_2$-colour 0 is related to $A'\circ(u\oplus\ldots\oplus u)$ where A' is obtained from A by changing the $\mathfrak{L}_2$-colour of all edges to 1.

For any $W\mathfrak{B}$-space (X,α) let $Q(X,\alpha) = (QX,\alpha')$ be the $\mathfrak{B}$-space with

$$QX_k = \bigcup_{\underline{i}\in\mathfrak{B}} \mathfrak{C}(\underline{i}^{\circ},k^2)\times X_{\underline{i}}/\sim$$

with the relation

$$(*) \qquad (A_1\circ A_2;\ x_1,\ldots,x_n) \sim (A_1\circ(v\oplus\ldots\oplus v);\ \alpha(A_2')(x_1,\ldots,x_n))$$

where A_1 and A_2 represent morphisms in $\mathfrak{C}$, and $A_2' = A_2$ and $(v\oplus\ldots\oplus v) = id$ if the twigs of A_1 have $\mathfrak{L}_2$-colour 0 while A_2' is obtained from A_2 by changing the $\mathfrak{L}_2$-colour of all edges to 0 and $(v\oplus\ldots\oplus v)=(u\oplus\ldots\oplus u)$ if the twigs of A_1 have $\mathfrak{L}_2$-colour 1. Note that $\mathfrak{B}$ is contained in $\mathfrak{C}$ as the subcategory of all objects $\underline{i}^2$, $\underline{i}\in\mathfrak{B}$ (as usually, $\underline{i}^2=\{(i_1,2),\ldots,(i_n,2)\} \in \mathfrak{B}\otimes\mathfrak{L}_2$ for $\underline{i} = \{i_1,\ldots,i_n\} \in \mathfrak{B}$). The $\mathfrak{B}$-structure on $Q(X,\alpha)$ is given by

$$b[(c_1;\ \underline{x}_1),\ldots,(c_n;\ \underline{x}_n)] = (b\circ(c_1\oplus\ldots\oplus c_n);\ \underline{x}_1,\ldots,\underline{x}_n)$$

Define $\mathfrak{B}$-homomorphisms $i_o, i_1 : M(X,\alpha) \longrightarrow Q(X,\alpha)$ and $r : Q(X,\alpha) \longrightarrow M(X,\alpha)$ on representatives by

$$i_o(A;\ x_1,\ldots,x_n) = (A';\ x_1,\ldots,x_n)$$

$$i_1(A;\ x_1,\ldots,x_n) = (A''\circ(u\oplus\ldots\oplus u);\ x_1,\ldots,x_n)$$

$$r(B;\ x_1,\ldots,x_n) = (B';\ x_1,\ldots,x_n)$$

where A' is obtained from A by changing the root colour from 1 to 2, and A" from A by changing the root colour from 1 to 2 and the colour of all other edges from 0 to 1. The tree B' is obtained from B by changing the colours 2 and 1 of the edges to 1 respectively 0. Then $r\circ i_o = id = r\circ i_1$. We have an inclusion $j : X \longrightarrow Q(X,\alpha)$ given by

$x \longmapsto (A_k;x)$, $x\in X_k$ with $A_k =$

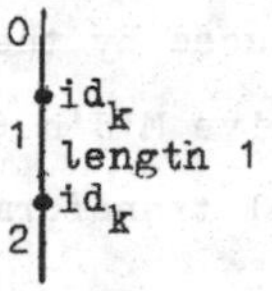

Then $j = i_1 \circ i_\alpha$. Now $j(X)$ is a SDR of $Q(X,\alpha)$. The deformation is given by substituting the roots of the representing trees of $Q(X,\alpha)$ at time t by

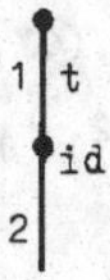

At $t = 0$, we have the identity, and at $t = 1$ relation (*) reduces the element to one in $j(X)$. Moreover $j(X)$ is kept pointwise fixed throughout the deformation. Since i_α is a homotopy equivalence, so is i_1 and hence r and i_o.

Now given two reduced $\mathfrak{B}$-maps $(f_l,\rho_l) : (X,\alpha) \longrightarrow (Y,\beta)$, $l = 0,1$, which are homotopic. Then there exists an action $\sigma : HW_r(\mathfrak{B} \otimes \mathfrak{L}_2) \rightarrow \mathfrak{Top}$ such that $d^o(\sigma) = \rho_1$, $d^1(\sigma) = \rho_o$ and $d^2(\sigma) = s^o(\alpha)$. This action σ induces a $\mathfrak{B}$-homomorphism $F : Q(X,\alpha) \longrightarrow (Y,\beta)$ by

$$F(A; x_1,\ldots,x_n) = \sigma(A)\,(x_1,\ldots,x_n)$$

If $h_l : M(X,\alpha) \longrightarrow (Y,\beta)$ is the $\mathfrak{B}$-homomorphism induced by (f_l,ρ_l), then $h_l = F \circ i_l$, $l = 0,1$. Now

$$P_\Sigma(h_o) = P_\Sigma(F) \circ P_\Sigma(i_o) = P_\Sigma(F) \circ P_\Sigma(i_1) = P_\Sigma(h_1)$$

because $P_\Sigma(i_o)$ and $P_\Sigma(i_1)$ are both inverses of $P_\Sigma(r)$. This proves the claim.

The natural equivalence $Id \longrightarrow F \circ (P_T \circ M') = F \circ H' \circ R = G \circ R$ of the previous proposition provides the first equivalence. We can choose representing homomorphisms $\varkappa_\beta : RG(Y,\beta) = M'J'(Y,\beta) \longrightarrow (Y,\beta)$ of the back adjunction $\mu_\beta : M'J'(Y,\beta) \longrightarrow (Y,\beta)$ of the adjoint pair (M',J') such that $Id_{(Y,\beta)}$ is the canonical composite $Id = J'\varkappa_\beta \circ \nu J'(Y,\beta)$. Given a $\mathfrak{B}$-homomorphism $h : (Y,\beta) \longrightarrow (Z,\gamma)$ define $M'J'h$ to be the $\mathfrak{B}$-homomorphism induced by the canonical composite $\nu J'(Z,\gamma) \circ h$. Using this representative $M'J'h \in \mathfrak{Mor}_{\mathfrak{B}}$ it is easy to check that the $\varkappa_\beta$ constitute a natural transformation $R \circ G \longrightarrow Id$. Since $\nu J'(Y,\beta)$ is a homotopy equivalence on the underlying spaces, so is $\varkappa_\beta$. Hence $R \circ G \longrightarrow Id$ is a natural equivalence. ∎

Corollary 4.55: Since $\mathfrak{Map}_{\mathfrak{B}}$ is isomorphic to $\mathfrak{Map}_{h\mathfrak{B}}$, the results (4.53) and (4.54) hold for $\mathfrak{Map}_{\mathfrak{B}}$, too. ∎

Proposition 4.54 to some extent generalizes a result of Malraison [31] who by slightly different means proved that the category $\mathfrak{H}$ of monoids and homotopy classes of A_∞-maps (see (1.14)) is isomorphic (and not just equivalent) to the category of fractions $\mathfrak{Mon}/\Sigma$, where $\mathfrak{Mon}$ is the category of monoids and homomorphisms and Σ is the class of homomorphisms which are homotopy equivalences. By (3.25), the category $\mathfrak{H}$ is essentially the full subcategory of $\mathfrak{Map}_{h\mathfrak{A}}$ of all $\mathfrak{A}$-spaces (= monoids).

7. HOMOTOPY INVARIANCE FOR GENERAL THEORIES

In §5 we showed that each $W\mathfrak{B}$-space can be embedded as a SDR into a $\mathfrak{B}$-space. The PROPs $W\mathfrak{B}$ and $\mathfrak{B}$ are related by the equivariant equivalence $\varepsilon : W\mathfrak{B} \longrightarrow \mathfrak{B}$. We want to generalize this result by substituting $W\mathfrak{B}$ and $\mathfrak{B}$ by general theories Θ_1 and Θ_2 and the functor ε by a theory functor which is a homotopy equivalence on each morphism space. So we are aiming towards a result of the nature that for reasonable theories Θ_1 and Θ_2 every reasonable Θ_1-space embeds as a SDR of a Θ_2-space. We restrict our attention to monochrome theories and leave it to the reader to make the necessary modifications for the general case. As usual, n denotes the unique object $[n] \longrightarrow *$ and S_k the group of permutations of $[k]$.

Definition 4.56: A morphism $b \in \Theta(k,1)$ is called non-degenerate if it does not have the form $b = c \bullet \sigma^*$ with $\sigma : [l] \longrightarrow [k]$ a proper monomorphism. The theory Θ is called <u>proper</u> if each morphism $b \in \Theta(k,1)$ has the form $b = c \bullet \sigma^*$ with c non-degenerate and σ a monomorphism, uniquely up to the equivalence $(c \bullet \pi^*) \bullet (\sigma \bullet \pi^{-1})^* = c \bullet \sigma^*$ for permutations π.

The following result shows that all interesting theories are proper ones.

<u>Lemma 4.57</u>: (a) Let Θ be a theory such that $b \bullet \sigma^* = c \bullet \sigma^*$ for $b,c \in \Theta(0,1)$ implies $b = c$. Then each morphism $f \in \Theta(k,1)$ is of the form $f = g \bullet \sigma^*$ with g non-degenerate and σ a monomorphism. If $g \bullet \sigma^* = g' \bullet \tau^*$ with g and g' non-degenerate, then there is a permutation π with $g = g' \bullet \pi^*$.
(b) If in addition $b \bullet \sigma^* = b \bullet \tau^*$ for $b \in \Theta(1,1)$ and σ,τ monomorphisms implies $\sigma=\tau$, then Θ is proper
(c) Let Θ be a theory such that for each composite $b \bullet \sigma^* = c \bullet \tau^*$ with b, c non-degenerate and $\sigma : p \longrightarrow n$, $\tau : q \longrightarrow n$ monic there exist monomorphisms $\mu : l \longrightarrow p$ and $\nu : l \longrightarrow q$ and a morphism $d \in \Theta(l,1)$ such that $b = d \bullet \mu^*$, $c = d \bullet \nu^*$ and $\sigma \bullet \mu = \tau \bullet \nu$. Then Θ is proper.

All interesting theories satisfy (c), because given an operation $a : A \times B \times C \longrightarrow X$ which factors through the projections $A \times B \times C \longrightarrow B \times C$ and $A \times B \times C \longrightarrow A \times B$, then it factors through the projection $A \times B \times C \to B$ in all interesting cases.

<u>Proof</u>: Obviously, any morphism $a \in \Theta(n,1)$ can be decomposed as $a = b \bullet \sigma^*$ with b non-degenerate and $\sigma : p \longrightarrow n$ a monomorphism. Let $a = c \bullet \tau^*$ be another such decomposition with $\tau : q \longrightarrow n$.
Proof (a): If $p = q = 0$, then $\sigma = \tau$ and $b = c$ by assumption. So suppose $0 < p \geq q$. Then there is an epimorphism $\mu : n \longrightarrow p$ such that

$\mu \circ \sigma = id$. Hence

$$b = b \circ \sigma^* \circ \mu^* = c \circ \tau^* \circ \mu^* = c \circ (\mu \circ \tau)^*$$

Since b is non-degenerate, $\mu \circ \tau : q \longrightarrow p$ is an epimorphism. Hence $p = q$ and $\mu \circ \tau$ is an isomorphism.

Proof (b): Using the notation of (a) we have to show that $\tau = \sigma \circ \mu \circ \tau$, because then $b = c \circ \pi^*$ and $\sigma = \tau \circ \pi^{-1}$ with $\pi = \mu \circ \tau$, which we have shown to be a permutation. If $p = q = 1$ this follows from

$$c \circ \tau^* = b \circ \sigma^* = c \circ (\sigma \circ \mu \circ \tau)^*$$

and the assumption. So suppose $p = q > 1$. If $\sigma([p]) = \tau([p])$, then there is a permutation $\rho : [p] \longrightarrow [p]$ such that $\tau = \sigma \circ \rho$. Hence

$$\tau = \sigma \circ \rho = \sigma \circ \mu \circ \sigma \circ \rho = \sigma \circ \mu \circ \tau$$

If $\sigma([p]) \neq \tau([p])$, there exists $i \in [p]$ such that $i \in \sigma([p]) - \tau([p])$. Since $p > 1$, we can choose the epimorphism μ such that $\mu^{-1}(\mu(i)) = \{i\}$, so that $\mu \circ \tau$ is not a permutation, which is a contradiction.

Proof (c): By assumption, there are monomorphisms $\mu : l \longrightarrow p$ and $\nu : l \longrightarrow q$ and a morphism $d \in \Theta(l,1)$ such that $b = d \circ \mu^*$, $c = d \circ \nu^*$ and $\tau \circ \nu = \sigma \circ \mu$. Since b and c are non-degenerate, μ and ν are isomorphisms. It follows that $b = c \circ \pi^*$ and $\sigma = \tau \circ \pi^{-1}$ with $\pi = \mu \circ \nu^{-1}$. ∎

Let $D\Theta(n,1) \subset \Theta(n,1)$ denote the subspace of degenerate morphisms and $\Delta_k X \subset X^k$ the diagonal

Theorem 4.58: Let Θ_1 and Θ_2 be monochrome proper theories and (X,α) a Θ_1-space. Suppose

(a) each finite product of spaces $\Theta_1(k,1)$, X and a single space $\Theta_2(l,1)$ is paracompact

(b) $(\Theta(1,1),\{id\})$ and $(\Theta_2(1,1),\{id\})$ are NDRs

(c) $(\Theta(r,1)^k, \Delta_k\Theta_1(r,1))$ are $S_k \times S_r$-NDRs where S_k acts by permuting the factors and S_r by componentwise composition on the right

(d) $(\Theta_i(r,1), D\Theta_i(r,1))$ are S_r-NDRs and for each fixed monomorphism

$\sigma : [k] \longrightarrow [r]$, the injective map $b \longmapsto b \circ \sigma^*$, $b \in \Theta_i(k,1)$ is a homeomorphism onto a closed subspace of $\Theta_i(r,1)$, $i = 1,2$.

(e) $(X^k, \Delta_k X)$ is a S_k-NDR

Then given a theory functor $F : \Theta_1 \longrightarrow \Theta_2$ which is a homotopy equivalence on each morphism space, there exists a Θ_2-space (Y,α^*) containing X as a SDR.

<u>Proof</u>: The augmentation $\varepsilon : W\Theta_1 \longrightarrow \Theta_1$ allows us to regard X as $W\Theta_1$-space. Replace X by the universal Θ_1-space $M(X, \alpha \circ \varepsilon)$, the analogue of the construction of §5 for theories. As shown in (4.49), we can embed X as a SDR into MX. The required Θ_2-space Y looks like MX, except that the label of the root vertex lies in Θ_2 instead of Θ_1. But let us give precise descriptions. As for PROPs, we can define theories $W_r(\Theta_1 \otimes \mathfrak{L}_1)$ by adding the extra relation that the length of an internal edge of $\mathfrak{L}_1$-colour 1 of a representing tree may be changed to 0 to the relations (3.1 a,b,c). Then Θ_1 is contained in $W_r(\Theta_1 \otimes \mathfrak{L}_1)$ as the full subcategory of all objects n^1. Now the construction of MX carries over. Let T_λ denote the space of all trees of shape λ whose edges all have $\mathfrak{L}_1$-colour 0 except of the root, which has $\mathfrak{L}_1$-colour 1. Let n_λ denote the source of λ. Then

$$MX = \bigcup_\lambda T_\lambda \times X^{n_\lambda}/\sim$$

with the relations

(a) = (3.1 a). It does not apply to the root vertex because of the change of colour

(b) = (3.1 b), but for cherry trees, i.e. the cherries are affected by the set operations in the same way as the twigs

(c) = (3.1 c)

(d) if the cherry tree A has an edge of length 1, then the subtree B with cherries $y_1,\ldots,y_r$ sitting on this edge may be replaced by the cherry $[\alpha \circ \varepsilon(B)]\,(y_1,\ldots,y_r)$

Let T'_λ be obtained from T_λ by replacing the space of root vertex labels $\Theta_1(k,1)$ by $\Theta_2(k,1)$. Then

$$Y = \bigcup_\lambda T'_\lambda \times X^{n_\lambda}/\sim$$

with the relations (a),(b),(d) above and (c) substituted by

(c') any edge of length 0 which does not meet the root vertex may be shrunk as in (c), while an incoming edge of length 0 to the root vertex may be shrunk as follows: We substitute the vertex label, a say, above that edge by F(a) and then shrink as described in (3.1 c).

The Θ_2-action on Y is defined on representing cherry trees by

$$b(A_1,\ldots,A_n) = B$$

where B is obtained from $A_1,\ldots,A_n$ by identifying their root vertices. Then the data of B is given by $A_1,\ldots,A_n$ except of the root vertex label which is $b\cdot(a_1 \oplus \ldots \oplus a_n)$ if $a_1,\ldots,a_n$ are the root vertex labels of $A_1,\ldots,A_n$. For example,

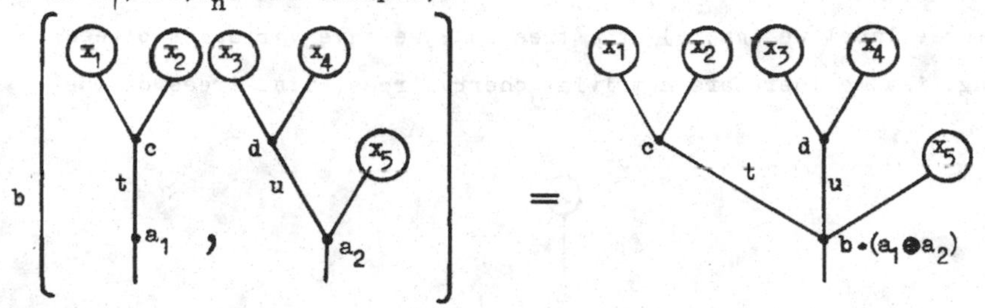

This definition coincides with the definition of the Θ_1-structure of MX if $\Theta_1 = \Theta_2$. There are again inclusions

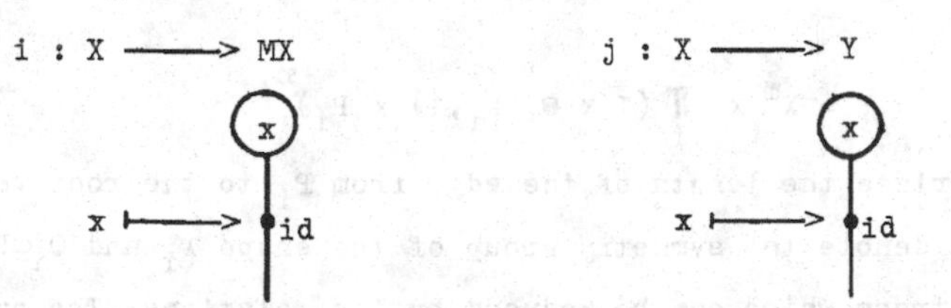

and as in (4.49) one shows that X is a SDR of MX. Define a map $f : (MX,X) \longrightarrow (Y,X)$ by substituting the vertex label b of a representing cherry tree of MX by F(b) to obtain a representing cherry tree of Y. Note that f is a homomorphism of Θ_1-spaces if we give Y the

Θ_1-structure induced by F. Suppose we know

Claim 1: j is a cofibration

Claim 2: $f : MX \longrightarrow Y$ is a homotopy equivalence

Then $j = f \circ i : X \longrightarrow Y$ is a homotopy equivalence and cofibration, and hence X a SDR of Y. (see [14; (3.7)]).

To prove the claims we filter the spaces MX and Y by the subspaces M_k and Y_k of cherry trees which are related to a cherry tree with at most k edges. The spaces of all cherry trees of MX and Y of a given shape λ with k edges are of the forms

$$Z_1 = \Theta_1(r,1) \times P \qquad \text{and} \qquad Z_2 = \Theta_2(r,1) \times P$$

where $\Theta_1(r,1)$ and $\Theta_2(r,1)$ are the spaces of root vertex labels and P is the space of all cherry trees of MX with shape λ, ignoring the root vertex label. Let $Q \subset P$ be the subspace of trees which can be reduced by the relations to a tree with less edges. On the root vertex (whose label we ignore) of a tree in P we have various subtrees sitting. Assume there are m trivial cherry trees, i.e. trees of the form

(perhaps m = 0), and n_i non-trivial trees of shape λ_i forming spaces $\Theta_1(k_i,1) \times P_i$, where $\Theta_1(k_i,1)$ again is the space of root vertex labels. Then

$$P = X^m \times \prod_i (I \times \Theta_1(k_i,1) \times P_i)^{n_i}$$

where I parameterizes the length of the edge from P_i to the root vertex of P. Let G_i denote the symmetry group of the shape λ_i and $Q_i \subset P_i$ the subspace of trees which can be reduced by the relations. The symmetry group G of P is $S_m \times \prod_i H_i$ where $H_i = G_i \wr S_{n_i}$, the wreath product of G_i and S_{n_i} (for a definition see Appendix II). A tree A of P lies in Q iff it satisfies one of the following conditions

(i) any two coordinates in X coincide (because then we can reduce A by (b))

(ii) any I-coordinate is 0 or 1 (because then (c),(c') or (d) applies)

(iii) any $\Theta_1(k_i,1)$-coordinate is degenerate or an identity (because then (a) or (b) applies)

(iv) any P_i-coordinate lies in Q_i

(v) two $(I\times\Theta_1(k_i,1)\times P_i)$-coordinates are in the same G_i-orbit (because then (b) applies)

Examples to (iii) and (v): Let $\sigma : \{1,2\} \longrightarrow \{1,2,3\}$ map 1 to 3 and 2 to 1 and let $\tau : \{1,2,3\} \longrightarrow \{1,2\}$ map 1,3 to 2 and 2 to 1. Then

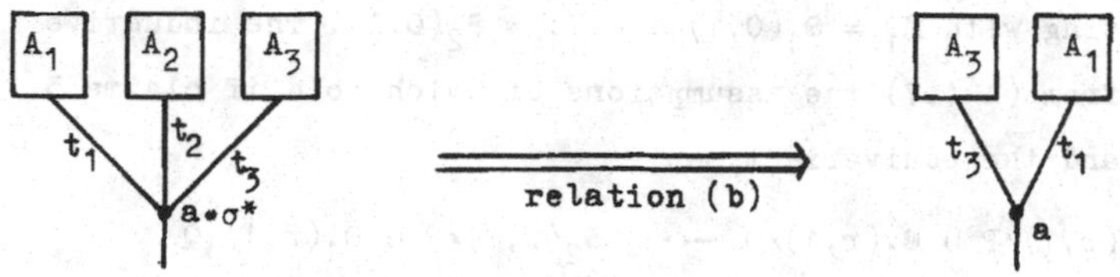

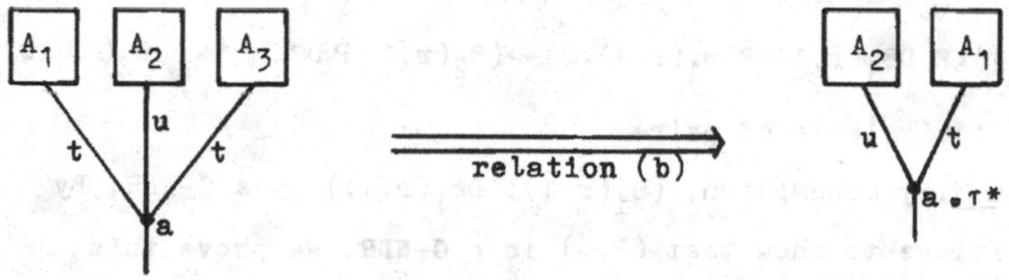

It is not difficult to show that each representing tree in the spaces Z_1 and Z_2 is related to a tree to which (i),...,(v) does not apply, uniquely up to relation (b) but for permutations only. This is precisely the situation we dealt with in chapter III.

The group G permutes the r incoming edges to the root vertex of P. Hence there is a homomorphism $G \longrightarrow S_r$ making $\Theta_1(r,1)$ and $\Theta_2(r,1)$ into G-spaces. We will show later

Claim 3: $(\Theta_i(r,1)\times P,\ D\Theta_i(r,1)\times P \cup \Theta_i(r,1)\times Q)$, $i = 1,2$, are G-NDRs

Claim 4: $\Theta_i(r,1)\times P - (D\Theta_i(r,1)\times P \cup \Theta_i(r,1)\times Q)$, $i = 1,2$, is a numerable principal G-space

<u>Claim 5</u>: $F : (\Theta_1(r,1), D\Theta_1(r,1)) \longrightarrow (\Theta_2(r,1), D\Theta_2(r,1))$ is an ordinary homotopy equivalence of pairs

Then MX and Y are proper iterated adjunction spaces because they are obtained by adjoining spaces

$$((\Theta_i(r,1)\times P)/G,\ (D\Theta_i(r,1)\times P \cup \Theta_i(r,1)\times Q)/G) \qquad i = 1,2$$

to M_{k-1} respectively Y_{k-1}, one for each shape orbit λ with k edges, and these pairs are NDRs by claim 3. Since $X \subset Y_2$ is a cofibration, claim 1 follows. In view of (A 4.4) it suffices to show that each map $f_k = f|M_k : M_k \longrightarrow Y_k$ is a homotopy equivalence. We prove this by induction, starting with $M_1 = \Theta_1(0,1)$ and $Y_1 = \Theta_2(0,1)$. The inductive step follows from (A 4.7) the assumptions of which hold if claims 3, 4,5 are true and the equivariant map

$$f' : D\Theta_1(r,1)\times P \cup \Theta_1(r,1)\times Q \longrightarrow D\Theta_2(r,1)\times P \cup \Theta_1(r,1)\times Q$$

induced by F is an ordinary homotopy equivalence. But by claim 5,

$$F\times id : (\Theta_1(r,1)\times P, D\Theta_1(r,1)\times P\cup\Theta_1(r,1)\times Q) \longrightarrow (\Theta_2(r,1)\times P, D\Theta_2(r,1)\times P\cup\Theta_2(r,1)\times Q)$$

is a homotopy equivalence of pairs.

<u>Proof of claim 3</u>: By assumption, $(\Theta_i(r,1), D\Theta_i(r,1))$ is a G-NDR. By (A 2.5) it suffices to show that (P,Q) is a G-NDR. We prove this by induction starting with a shape λ with one edge only, the root, i.e. $P = Q = \emptyset$. For the inductive step, consider for a given shape λ with k edges the pair

$$(P,Q) = (X^m, \Delta' X^m) \times \prod_i ((I\times\Theta_1(k_i,1)\times P_i)^{n_i}, R_i)$$

where $\Delta' X^m = \{(x_1,\ldots,x_m) \in X^m | x_i = x_j \text{ for some } i \neq j\}$ and $R_i \subset (I\times\Theta_1(k_i,1)\times P_i)^{n_i}$ is the subspace of all elements satisfying (ii),(iii),(iv) or (v). By (A 2.8) and assumption (e), $(X^m, \Delta' X^m)$ is a S_m-NDR. So by (A 2.4) it suffices to show that $((I\times\Theta_1(k_i,1)\times P_i)^{n_i}, R_i)$ is a $G_i \wr S_{n_i}$-NDR. We use the following

<u>Observation</u>: G acts freely on P-Q

Proof: Suppose $gA = A$ for $A \in P-Q$ and $g \in G$, $g \neq$ identity. Then two $(I \times \Theta_1(k_i,1) \times P_i)$-coordinates of A are in the same G_i-orbit for some i or two X-coordinates of A coincide, which is a contradiction.

Let $V_i \subset I \times \Theta_1(k_i,1) \times P_i$ be the subspace of all elements satisfying (ii),(iii) or (iv). Then G_i acts freely on $I \times \Theta_1(k_i,1) \times P_i - V_i$. By induction hypothesis, (P_i, Q_i) is a G_i-NDR, by assumption (b),(d) and (A 2.7) the pair $(\Theta_1(k_i,1), D'\Theta_1(k_i,1))$ is a G_i-NDR where $D'\Theta_1(k_i,1) = D\Theta(k_i,1) \cup \{id_1\}$ if $k_i = 1$ and $D\Theta(k_i,1)$ otherwise. Hence, by the product theorem for cofibrations, $(I \times \Theta_1(k_i,1) \times P_i, V_i)$ is a G_i-NDR. By assumption (c) and the product theorem $((I \times \Theta_1(k_i,1) \times P_i)^l, \Delta_l(I \times \Theta_1(k_i,1) \times P_i))$ is a $(S_l \times G_i)$-NDR. Hence, by (A 2.10), the pair $((I \times \Theta_1(k_i,1) \times P_i)^{n_i}, R_i)$ is a $G_i \wr S_{n_i}$-NDR.

<u>Proof of claim 4</u>: We know that G is finite and operates freely on $\Theta_i(r,1) \times P - (D\Theta_i(r,1) \times P \cup \Theta_i(r,1) \times Q)$. Moreover, there is a map $u : \Theta_i(r,1) \times P \longrightarrow I$ such that $u^{-1}(0) = D\Theta_i(r,1) \times P \cup \Theta_i(r,1) \times Q$ because of claim 3. By assumption (a), the space $\Theta_i(r,1) \times P$ is paracompact. So claim 4 follows from (A 3.8).

<u>Proof of claim 5</u>: The proof proceeds by induction starting with $r = 0$ where $D\Theta_1(r,1) = D\Theta_2(r,1) = \emptyset$. For each subset $A \subset [r]$ of k elements, $k<r$, let $\sigma_A : [k] \longrightarrow [r]$ be the order preserving monomorphism with image A. Let

$$D_k = \{b \cdot \sigma^* \in \Theta_1(r,1) \mid \sigma : [k] \longrightarrow [r] \text{ injective}\} \subset \Theta_1(r,1)$$

Denote the subspace of $\Theta_1(r,1)$ of all elements of the form $b \cdot \sigma_A^*$ by B_A. Then

$$D_k = \bigcup_{|A|=k} B_A \quad \text{and} \quad B_A \cap B_{A'} \subset D_{k-1} \text{ for } A \neq A',\ |A| = |A'| = k$$

($|A|$ = cardinality of A) because Θ_1 is proper. Hence D_k can be obtained from D_{k-1} by successively adjoining spaces $(B_A, B_A \cap D_{k-1})$ one for each subset $A \subset [r]$ with k elements. The same holds for Θ_2;

we denote the corresponding spaces by the same symbol with dash. Obviously $F : D_0 \cong \Theta_1(0,1) \cong \Theta_2(0,1) \cong D'_0$. Assume inductively, that $F : D_{k-1} \simeq D'_{k-1}$. By assumption (d), composition with σ^*_A induces homeomorphisms

$$(\Theta_1(k,1), D\Theta_1(k,1)) \cong (B_A, B_A \cap D_{k-1}),\ (\Theta_2(k,1), D\Theta_2(k,1)) \cong (B'_A, B'_A \cap D'_{k-1})$$

Hence, by assumption (d) and induction hypothesis,

$$F : (B_A, B_A \cap D_{k-1}) \simeq (B'_A, B'_A \cap D'_{k-1})$$

and $(B_A, B_A \cap D_{k-1})$, $(B'_A, B'_A \cap D'_{k-1})$ are NDRs. By (A 4.6), we find that $F : D_k \simeq D'_k$. Hence $F : D\Theta_1(r,1) = D_{r-1} \simeq D'_{r-1} = D\Theta_2(r,1)$ and hence

$$F : (\Theta_1(r,1), D\Theta_1(r,1)) \simeq (\Theta_2(r,1), D\Theta_2(r,1))$$

is a homotopy equivalence of pairs by (A 4.3). ∎

Remark: Actually, we have proved a little more than stated in the theorem. We have constructed a Θ_2-space Y which we can consider as a Θ_1-space because of the functor F, and the inclusion $j : X \longrightarrow Y$ carries the structure of a reduced Θ_1-map.

Presumably most of the results of chapter IV can be generalized to arbitrary theories under assumptions similar to those of Theorem 4.58. But as we see from the proof of (4.58), the details are formidable.

V. Chapter

STRUCTURES ON BASED SPACES

One of the main applications of our theory will be the classification of the algebraic topological structures of iterated loop spaces. These spaces live naturally in the category $\mathfrak{Top}^o$ of based topological spaces and based maps. Therefore we have to modify our constructions to cover this case.

1. BASED THEORIES

Let X be a topological space and K a set. Then X^+ is X with a disjoint point $\{*\}$ attached, which serves as base point, and X+K the disjoint union of X and K. Let $\mathfrak{S}_K^o$ denote the category whose objects are functions $\underline{i} = (\underline{i}', id) : [n]+K \longrightarrow K$ and whose morphisms from $\underline{i}$ to $\underline{j}$ are all functions σ making

$$\begin{array}{ccc} [n]+K & \xrightarrow{\ \sigma\ } & [m]+K \\ & \underline{i}\searrow \quad \swarrow\underline{j} & \\ & K & \end{array}$$

commute. A <u>basic</u> <u>object</u> in $\mathfrak{S}_K^o$ is a function $[1]+K \longrightarrow K$. Since this function is the identity on the second summand it is uniquely determined by the image k of $1 \in [1]$. Hence we often denote it by k. Then $\underline{i} : [n]+K \longrightarrow K$ is the categorical sum of the basic objects $\underline{i}(1), \ldots, \underline{i}(n)$.

Let $\mathfrak{Top}^o_K$ denote the category of K-graded spaces in $\mathfrak{Top}^o$. An object $X \in \mathfrak{Top}^o_K$ determines a product preserving functor $(\mathfrak{S}^o_K)^{op} \longrightarrow \mathfrak{Top}^o$ sending the object $\underline{i}$ to $X_{\underline{i}} = X_{\underline{i}(1)} \times \ldots \times X_{\underline{i}(n)}$ and the morphism $\sigma : \underline{i} \longrightarrow \underline{j}$ in $\mathfrak{S}^o_K$ to $\sigma^* : X_{\underline{j}} \longrightarrow X_{\underline{i}}$

$$\sigma^*(x_1, \ldots, x_m) = (x_{\sigma(1)}, \ldots, x_{\sigma(n)})$$

where $x_{\sigma(r)} = * \in X_k$ if $\sigma(r) = k \in K \subset [m]+K$ (we always denote base points by $*$). We see that there are additional set operations.

The sets $\mathfrak{S}^o_K(\underline{i},\underline{j})$ are canonically based, the base point being the function $\underline{i} : [n]+K \longrightarrow K \subset [m]+K$. The corresponding set operation is the constant map in $\mathfrak{Top}^o(X_{\underline{j}}, X_{\underline{i}})$.

A function $f : K \longrightarrow L$ induces a functor $f_* : \mathfrak{S}^o_K \longrightarrow \mathfrak{S}^o_L$ and hence a functor $f^{op}_* : (\mathfrak{S}^o_K)^{op} \longrightarrow (\mathfrak{S}^o_L)^{op}$. If $\underline{i} = (\underline{i}', id) : [n]+K \longrightarrow K$ is an object and $\sigma = (\sigma', id) : [n]+K \longrightarrow [m]+K$ a morphism from $\underline{i}$ to $\underline{j}$ in $\mathfrak{S}^o_K$, then $f_*(\underline{i}) = (f \cdot \underline{i}', id_L)$ and $f_*(\sigma) = ((id_{[m]}+f) \cdot \sigma', id_L)$.

<u>Definition 5.1</u>: A (finitary) <u>based</u> K-<u>coloured</u> <u>topological</u>-<u>algebraic</u> <u>theory</u> is a category Θ with ob Θ = ob $\mathfrak{S}^o_K$ together with a faithful functor $(\mathfrak{S}^o_K)^{op} \longrightarrow \Theta$ preserving objects and products. The latter means that $\Theta(\underline{i},\underline{j}) \cong \Theta(\underline{i},\underline{j}(1)) \times \ldots \times \Theta(\underline{i},\underline{j}(m))$ is a based homeomorphism where the base points are the images of the base points of $(\mathfrak{S}^o_K)^{op}$.
A Θ-<u>space</u> is a continuous functor $\Theta \longrightarrow \mathfrak{Top}^o$ such that $(\mathfrak{S}^o_K)^{op} \rightarrow \Theta \rightarrow \mathfrak{Top}^o$ preserves products. The images of the basic objects determine an object in $\mathfrak{Top}^o_K$, the <u>underlying</u> <u>space</u>.
A <u>homomorphism</u> between Θ-spaces is a natural transformation of such functors.
A <u>theory</u> <u>functor</u> from a K-coloured theory Θ_1 to an L-coloured theory Θ_2 is a continuous functor $F : \Theta_1 \longrightarrow \Theta_2$ together with a function $f : K \longrightarrow L$ such that

$$\begin{array}{ccc} (\mathfrak{S}^o_K)^{op} & \xrightarrow{f^{op}_*} & (\mathfrak{S}^o_L)^{op} \\ \downarrow & & \downarrow \\ \Theta_1 & \xrightarrow{F} & \Theta_2 \end{array}$$

commutes.

2. BASED PROs AND PROPs

Denote the 0-ary set operation in $\mathfrak{S}_K^o(k,0)$ by ω_k.

In contrary to the unbased case we consider four types of spines: Besides the PROs and PROPs, a notion which makes sense for based theories too, we consider "based" PROs and PROPs which are spines with respect to the subcategory of $\mathfrak{S}_K^o$ generated under $\oplus$ and composition by the ω_k respectively the ω_k and the isomorphisms. The reason is that for a based monoid and a based abelian monoid one usually assumes that the identity is the base point, i.e. the 0-ary operation λ_o including the unit coincides with the 0-ary set operation ω. Hence a based PROP is a category $\mathfrak{B}$ with bifunctor $\oplus : \mathfrak{B}\times\mathfrak{B} \longrightarrow \mathfrak{B}$ as defined in (2.43) with G substituted by the subcategory of $\mathfrak{S}_K^o$ generated under $\oplus$ and composition by isomorphisms and the ω_k. The axioms (2.43) (a),...,(d) hold too with the appropriate modification of (2.43 d). The definition of a based PRO is analogous.

Let us give an example where the additional set operation $\omega : [1]^+ \longrightarrow [0]^+$ causes changes. Since $\mathfrak{A}$ does not contain set operations the unbased free monoid on a space X is the disjoint union $\bigcup_{n=0}^{\infty} X^n$. In the based case, we have a single set operation left. Hence the based free monoid on a based space X is

$$\dot{\bigcup_{n=0}^{\infty}} X^n/\!\sim$$

where $(x_1,\ldots,x_n)\sim(x_1,\ldots,\hat{x}_i,\ldots,x_n)$ if $x_i = *$ and $(x) \sim *$ if $x = *$, because if $x_i = *$, then $(x_1,\ldots,x_n)=(\mathrm{id}\oplus\omega\oplus\mathrm{id})^*(x_1,\ldots,\hat{x}_i,\ldots,x_n)$. (Here $\hat{x}$ means delete x).

3. THE BASED BAR CONSTRUCTION

The bar construction for unbased theories carries over to based theories with the following modifications: The trees have again edges with colours in K and lengths in I, and vertex labels in the appropriate morphism spaces of Θ, but its twigs are labelled by elements in $[n]+K$, if it represents a morphism with source $\underline{i} : [n]+K \longrightarrow K$, instead of elements in $[n]$ only. If we have m twigs, the twig labels then define a morphism $[m]+K \longrightarrow [n]+K$ in $\mathfrak{S}_K^o$, which stands for a set operation. The relations among the trees of $W^b\Theta$ (b for "based") are the same as relations (3.1) with the following modification of (3.1 b):

(3.2) Let $\sigma : [n]+K \longrightarrow [m]+K$ be a morphism in $\mathfrak{S}_K^o$. We may replace any vertex label $a \bullet \sigma^*$ by a, by changing the part of the tree above this vertex:

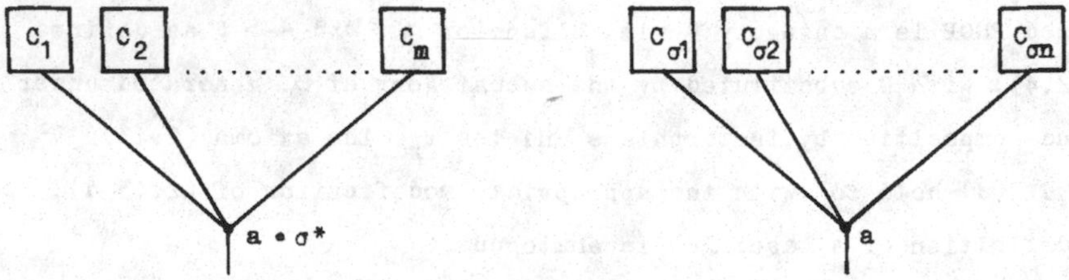

where $C_{\sigma r}$ is a single twig with label (and colour) k if $\sigma(r)=k\in K\subset[m]+K$.

If each space $\Theta(0,k)$ has only one element, i.e. the 0-ary operations are set operations, then $W^b\Theta(0,k)$ also has only one element. Moreover, all stumps may be pruned away by shrinking their outgoing edge. We use relations (b) and (a) for this. In terms of operations this implies that in any $W^b\Theta$-space the base points behave like strict identities.

We again specialize to PROs and PROPs. In the case of unbased PROs and PROPs, there are, of course, no changes. So let $\mathfrak{B}$ be a K-coloured based PROP (for based PROs the construction is similar) and denote the 0-ary set operation in $\mathfrak{B}(0,k)$ by ω_k^*. Then $W^b\mathfrak{B}$ is the quotient of $W\mathfrak{B}$ (here we consider ω_k^* as an ordinary 0-ary operation) modulo the relation

(5.3) a stump 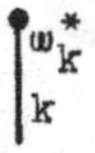 may be shrunk

$W^b\mathfrak{B}$ is the correct spine of the theory $W^b\Theta$ obtained from the based theory Θ associated to $\mathfrak{B}$.

Since all 0-ary operations on a based $\mathfrak{B}$-space coincide it seems to be reasonable to restrict the attention to PROs and PROPs having at most one 0-ary operation $0 \longrightarrow k$ for each $k \in K$. If $\mathfrak{B}$ is an unbased PROP satisfying this condition, then $W\mathfrak{B}(0,k)$ nevertheless may have many 0-ary operations. We introduce two modifications of the construction W that correct this. Define $W'\mathfrak{B}$ to be the quotient of $W\mathfrak{B}$ modulo the relation

(5.4) two trees without twigs having the same root colour coincide,

and define $W''\mathfrak{B}$ to be the quotient of $W\mathfrak{B}$ modulo the relation

(5.5) any stump may be shrunk.

Then $W'\mathfrak{B}(0,k)$ has at most one element and $W''\mathfrak{B}(0,k) = \mathfrak{B}(0,k)$.

The basic difference between W' and W" is best illustrated with an example. Let $\mathfrak{B} = \mathfrak{A}$. In $W'\mathfrak{A}(1,1)$ we have a representing tree

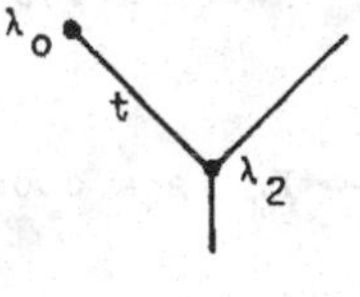

giving a path from $e \cdot x$ to x, i.e. the base point e of a $W'\mathfrak{A}$-space X is a homotopy unit, while in $W''\mathfrak{A}$ this tree coincides with

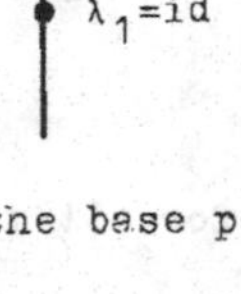

which means that in a $W''\mathfrak{A}$-space the base point e is a strict unit.

Convention: When dealing with constructions W' and W", we consider the 0-ary set operation of a based PROP as ordinary 0-ary operation, thus obtaining an unbased PROP. For actions on based spaces this does not make any difference.

With this convention, we find that for a based PROP with $\mathfrak{B}(0,k)=\{\omega_k\}$ for all $k\in K$ the categories $W^b\mathfrak{B}$ and $W''\mathfrak{B}$ coincide. But $W''\mathfrak{B}$ has also some relevance in the unbased case.

Let X be an object in $\mathfrak{Top}_K^{\circ}$ and let α be an action of an unbased K-coloured PROP $\mathfrak{B}$. We have $\mathfrak{B}(0,k) \subset W\mathfrak{B}(0,k)$ by the standard section, i.e. $\mathfrak{B}(0,k)$ is the subspace of all stumps. The action α induces maps

$$\hat{\alpha}_k : \mathfrak{B}(0,k) \subset W\mathfrak{B}(0,k) \longrightarrow X_k$$

Theorem 5.6: (a) If the maps $\hat{\alpha}_k$ are homotopic to the constant maps to $* \in X_k$, then the action α is homotopic through actions to an action β of $W\mathfrak{B}$ on X such that $\beta(\mathfrak{B}(0,k)) = *$ for all $k\in K$

(b) Let $\mathfrak{B}$ be a K-coloured PROP. (As always, we assume that $(\mathfrak{B}(k,k),\{id_k\})$ is a NDR for all k.) Let $\pi : W\mathfrak{B} \longrightarrow W''\mathfrak{B}$ be the projection and $\mathfrak{C} \subset W''\mathfrak{B}$ a subcategory such that $\mathfrak{C}(0,k) = W''\mathfrak{B}(0,k)$ for all $k \in$ ob $\mathfrak{C}$ and $\pi^{-1}(\mathfrak{C}) \subset W\mathfrak{B}$ is an admissible subcategory. Assume we are given a homotopy of actions $\alpha(t) : W\mathfrak{B} \longrightarrow \mathfrak{Top}$ on $X \in \mathfrak{Top}_K$ satisfying

(i) $\hat{\alpha}_k(t) = \hat{\alpha}_k(0)$ for all $t\in I$

(ii) $\hat{\alpha}_k(0) : \mathfrak{B}(0,k) \subset W\mathfrak{B}(0,k) \longrightarrow X_k$ is a closed cofibration for all $k\in K$ and $t\in I$

(iii) $\alpha(t)|\pi^{-1}(\mathfrak{C})$ factors through π

(iv) $\alpha(t)$ factors through π for $t = 0$ [or $t\in\partial I$, or (iv) is empty]

Then there exists a homotopy $\beta(t_1,t_2) : W\mathfrak{B} \longrightarrow \mathfrak{Top}_K$ through homotopies of actions such that

(i) $\beta(t,0) = \alpha(t)$

(ii) $\beta(t_1,t_2)|\pi^{-1}(\mathfrak{C}) = \alpha(t_1)$ for all t_2

(iii) $\beta(t_1,t_2) = \alpha(t_1)$ for $t_1 = 0$ [or $t_1 \in \partial I$, or (iii) is empty]

(iv) $\beta(t,1)$ factors through π

(v) $\hat{\beta}_k(t_1,t_2) = \hat{\alpha}_k(0)$ for all $t_1,t_2 \in I$

Under suitable hypotheses this result makes unbased W$\mathfrak{B}$-actions to based W"$\mathfrak{B}$-actions. We have stated it in greater generality than we need for this purpose, but we will use the full generality later.

Proof: Part (a) follows immediately from (3.14) with the subcategory $\mathfrak{D}$ generated by the morphisms of $\mathfrak{B}(0,k)$, $k \in K$.

The homotopy $\beta(t_1,t_2)$ of part (b) is constructed inductively. We use the cherry tree definition of an action (cf.3.24) and induct on the number of vertices and the number of cherries. The induction starts from the requirement that the trivial cherry tree with cherry x has value x and a stump b has value $\hat{\alpha}(0)(b)$. For the inductive step from r-1 to r we consider a space P of trees of a given shape λ with n twigs, $n \le r$, and r-n vertices. We convert P into a space of trees of filtration r by attaching to each twig a cherry or convert it into a stump of length t and some label in $\mathfrak{B}(0,k)$, k = twig colour. By (3.24 e) a stump b of length 1 must be equivalent to the appropriate cherry $\hat{\alpha}(0)(b)$. To take care of this we introduce

$$Y_k = \mathfrak{B}(0,k)\times I \cup X_k/((b,1)\sim\hat{\alpha}_k(0)(b))$$

Let Y_k' be the image of $\mathfrak{B}(0,k)\times I$ and $Y_k'' = \mathfrak{B}(0,k)\times\{0\}$. What we look for is a homotopy

$$H : P\times Y_{k_1}\times\ldots\times Y_{k_n}\times I\times I \longrightarrow X_l$$

where $k_1,\ldots,k_n$ are the n colours of the twigs of λ and l the colour of its root. By induction, $H(A,y_1,\ldots,y_n,u_1,u_2)$ is already determined iff one of the following holds

(i) A lies in the subspace $Q \subset P$ of trees that simplify by (3.24 (a) or (c)), or decompose and hence simplify by (3.24 e), or that represent an element of $\pi^{-1}(\mathfrak{B})$. (Note that A represents a tree of $\pi^{-1}(\mathfrak{B})$ iff any tree obtained from A by converting a twig to a stump of arbitrary length represents a tree of $\pi^{-1}(V)$, because $\mathfrak{B}(0,k) = W''\mathfrak{B}(0,k)$ for $k\in\mathfrak{B}$)

(ii) some y_i lies in some Y_k'', because then we have a stump of length 0 which can be shrunk

(iii) some y_i lies in some Y_k' and $u_2 = 1$, by the requirement on $\beta(u_1,1)$

(iv) $u_2 = 0$

(v) $u_1 = 0$ [or $u_1\in\partial I$, or (v) is empty] (Denote this subspace of I by I', i.e. $I' = \{0\}$ or ∂I or $\emptyset$)

Let G be the symmetry group of the shape λ. The action of G on the twigs makes $Y = Y_{k_1}\times\ldots\times Y_{k_n}$ into a G-space. We will show that there is an equivariant retraction of $P\times Y\times I\times I$ onto the subspace of elements satisfying one of the conditions (i),...,(v) so that the induction can proceed. We already know that (P,Q) is a G-NDR. Let Z' and Z" be the subspace of $y\in Y$ with some y_i in some Y_k' respectively Y_k''. Now (Y_k',Y_k'') is homeomorphic to $(\mathfrak{B}(0,k)\times I,\mathfrak{B}(0,k)\times 0)$ so that Y_k'' is a SDR of Y_k'. We prove later

Claim: Z" is an equivariant SDR of Z'

From the product formula for equivariant SDRs we obtain that $Z'\times I'\times I \cup Z'\times I\times\partial I \cup Z''\times I\times I$ is an equivariant SDR of $Z'\times I\times I$ and that $Z'\times I\times I \cup Y\times I'\times I \cup Y\times I\times 0$ is an equivariant SDR of $Y\times I\times I$, the latter because $\{0\}\subset I$ is a SDR and $(Y\times I, Z'\times I \cup Y\times I')$ is a G-NDR. Hence $Z''\times I\times I \cup Z'\times I\times 1 \cup Y\times I'\times I \cup Y\times I\times 0$ is an equivariant SDR of $Y\times I\times I$ and hence $P\times(Z''\times I\times I \cup Z'\times I\times 1 \cup Y\times I'\times I \cup Y\times I\times 0)\cup Q\times Y\times I\times I$ of $P\times Y\times I\times I$, what we had to show.

It remains to prove the claim: Let V_r be the subspace of points in Y with at least n-r coordinates in some Y_k'. Consider the pair $(V_r\cup Z'',V_{r-1}\cup Z'')$. Then $(V_r\cup Z'')-(V_{r-1}\cup Z'')$ consists of a collection of spaces homeomorphic to copies

$$(Y_{l_1}'-Y_{l_1}'')\times\ldots\times(Y_{l_{n-r}}'-Y_{l_{n-r}}'')\times(Y_{m_1}-Y_{m_1}')\times\ldots\times(Y_{m_r}-Y_{m_r}')$$

after a suitable shuffle of coordinates, $\{l_1,\ldots,l_{n-r}\}\cup\{m_1,\ldots,m_r\} = \{k_1,\ldots,k_n\}$, on which the homomorphic image G' (determined by the coordinate shuffle) of the subgroup of G acts which keeps this space invariant. Let $U \subset Y'_{l_1}\times\ldots\times Y'_{l_{n-r}}$ be the subspace of all points having some coordinate in some Y''_{l_i} and $W \subset Y_{m_1}\times\ldots\times Y_{m_r}$ the subspace of all points having some coordinate in some Y'_{m_i}. Then

$$U\times Y_{m_1}\times\ldots\times Y_{m_r} \cup Y'_{l_1}\times\ldots\times Y'_{l_{n-r}}\times W = Y'_{l_1}\times\ldots\times Y'_{l_{n-r}}\times Y_{m_1}\times\ldots\times Y_{m_r}\cap(V_{r-1}\cup Z'')$$

is a G'-equivariant SDR of $Y'_{l_1}\times\ldots\times Y'_{l_{n-r}}\times Y_{m_1}\times\ldots\times Y_{m_r}$ by (A 2.4). By induction, $V_o\cup Z''$ is an equivariant SDR of $Z' = V_{n-1}\cup Z''$. But $V_o\cap Z''$ is an equivariant SDR of V_o, again by (A 2.4). Hence Z" is an equivariant SDR of $V_o\cup Z''$. ∎

By a similar argument, one can also prove a partly homogeneous version of this theorem.

The following result, which can be proved in the same manner as the first part of (4.13), indicates how we will apply Theorem 5.6.

<u>Lemma 5.7</u>: Let $\alpha_t : W\mathfrak{B} \longrightarrow \mathfrak{Top}$ be a homotopy of actions on $X \in \mathfrak{Top}_K$. Then id_X carries the structure of a $\mathfrak{B}$-map from (X,α_o) to (X,α_1). The same holds for the partly homogeneous case. ∎

4. LIFTING AND EXTENSION THEOREMS

When dealing with $W'\mathfrak{B}$ we usually have to assume

<u>Assumption 5.8</u>: $\mathfrak{B}(0,k)$ has at most one element for all k.

For based actions this is no restriction, as we have seen before. Since $W^b\mathfrak{B}$ coincides with $W''\mathfrak{B}$ for a based PROP $\mathfrak{B}$ satisfying (5.8) and since it suffices to study PROPs satisfying (5.8) when we consider based actions, the case $W^b\mathfrak{B}$ is covered by the case $W''\mathfrak{B}$.

Recall that we have filtered each space $W\mathfrak{B}(\underline{i},k)$ by subspaces F_r of morphisms represented by trees with at most r internal edges. This filtration induces a filtration F'_r of $W'\mathfrak{B}(\underline{i},k)$ and F''_r of $W''\mathfrak{B}(\underline{i},k)$. Let P be the space of trees of a given shape λ with r internal edges and G its symmetry group. Then a tree A of P represents an element of F'_{r-1} iff one of the following holds

(5.9)(a) Some vertex label is an identity

(b) some internal edge has length 0

(c) there is an internal edge of length 1 having a subtree with at least one internal edge and with source 0 above it (because then the additional relation applies)

The tree A of P represents an element of F''_{r-1} iff one of the following holds

(5.10)(a) some vertex label is an identity

(b) some internal edge has length 0

(c) A has a stump and $r\geq 1$.

Let $Q'\subset P$ and $Q''\subset P$ be the subspace of those elements satisfying (5.9) respectively (5.10). Then (P,Q') and (P,Q'') are equivariant NDRs.

Proposition 5.11: The augmentation $\varepsilon : W\mathfrak{B} \longrightarrow \mathfrak{B}$ induces an equivariant equivalence $\varepsilon'' : W''\mathfrak{B} \longrightarrow \mathfrak{B}$ and, provided $\mathfrak{B}$ satisfies (5.8), an equivariant equivalence $\varepsilon' : W'\mathfrak{B} \longrightarrow \mathfrak{B}$.

Proof: We obtain F'_r and F''_r from F'_{r-1} and F''_{r-1} by attaching spaces (P,Q') and (P,Q'') (cf. 3.15). As in (3.15) it suffices to show that $Q'\subset P$ and $Q''\subset P$ are equivariant SDRs. Since $\{0\}\subset I$ is an SDR, this is clear for (P,Q'') by (A 2.4). For (P,Q') this holds for the same reason if the shape λ has no stumps. If it has a stump and $r\geq 1$ then the stump is an internal edge. Let $\overline{P}$ be the space obtained from P by deleting the coordinate giving the length of the stump, so that $P = \overline{P}\times I$. Then Q' is of the form $Q' = \overline{P}\times 0 \cup \overline{Q}\times I$, where $\overline{Q}$ is a suitable subspace of $\overline{P}$. Hence $(P,Q') = (\overline{P},\overline{Q})\times(I,0)$ so that Q' is an equivariant SDR of P by (A 2.4). ∎

For $W'\mathfrak{B}$ and any partly homogeneous version we have a lifting theorem

Theorem 5.12: The lifting theorems (3.17) and (3.20) also hold if we substitute $H_L W\mathfrak{B}$ by $H_L W'\mathfrak{B}$ provided all PROPs involved satisfy (5.8).

Proof: Let $P : H_L W\mathfrak{B} \longrightarrow H_L W'\mathfrak{B}$ be the projection functor. Then there exist extensions $H : H_L W\mathfrak{B} \longrightarrow \mathfrak{D}$ of $H' : P^{-1}(\mathfrak{B}) \longrightarrow \mathfrak{D}$ and $K_t : H_L W\mathfrak{B} \longrightarrow \mathfrak{C}$ of $K_t' : P^{-1}(\mathfrak{B}) \longrightarrow \mathfrak{C}$ which factor through $H_L W'\mathfrak{B}$, because $\mathfrak{C}$ and $\mathfrak{D}$ have at most one 0-ary operation. ∎

For the same reasons we have a homotopy extension theorem for $W'\mathfrak{B}$.

Theorem 5.13: Proposition 3.14 also holds if we substitute $H_L W\mathfrak{B}$ by $H_L W'\mathfrak{B}$ and $\mathfrak{C}$ by $\mathfrak{Top}^o$. ∎

Since we work with functors into $\mathfrak{Top}^o$ the actions are automatically based. The Theorems 5.12 and 5.13 show that we may well restrict to PROPs and PROs satisfying (5.8) and substitute $W\mathfrak{B}$ by $W'\mathfrak{B}$ when we work with based actions. The results of IV, §1,2,3 carry over with exception of (4.16), on which the homotopy invariance results rely. The based analogue of (4.16) can be proved for <u>well-pointed</u> spaces X, i.e. each pair $(X_k,*)$ is a NDR (see (5.16)), so that we obtain the homotopy invariance results for well-pointed spaces. Since there is nothing spectacular about $W'\mathfrak{B}$ we concentrate on $W''\mathfrak{B}$ from now on.

(3.17) does not imply a lifting theorem for $W''\mathfrak{B}$, because indecomposable trees in $W\mathfrak{B}$ may well represent a morphism in $W''\mathfrak{B}$ which can be decomposed into some morphism of $W\mathfrak{B}''$ and a sum of 0-ary operations and identities. Fortunately, we have an adequate substitute for (3.17).

<u>Theorem 5.14</u>: Suppose given a diagram

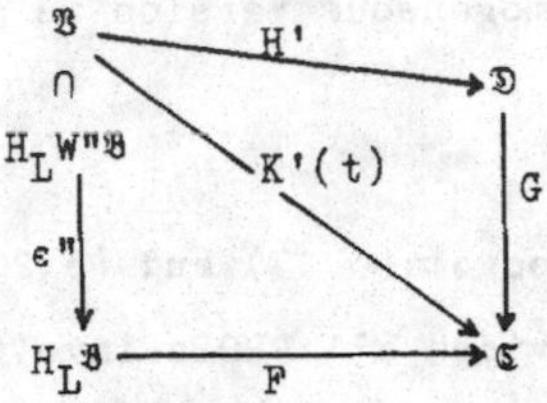

of categories and functors satisfying the assumptions of (3.17). We suppose further that $\mathfrak{A}(0,k)$ contains $\mathfrak{B}(0,k)$ and that $H' : \mathfrak{B}(0,k) \subset \mathfrak{A}(0,k) \to \mathfrak{D}(0,k)$ is a closed cofibration for all $k \in K$. Let $\rho : \mathfrak{D} \longrightarrow \mathfrak{Top}$ be an action on $X \in \mathfrak{Top}_{K \times L}$ such that each $\hat{\rho}_k : \mathfrak{D}(0,k) \longrightarrow X_k$ is a closed cofibration. Then there is an action $\varkappa : H_L W''\mathfrak{B} \longrightarrow \mathfrak{Top}$ on X extending the multiplicative functor $\rho \circ H' : \mathfrak{A} \longrightarrow \mathfrak{Top}$.

<u>Proof</u>: We prove the homogeneous version. Let $\pi : W\mathfrak{B} \longrightarrow W''\mathfrak{B}$ be the projection. Then $\mathfrak{A}' = \pi^{-1}\mathfrak{A}$ satisfies the assumptions of (3.17) so that there is an extension $H : W\mathfrak{B} \longrightarrow \mathfrak{D}$ of $H' \circ \pi : \mathfrak{A}' \longrightarrow \mathfrak{D}$. Now apply (5.6) with $\alpha(t) = \rho \circ H$ to obtain the required action $\varkappa$.

The partly homogeneous version follows from the partly homogeneous analogue of (5.6). ▌

In a similar manner we can also prove a homotopy extension result.

<u>Theorem 5.15</u>: Let $\mathfrak{A} \subset H_L W''\mathfrak{B}$ be an admissible subcategory of $H_L W''\mathfrak{B}$ such that $\mathfrak{B}(0,k) \subset \mathfrak{A}(0,k)$ for all $k \in K$. Let $\rho : H_L W''\mathfrak{B} \longrightarrow \mathfrak{Top}$ be an action on $X \in \mathfrak{Top}_{K \times L}$ and $\alpha(t) : \mathfrak{A} \longrightarrow \mathfrak{Top}$ a homotopy of multiplicative functors such that $\alpha(0) = \rho|\mathfrak{A}$, $\hat{\alpha}_k(t) = \hat{\alpha}_k(0) : \mathfrak{B}(0,k) \subset \mathfrak{A}(0,k) \longrightarrow X_k$ for

all $t\in I$, and $\hat{\alpha}_k(0)$ is a closed cofibration. Then there is a homotopy of actions of $H_L W''\mathfrak{B}$ on X extending ρ and $\alpha(t)$.

Proof: Using the notation of the proof of (5.6), we have an extension $\beta(t)$ of $\alpha(t)\circ\pi : \pi^{-1}(\mathfrak{B}) \longrightarrow \mathfrak{Top}$ to $H_L W\mathfrak{B}$ such that $\beta(0) = \rho\circ\pi$. We deform $\beta(t)$ relative to $\pi^{-1}(\mathfrak{B})$ by (5.6) keeping $\beta(0)$ fixed and obtain an action which factors through $H_L W''\mathfrak{B}$. ▮

5. BASED HOMOTOPY HOMOMORPHISMS

If we deal with based actions of unbased PROs and PROPs, we can define based $\mathfrak{B}$-maps and $h\mathfrak{B}$-maps as in chapter IV. Since the lifting theorem (3.17) and the homotopy extension theorem (3.14) can be used for based actions of unbased PROPs and PROs, the results of chapter IV, §1,2,3 carry over as long as they do not rely on Lemma 4.16. This lemma is substituted by

Lemma 5.16: Let $p : X \longrightarrow Y$ be a morphism of $\mathfrak{Top}_K^o$ which is an unbased homotopy equivalence and suppose that X and Y are well-pointed. Then p admits a based $W(K\otimes\mathfrak{J})$-action.

The proof proceeds as the proof of (4.16) with the exception that the maps

$$f_k : I^{2n}\times Y_k \longrightarrow X_k \qquad h_k : I\times I^{2n}\times Y_k \longrightarrow Y_k$$

to be constructed in the inductive step are not only given on $\partial I^{2n}\times Y_k$ respectively $(0\times I^{2n}\cup I\times\partial I^{2n})\times Y_k$ in advance, but also on $I^{2n}\times\{*\}\subset I^{2n}\times Y_k$ and $I\times I^{2n}\times\{*\}\subset I\times I^{2n}\times Y_k$. Again (A 3.5) provides the required maps, because $(I^{2n}\times Y_k, \partial I^{2n}\times Y_k \cup I^{2n}\times\{*\})$ is a NDR. ▮

Hence for an unbased PROP $\mathfrak{B}$ we explicitly obtain

Theorem 5.17: The simplicial class $\mathfrak{K}^o_{\mathfrak{B}}$ [$\mathfrak{K}^o_{h\mathfrak{B}}$] whose n-simplexes are based $W(\mathfrak{B}\otimes\mathfrak{L}_n)$-actions [$H_{\mathfrak{L}_n}(\mathfrak{B}\otimes\mathfrak{L}_n)$-actions] satisfies the restricted Kan condition. Hence the category $\mathfrak{Map}^o_{\mathfrak{B}}$ [$\mathfrak{Map}^o_{h\mathfrak{B}}$] of based $W\mathfrak{B}$-spaces and simplicial homotopy classes of based $\mathfrak{B}$-maps [$h\mathfrak{B}$-maps] exists. ∎

Proposition 5.18: Two based $\mathfrak{B}$-maps [$h\mathfrak{B}$-maps] $(f_i,\rho_i):(X,\alpha)\longrightarrow(Y,\beta)$, $i = 0,1$, are simplicially homotopic iff there is homotopy through based $\mathfrak{B}$-maps [$h\mathfrak{B}$-maps] $(f_t,\rho_t):(X,\alpha)\longrightarrow(Y,\beta)$ from (f_o,ρ_o) to (f_1,ρ_1). ∎

Proposition 5.19: Let $(f,\rho):(X,\alpha)\longrightarrow(Y,\beta)$ be a based $\mathfrak{B}$-map [$h\mathfrak{B}$-map] and $f\simeq g$. Then g admits a structure $\varkappa$ of a based $\mathfrak{B}$-map [$h\mathfrak{B}$-map] such that (f,ρ) and $(g,\varkappa)$ are homotopic. ∎

Theorem 5.20: Let $\mathfrak{D}\subset\mathfrak{B}$ be a sub-PROP such that $(\mathfrak{B}(\underline{i},k),\mathfrak{D}(\underline{i},k))$ is a $S_{\underline{i}}$-NDR for all $\underline{i}$ and k in $\mathfrak{D}$, and let $p : X \longrightarrow Y$ be a homotopy equivalence of well-pointed spaces in $\mathfrak{Top}^o_K$. Let $i : W\mathfrak{D}\subset W\mathfrak{B}$ and $j : W(\mathfrak{D}\otimes\mathfrak{L}_1)\subset W(\mathfrak{B}\otimes\mathfrak{L}_1)$ [$HW(\mathfrak{D}\otimes\mathfrak{L}_1)\subset HW(\mathfrak{B}\otimes\mathfrak{L}_1)$] be the inclusion functors

(a) If (X,α') is a based $W\mathfrak{D}$-space, (Y,β) a based $W\mathfrak{B}$-space, and $(p,\rho'):(X,\alpha)\longrightarrow(Y,\beta\circ i)$ a based $\mathfrak{D}$-map [$h\mathfrak{D}$-map], then we can extend α' to a based $W\mathfrak{B}$-action α on X and ρ' to a based $\mathfrak{B}$-map [$h\mathfrak{B}$-map] $(p,\rho):(X,\alpha)\longrightarrow(Y,\beta)$

(b) If $(p,\rho):(X,\alpha)\longrightarrow(Y,\beta)$ is a based $\mathfrak{B}$-map [$h\mathfrak{B}$-map], then any homotopy inverse based $\mathfrak{D}$-map [$h\mathfrak{D}$-map] $(q,\varkappa'):(Y,\beta\circ i)\to(X,\alpha\circ i)$ of $(p,\rho\circ j)$ can be extended to a homotopy inverse $(q,\varkappa):(Y,\beta)\longrightarrow(X,\alpha)$ of (p,ρ). ∎

To cover the case of based PROPs, we also study maps of $W''\mathfrak{B}$-spaces. We define a $\mathfrak{B}''$-map from a $W''\mathfrak{B}$-space (X,α) to a $W''\mathfrak{B}$-space (Y,β) to be an action $\varkappa : W''(\mathfrak{B}\otimes\mathfrak{L}_1)\longrightarrow\mathfrak{Top}$ with $d^o(\varkappa)=\beta$ and $d^1(\varkappa)=\alpha$. The homogeneous version is defined analogously. The results of IV, §1,2 carry over provided we only consider actions $\alpha : W''\mathfrak{C} \longrightarrow \mathfrak{Top}$ respectively their partly homogeneous analogues, $\mathfrak{C}$ an arbitrary unbased PROP, on objects $X \in \mathfrak{Top}_K$ satisfying

Assumption 5.21: The induced maps $\hat{\alpha}_k : \mathfrak{C}(0,k) \subset W''\mathfrak{C}(0,k) \longrightarrow X_k$ are closed cofibrations.

This assumption is in particular satisfied if each space $\mathfrak{C}(0,k)$ has at most one element b_k and α is an action on a well-pointed object $X \in \mathfrak{Top}_K^o$ whose base points are the images of

$$\hat{\alpha}_k(b_k) : * \longrightarrow X_k$$

if $\mathfrak{C}(0,k) \neq \emptyset$. Note that α is then automatically a based action, provided each $\mathfrak{C}(0,k)$ has exactly one element.

Explicitly, we have

Theorem 5.22: Let $\mathfrak{B}$ be an arbitrary PROP. The restricted Kan condition holds for the simplicial class $\mathfrak{K}''_{\mathfrak{B}}$ [$\mathfrak{K}''_{h\mathfrak{B}}$] whose n-simplexes are $W''(\mathfrak{B}\otimes\mathfrak{L}_n)$-actions [$H_{\mathfrak{L}_n} W''(\mathfrak{B}\otimes\mathfrak{L}_n)$-actions] satisfying (5.21). Hence the category $\mathfrak{Map}''_{\mathfrak{B}}$ [$\mathfrak{Map}''_{h\mathfrak{B}}$] of $W''\mathfrak{B}$-spaces satisfying (5.21) and simplicial homotopy classes of [homogeneous] $\mathfrak{B}''$-maps satisfying (5.21) exists. ▌

Proposition 5.23: Two $\mathfrak{B}''$-maps [$h\mathfrak{B}''$-maps] $(f_i,\rho_i) : (X,\alpha) \longrightarrow (Y,\beta)$, $i = 0,1$, are simplicially homotopic iff there is a homotopy $(f_t,\rho_t) : (X,\alpha) \longrightarrow (Y,\beta)$ through $\mathfrak{B}''$-maps [$h\mathfrak{B}''$-maps] from (f_o,ρ_o) to (f_1,ρ_1), provided all actions satisfy (5.21).

The proof is not a direct translation of (4.13) to our case, because we do not have a uniqueness part in (5.14). If $\pi : W(\mathfrak{B}\otimes\mathfrak{L}_1) \longrightarrow W''(\mathfrak{B}\otimes\mathfrak{L}_1)$ is the projection, then we know that the $\mathfrak{B}$-maps $(f_o,\rho_o\circ\pi)$ and $(f_1,\rho_1\circ\pi)$ are homotopic by a homotopy through $\mathfrak{B}$-maps $(f_t,\varkappa_t)$. By (5.6) we can deform $\varkappa_t$ to a homotopy which factors through π inducing the required homotopy through $\mathfrak{B}''$-maps from (f_o,ρ_o) to (f_1,ρ_1). The homogeneous version is proved in the same manner. ▌

Theorem 5.24: Let $(p,\rho) : (X,\alpha) \longrightarrow (Y,\beta)$ be a $\mathfrak{B}''$-map [$h\mathfrak{B}''$-map] whose underlying map p is a homotopy equivalence in $\mathfrak{Top}_K$. Let $\mathfrak{D}\subset\mathfrak{B}$ be a sub-PROP such that $(\mathfrak{B}(\underline{i},k),\mathfrak{D}(\underline{i},k))$ is an $S_{\underline{i}}$-NDR for all $\underline{i}$ and k. Denote the inclusions $W''\mathfrak{D} \subset W''\mathfrak{B}$ and $W''(\mathfrak{D}\otimes\mathfrak{L}_1) \subset W''(\mathfrak{B}\otimes\mathfrak{L}_1)$ $[H_{\mathfrak{L}_1}W''(\mathfrak{D}\otimes\mathfrak{L}_1)\subset H_{\mathfrak{L}_1}W''(\mathfrak{B}\otimes\mathfrak{L}_1)]$ by i and j and let $(q,\varkappa') : (Y,\beta\circ i) \longrightarrow (X,\alpha\circ i)$ be a homotopy inverse $\mathfrak{D}''$-map [$h\mathfrak{D}''$-map] of $(p,\rho\circ j)$. Assume all actions satisfy (5.21). Then $(q,\varkappa')$ can be extended to a $\mathfrak{B}''$-map [$h\mathfrak{B}''$-map] $(q,\varkappa) : (Y,\beta) \longrightarrow (X,\alpha)$ which is homotopy inverse to (p,ρ) and satisfies (5.21). ▌

Unfortunately the proof of (4.18) has to be changed to work for $\mathfrak{B}''$-maps because the category $\mathfrak{C}$ used in the proof does not necessarily contain the spaces $\mathfrak{B}(0,k)$. We indicate the modifications for the special case we need.

Theorem 5.25: Let $\mathfrak{B}$ be a PROP such that each $\mathfrak{B}(0,k)$ has exactly one element. Let $\mathfrak{D}\subset\mathfrak{B}$ be a sub-PROP as in (5.24) and let $p : X \longrightarrow Y$ be a homotopy equivalence of well-pointed spaces in $\mathfrak{Top}^o_K$. If (X,α') is a based $W''\mathfrak{D}$-space, (Y,β) a based $W''\mathfrak{B}$-space, and $(p,\rho'):(X,\alpha')\to(Y,\beta\circ i)$ a based $\mathfrak{D}''$-map [$h\mathfrak{D}''$-map], we then can extend α' to a based $W\mathfrak{B}''$-action α on X and ρ' to a based $\mathfrak{B}''$-map [$h\mathfrak{B}''$-map] $(p,\rho) : (X,\alpha) \longrightarrow (Y,\beta)$.

Proof: By (5.20), there is a based $\mathfrak{B}$-map $(p,\rho'') : (X,\alpha'') \longrightarrow (Y,\beta\circ\pi_{\mathfrak{B}})$ where $\pi_{\mathfrak{B}} : W\mathfrak{B} \longrightarrow W''\mathfrak{B}$ is the projection, extending the based $\mathfrak{D}$-map $(X,\alpha'\circ\pi_{\mathfrak{D}}) \longrightarrow (Y,\beta\circ i\circ\pi_{\mathfrak{D}})$. Since $\hat{\rho}''_{(k,0)}$ and $\hat{\rho}''_{(k,1)}$ are the inclusions of the base point, which are closed cofibrations, we can deform ρ'' to a based $\mathfrak{B}''$-map $(X,\alpha) \longrightarrow (Y,\beta)$ by (5.6). The homogeneous version is proved analogously. ▌

6. THE BASED CONSTRUCTION M

We define reduced based $\mathfrak{B}$-maps and reduced $\mathfrak{B}''$-maps in a fashion analogous to chapter IV. They enjoy the same properties as in the unbased case with the modifications listed in the previous section. Therefore, one might expect that the based equivalent of the construction M of (4.49) can easily be obtained. This is not the case! Let $\mathfrak{B}$ be an unbased PROP and (X,α) a based $\mathfrak{B}$-space, then $M(X,\alpha)$ as defined in (4.49) is a $\mathfrak{B}$-space but it is not based. If we take the base point of X as base point of $M(X,\alpha)$ then the $\mathfrak{B}$-action does not preserve base points. By imposing new relations we can make $M(X,\alpha)$ into a based $\mathfrak{B}$-space having the correct universal property, but then it is not of the homotopy type of X any more, which is insufficient for us.

The situation is different for based PROPs (or PROs). Since we may restrict our attention to based PROPs having exactly one 0-ary operation, we may as well treat the case $W''\mathfrak{B}$.

Let $W_r''(\mathfrak{B}\otimes\mathfrak{L}_1)$ be the quotient of $W_r(\mathfrak{B}\otimes\mathfrak{L}_1)$ obtained by shrinking all stumps, and let (X,α) be a $W''\mathfrak{B}$-space. Define $M''(X,\alpha)=\{M''X_k\}$ by

$$M''X_k = \bigcup_{\underline{i}\in\mathfrak{B}} W_r''(\mathfrak{B}\otimes\mathfrak{L}_1)(\underline{i}^o,k^1)\times X_{\underline{i}}/\sim \tag{5.26}$$

$$(c\circ a;x_1,\ldots,x_n)\sim(c;\alpha(a)(x_1,\ldots,x_n))$$

$c\in W_r''(\mathfrak{B}\otimes\mathfrak{L}_1)(\underline{j}^o,k^1)$ and $a\in W_r''(\mathfrak{B}\otimes\mathfrak{L}_1)(\underline{i}^o,j^o)$ (the upper index specifies the $\mathfrak{L}_1$-colour).

In terms of cherry trees $M''(X,\alpha)$ is the quotient of $M(X,\alpha)$ by adding the relation

$$(A;y_1,\ldots,y_n)\sim(A';y_1,\ldots,y_n) \tag{5.27}$$

if A' is obtained from A by shrinking stumps.
As a consequence of this relation we obtain

$$(A;y_1,\ldots,y_n)\sim(A';y_1,\ldots,y_i,b(*),y_{i+1},\ldots,y_n)$$

if A has a stump b between its i-th and (i+1)-st twig and A' is obtained by converting this stump to a twig.

Theorem 5.28: Let $\mathfrak{B}$ be a PROP and (X,α) a $W''\mathfrak{B}$-space. Then there is a reduced $\mathfrak{B}''$-map $(i_\alpha,\nu_\alpha) : (X,\alpha) \longrightarrow M''(X,\alpha)$ such that

(a) i_α embeds X as SDR into $M''(X,\alpha)$

(b) any reduced $\mathfrak{B}''$-map $(f,\rho) : (X,\alpha) \longrightarrow (Y,\beta)$ is the canonical composite of (i_α,ν_α) and a unique $\mathfrak{B}$-homomorphism $M''(X,\alpha) \longrightarrow (Y,\beta)$

(c) if $(f,\rho) : (X,\alpha) \longrightarrow (Y,\beta)$ is a reduced $\mathfrak{B}''$-map and each $\hat{\alpha}_k$ and $\hat{\beta}_k$ is a closed cofibration, the homotopy class of the induced homomorphism $h : M''(X,\alpha) \longrightarrow (Y,\beta)$ depends only on the simplicial homotopy class of (f,ρ).

The proof is as for Theorem 4.49. ∎

The construction M'' gives actually a little more. If (X,α) is a $W''\mathfrak{B}$-space such that each $\hat{\alpha}_k : \mathfrak{B}(0,k) \longrightarrow X_k$ is an inclusion and $A \in \mathfrak{Top}_K$ is the collection of images of the $\hat{\alpha}_k$, the action α on X makes A into a $\mathfrak{B}$-space and the inclusion i_α restricted to A is a $\mathfrak{B}$-homomorphism. This has relevance for the based case:

Corollary 5.29: Let $\mathfrak{B}$ be a K-coloured PROP such that each $\mathfrak{B}(0,k)$ has exactly one element and let (X,α) be a $W\mathfrak{B}$-space such that each $\hat{\alpha}_k : \mathfrak{B}(0,k) \longrightarrow X_k$ is a closed cofibration. Denote the image of $\hat{\alpha}_k$ by x_k. Then there is a based $\mathfrak{B}$-space (Y,β) and an unbased reduced $\mathfrak{B}$-map $(f,\rho) : (X,\alpha) \longrightarrow (Y,\beta)$ whose underlying map consists of maps of pairs $f_k : (X_k,x_k) \longrightarrow (Y_k,*)$, which embeds X_k into Y_k as SDR.

Proof: By (5.6) we can homotop α to a $W\mathfrak{B}''$-action $\bar{\alpha}$ keeping $\hat{\alpha}_k(*)$ fixed. Take $Y = M''(X,\bar{\alpha})$ with base points given by $\mathfrak{B}(0,k) \subset Y_k$. By (5.7) and (5.28) there is an unbased reduced $\mathfrak{B}$-map

$$(f,\rho) : (X,\alpha) \xrightarrow{(id_X,\varkappa)} (X,\bar{\alpha}) \xrightarrow{(i_{\bar{\alpha}},\nu_{\bar{\alpha}})} M''(X,\bar{\alpha}) ,$$

$f = i_\alpha$, with the required properties. ∎

In view of (5.28 c) we also show

Lemma 5.30: Let $\mathfrak{B}$ be a PROP and (X,α) a $W''\mathfrak{B}$-space such that each $\hat{\alpha}_k$ is a closed cofibration. Then each inclusion $\mathfrak{B}(0,k) \subset M''X_k$ is a closed cofibration.

Proof: We filter $M''(X,\alpha)$ by subspaces M''_r of elements represented by a reduced cherry tree with m vertices and n cherries, $m+n \le r$. An element of the space P of all cherry trees with m vertices and n cherries, $m+n=r$, represents an element of filtration $r-1$ iff a vertex is labelled by an identity, or an internal edge has length 0 or 1 or supports a stump, or a cherry lies in the image of some $\hat{\alpha}_k$. If Q is the space of these cherry trees, then (P,Q) is a NDR so that (M''_r, M''_{r-1}) is a NDR (A 4.1). Since $M''_1 = \bigcup_k \mathfrak{B}(0,k)$, the result follows (A 4.1). ∎

VI. Chapter

ITERATED LOOP SPACES AND ACTIONS ON CLASSIFYING SPACES

It is the aim of this chapter to show that E-spaces (see 2.46) coincide with infinite loop spaces. As an application we prove that the stable groups O,U,SO,SU,Top,F and their classifying spaces are infinite loop spaces. For this purpose we investigate how much structure on a space X can be transferred to its classifying space BX if there is any.

1. THE CLASSIFYING SPACE CONSTRUCTION

In this chapter $\mathfrak{A}$ denotes the PRO associated with the theory Θ_m of monoids. Recall that $\mathfrak{Mor}_{\mathfrak{A}}$ is the category of monoids and homomorphisms and $\mathfrak{Hom}_{\mathfrak{A}}$ the category of moinoids and homotopy classes of homomorphisms.

Let G be any monoid with unit e. Consider G not as an $\mathfrak{A}$-space but as a monochrome PRO $\mathfrak{G}$ by putting $\mathfrak{G}(1,1) = G$ and $\mathfrak{G}(n,1) = \emptyset$ for $n \neq 1$. Composition is given by the multiplication in G. Take a single point P with its unique $W\mathfrak{G}$-structure and apply the construction M of (4.49). Note that M is defined even if (G,e) is not a NDR so that condition (3.7) is not satisfied for the PRO $\mathfrak{G}$. The space EG=MP is a contractible free G-space. Contractibility follows from (4.49 a) and freeness from the definition of the $\mathfrak{G}$-action on MP.

Definition 6.1: We call EG the universal space and BG=EG/G the classifying space of the monoid G.

Let us give direct descriptions of EG and BG. The representing cherry trees of EG are linear and vertical, and may be specified by giving in order, going up the tree, the vertex labels and edge lengths as $(g_o,t_1,g_1,t_2,g_2,\dots,g_k)$, $g_i \in G$, $t_i \in I$. We have the relations

(6.2)

$$\text{(a)}\quad (g_o,t_1,g_1,\dots,g_k) = \begin{cases} (g_o,t_1,g_1,\dots,t_{k-1},g_{k-1}) & g_k=e,\ k>0 \\ (g_o,t_1,\dots,g_{i-1},t_i*t_{i+1},g_{i+1},\dots,g_k) & g_i=e,\ 0<i<k \end{cases}$$

$$\text{(b)}\quad (g_o,t_1,g_1,\dots,g_k) = (g_o,t_1,\dots,t_{i-1},g_{i-1}\circ g_i,t_{i+1},\dots,g_k)\quad t_i=0$$

$$\text{(c)}\quad (g_o,t_1,g_1,\dots,g_k) = (g_o,t_1,\dots,g_{i-1})\qquad t_i=1$$

Here we use $t*t' = t+t'-tt'$ and <u>not</u> $t*t' = \max(t,t')$. Hence

$$EG = \bigcup_{k=0}^{\infty} G^{k+1}\times I^k/\sim$$

Relation (a),(b), and (c) stand for relations (3.1 a),(3.1 c), and (4.49 d). Note that $(g_o,t_1,\dots,g_k)$ stands for the representative $[(g_o,j),t_1,(g_1,id_o),t_2,\dots,(g_k,id_o);*]$ with $P=\{*\}$, $(g_i,h_i)\in G \circledast \Omega_1$, and $j : 0 \longrightarrow 1$ in Ω_1.

The contraction H_t : EG ⟶ EG is given by

$$H_t(g_o,t_1,g_1,\dots,g_k) = (e,t,g_o,t_1,g_1,\dots,g_k)$$

t running from 0 to 1, and the G-action G×EG ⟶ EG by

$$[g,(g_o,t_1,g_1,\dots,g_k)] \longmapsto (g\circ g_o,t_1,g_1,\dots,g_k)$$

Consequently

$$BG = \bigcup_{k=0}^{\infty} G^k\times I^k/\sim$$

with $G^o=\{e\}$ and the relations

(6.3)

$$\text{(a)}\quad (t_1,g_1,t_2,\dots,g_k) = \begin{cases} (t_1,g_1,t_2,\dots,g_{k-1}) & g_k=e \\ (t_1,g_1,\dots,g_{i-1},t_i*t_{i+1},g_{i+1},\dots,g_k) & g_i=e,\ i<k \end{cases}$$

$$\text{(b)}\quad (t_1,g_1,t_2,\dots,g_k) = \begin{cases} (t_2,g_2,t_3,\dots,g_k) & t_1=0 \\ (t_1,g_1,\dots,t_{i-1},g_{i-1}\circ g_i,t_{i+1},\dots,g_k) & t_i=0,\ i>0 \end{cases}$$

(c) $(t_1,g_1,t_2,\dots,g_k) = (t_1,g_1,\dots,g_{i-1}) \qquad t_i=1$

We use the obvious convention, that ( ) = (e). The projection $pG : EG \longrightarrow BG$ is given by

$$pG(g_o,t_1,g_1,\dots,g_k) = (t_1,g_1,\dots,g_k)$$

A homomorphism $f : G \longrightarrow H$ of monoids induces maps $Ef : EG \longrightarrow EH$ and $Bf : BG \longrightarrow BH$ by $Ef(g_o,t_1,\dots,g_k) = (f(g_o),t_1,\dots,f(g_k))$ and $Bf(t_1,g_1,\dots,g_k) = (t_1,f(g_1),\dots,f(g_k))$, which makes E and B into functors

$$E,B : \mathfrak{Mor}_{\mathfrak{A}} \longrightarrow \mathfrak{Top}$$

and p into a natural transformation of functors. If $f_t : G \longrightarrow H$ is a homotopy through homomorphisms, then Ef_t and Bf_t are homotopies $Ef_o \simeq Ef_1$ respectively $Bf_o \simeq Bf_1$. Hence we can pass to the homotopy categories and obtain functors

$$\overline{E},\overline{B} : \mathfrak{Hom}_{\mathfrak{A}} \longrightarrow \mathfrak{Top}_h$$

Our functors E and B coincide with Milgram's classifying space functors [37]: Let $\Delta^n = \{(u_1,\dots,u_n)\in\mathbb{R}^n \mid 0\le u_1\le\dots\le u_n\le 1\}$, the Euclidean n-simplex. Define

$$E'G = \bigcup_{k=0}^{\infty} G^{k+1}\times\Delta^k/\sim$$

with the relations

(6.4)

(a) $$(g_o,u_1,g_1,\dots,g_k) = \begin{cases}(g_o,u_1,g_1,\dots,g_{k-1}) & \text{if } g_k=e,\ k>0\\ (g_o,u_1,\dots,g_{i-1},u_{i+1},g_{i+1},\dots,g_k) & \text{if } g_i=e, 0<i<k\end{cases}$$

(b) $$(g_o,u_1,g_1,\dots,g_k) = \begin{cases}(g_o\cdot g_1,u_2,g_2,\dots,g_k) & \text{if } u_1=0\\ (g_o,u_1,\dots,g_{i-1},u_{i+1},g_i\cdot g_{i+1},u_{i+2},\dots,g_k) & \text{if } u_i=u_{i+1}\\ (g_o,u_1,\dots,u_{k-1},g_{k-1}) & \text{if } u_k=1\end{cases}$$

There is a G-action on E'G given by $g\cdot(g_o,u_1,\dots,g_k)=(g\cdot g_o,u_1,\dots,g_k)$

<u>Definition 6.5</u>: We call E'G <u>Milgram's</u> <u>universal</u> <u>space</u> of G and B'G=E'G/G <u>Milgram's</u> <u>classifying</u> <u>space</u> of G.

We can extend E' and B' to functors $\mathfrak{Mor}_{\mathfrak{U}} \longrightarrow \mathfrak{Top}$ in the same manner as E and B.

<u>Proposition 6.6</u>: There is a natural equivariant homemorphism EG -> E'G. Consequently, the functors E and E' and the functors B and B' are naturally isomorphic.

<u>Proof</u>: Define a map h : EG $\longrightarrow$ E'G on representatives by $h(g_0,t_1,g_1,\ldots,g_k) = (g_0,u_1,g_1,\ldots,g_k)$ where $u_i = t_1 * t_2 * \ldots * t_i$. Then relation (6.2 a) corresponds to (6.4 a), and relation (6.2 b) and (6.2 c) correspond to (6.4 b). The inverse of h is given by

$$(g_0,u_1,g_1, \ldots,g_k) \longmapsto (g_0,t_1,g_1,\ldots,g_k)$$

with $t_1=u_1$ and $t_i = (u_i-u_{i-1})/(1-u_{i-1})$ for i>1, with the convention that 0/0=1. ∎

<u>Remark</u>: If we use $t_1 * t_2 = \max(t_1,t_2)$ instead, there is no such homeomorphism.

<u>Proposition 6.7</u>: The functors E and B preserve products.

<u>Proof</u>: It is well-known (see [30] or [50]) that the functors E' and B' preserve products. ∎

Next we will show that the functor B preserves homotopy equivalences. Since we later on need this result for monoids in the category of H-spaces, where H is a discrete group, we work in the category of H-spaces. An H-monoid is a monoid G together with an action of H on G such that $g \longmapsto h\cdot g$, $g \in G$, $h \in H$ is a monoid homomorphism for all $h \in H$.

Then EG and BG admit H-actions

$$h\cdot(g_o,t_1,g_1,\ldots,g_k) = (h\cdot g_o,t_1,h\cdot g_1,\ldots,h\cdot g_k) \quad \text{and}$$

$$h\cdot(t_1,g_1,t_2,\ldots,g_k) = (t_1,h\cdot g_1,t_2,\ldots,h\cdot g_k)$$

and $p : EG \longrightarrow BG$ is H-equivariant.

The functors E and B and accordingly $\overline{E}$ and $\overline{B}$ are canonically filtered: Put $E_iG = \bigcup_{n=0}^{i} G^{n+1}\times I^n/\sim$ and $B_iG = \bigcup_{n=0}^{i} G^n\times I^n/\sim$ with the relations (6.2) and (6.3). Let $p_iG : E_iG \longrightarrow B_iG$ be the projections.

<u>Lemma 6.8</u>: (a) $EG = \varinjlim E_iG$ and $BG = \varinjlim B_iG$

(b) If (G,e) is an H-NDR, then $E_iG\subset E_{i+1}G\subset EG$ and $B_iG\subset B_{i+1}G\subset BG$ are closed H-equivariant cofibrations. Hence, since $E_oG=G$ and $B_oG=(e)=*$, the pairs (EG,G) and $(BG,*)$ are NDRs.

<u>Proof</u>: By now standard. ∎

<u>Lemma 6.9</u>: Let G_1 and G_2 be H-monoids such that (G_1,e) and (G_2,e) are H-NDRs. Let $f : G_1 \longrightarrow G_2$ be an equivariant homomorphism. Then

(a) if f is an equivariant homotopy equivalence, so is Bf

(b) if f is an equivariant closed cofibration, so is Bf

(c) if f is a closed equivariant cofibration and equivariant homotopy equivalence, then Bf embeds BG_1 as equivariant SDR in BG_2

(d) if G_1 and G_2 are Hausdorff and f is a weak homotopy equivalence, then BG_1 and BG_2 are Hausdorff and Bf is a weak homotopy equivalence.

<u>Proof</u>: BG is an iterated adjunction space in the category of H-spaces obtained by adjoining spaces $(G^n\times I^n, D^nG\times I^n \cup G^n\times\partial I^n)$ where $D^nG \subset G^n$ is the subspace of points having a coordinate e. Part (b) follows from (A 4.9). For part (a) the map $f : (G_1,e) \longrightarrow (G_2,e)$ is an equivariant homotopy equivalence of pairs inducing a homotopy equivalence of pairs

$$(G_1^n\times I^n, G_1^n\times\partial I^n \cup D^nG_1\times I^n) \longrightarrow (G_2^n\times I^n, G_2^n\times\partial I^n \cup D^nG_2\times I^n)$$

Hence (a) follows from (A 4.6) and (A 4.4). Part (c) follows from (a) and (b) and the equivariant version of [14;(3.7)]. For part (d) it suffices to show that the map

$$r : G_1^n \times \partial I^n \cup D^n G_1 \times I^n \longrightarrow G_2^n \times \partial I^n \cup D^n G_2 \times I^n$$

is a weak homotopy equivalence and then apply (A 4.8). Again, by (A 4.8), it suffices tp prove that f induces a weak homotopy equivalence $D^n G_1 \longrightarrow D^n G_2$ because $G^n \times \partial I^n \cup D^n G \times I^n$ is obtained from $G^n \times \partial I^n$ by attaching $(D^n G \times I^n, D^n G \times \partial I^n)$. Since $D^n G$ is obtained from $G^{n-1} \times \{e\}$ by attaching $(D^{n-1} G \times G, D^{n-1} G \times \{e\})$ this follows from (A 4.8) by induction on n. ▮

Let $\Omega : \mathfrak{Top}^o \longrightarrow \mathfrak{Top}^o$ be the loop space functor and $L : \mathfrak{Top}^o \longrightarrow \mathfrak{Top}^o$ the based path space functor, i.e. $L(X) = \{\omega : I \longrightarrow X \mid \omega(0) = *\}$. For a monoid G we take $e \in G$ as base point and (e) as base point for EG and BG. Then E and B may be considered as functors

$$E, B : \mathfrak{Mor}_{\mathfrak{A}} \longrightarrow \mathfrak{Top}^o$$

By (2.53) and (3.25), we may interpret Ω as functor

$$\Omega : \mathfrak{Top}^o \longrightarrow \mathfrak{Map}^o_{\mathfrak{A}}$$

into the category of based W𝔄-spaces and based 𝔄-maps. The end point projection $\pi : LX \longrightarrow X$ is a fibration with fibre ΩX. There is a natural map of pairs

$$jG : (EG, G) \longrightarrow (LBG, \Omega BG)$$

defined by $[jG(g_o, t_1, g_1, \ldots, g_k)](t) = (1-t, g_o, t_1, \ldots, g_k)$ making the diagram

$$\begin{array}{ccc} EG & \xrightarrow{\;jG\;} & LBG \\ & {\scriptstyle pG}\searrow \quad \swarrow{\scriptstyle \pi} & \\ & BG & \end{array}$$

commute.

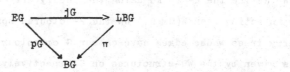

Proposition 6.10: The maps $jG : G \longrightarrow \Omega BG$ carry a natural $\mathfrak{A}$-map structure defining a natural transformation from the functor $J:\mathfrak{Mor}_{\mathfrak{A}} \to \mathfrak{Map}_{\mathfrak{A}}$ induced by the augmentation $\varepsilon : W\mathfrak{A} \longrightarrow \mathfrak{A}$ to the functor $\Omega B:\mathfrak{Mor}_{\mathfrak{A}} \to \mathfrak{Map}_{\mathfrak{A}}$. If (G,e) is a NDR, this $\mathfrak{A}$-map structure is homotopic through $\mathfrak{A}$-maps with carrier jG to a based $\mathfrak{A}$-map structure.

Proof: Let $\mathfrak{D}$ be the PRO of example (2.53) and $\rho : W\mathfrak{A} \longrightarrow \mathfrak{D}$ a PRO functor, which exists by (2.54) and the lifting theorem. We know that there is a based action of $\mathfrak{D}$ and hence of $W\mathfrak{A}$ on loop spaces. Let L^oEG be the subspace of LEG of all paths ω with $\omega(1) \in E_oG = G$. Recall that

$$\mathfrak{D}(n,1) = \{(x_1,y_1,\ldots,x_n,y_n) \in \mathbb{R}^{2n} \mid 0 \le x_1 < y_1 \le x_2 < y_2 \le \ldots \le x_n < y_n \le 1\}$$

Using the G-action on EG we define a based $\mathfrak{D}$-action and hence a based $W\mathfrak{A}$-action on L^oEG by

$$[(x_1,y_1,\ldots,x_n,y_n)(\omega_1,\ldots,\omega_n)](t) = \begin{cases} g_1g_2\ldots g_{i-1}\omega_i\left(\dfrac{t-x_i}{y_i-x_i}\right) & t\in[x_i,y_i] \\ g_1g_2\ldots g_i(e) & t\in[y_i,x_{i+1}] \end{cases}$$

where $g_i = \omega_i(1)$, $y_o=0$ and $x_{n+1}=1$.

We give jG an $\mathfrak{A}$-map structure by exhibiting it as composite of an $\mathfrak{A}$-map $kG : G \longrightarrow L^oEG$ and a $W\mathfrak{A}$-homomorphism $rG : L^oEG \longrightarrow \Omega BG$. The projection $p : EG \longrightarrow BG$ induces a based map $LEG \longrightarrow LBG$ which maps L^oEG into ΩBG defining rG. Since the $\mathfrak{D}$-action on ΩBG is given by

$$[(x_1,y_1,\ldots,x_n,y_n)(\omega_1,\ldots,\omega_n)](t) = \begin{cases} \omega_i\left(\dfrac{t-x_i}{y_i-x_i}\right) & t\in[x_i,y_i] \\ (e) & t\in[y_i,x_{i+1}] \end{cases}$$

it is clear that rG is a based $\mathfrak{D}$-homomorphism and hence a based $W\mathfrak{A}$-homomorphism.

We define the $\mathfrak{A}$-map kG using the cherry tree description (3.24) for an action α of $W(\mathfrak{A} \otimes \mathfrak{L}_1)$ on the 2-coloured space $\{G,L^oEG\}$. On cherry trees whose edges have colour G or colour L^oEG only, the action α is given by the $W\mathfrak{A}$-structures on G respectively L^oEG. On cherry trees having edges of both colours we define α by induction on the number k of internal edges. Since $\mathfrak{A}(m,1)$ consists of a single point for each m, the space of all cherry trees of a given shape with n twigs and k internal edges is of the form $X_1\times\ldots\times X_n\times I^k$, where X_i is a copy of G or L^oEG and $X_1\times\ldots\times X_n$ is the space of cherries. Moreover, we may restrict our attention to cherry trees containing no identity. For k=0, we define

$$f : X_1\times\ldots\times X_n\times I^o\times I \longrightarrow EG$$

by $f(x_1,\ldots,x_n,t)=(e,1-t,q(x_1)\cdot\ldots\cdot q(x_n))$ with $q(x_i)=x_i$ if $X_i=G$ and $q(x_i)=\omega(1)$ if $x_i=\omega\in L^oEG$. This defines the based action kG on cherry trees with no internal edge. Inductively, we have to find a map $f : X_1\times\ldots\times X_n\times I^k\times I \longrightarrow EG$ which is already given on $X_1\times\ldots\times X_n\times\partial I^k\times I$ and which satisfies

(a) $f(x_1,\ldots,x_n,s,0) = (e)$

(b) $f(x_1,\ldots,x_n,s,1) = (q(x_1)\cdot\ldots\cdot q(x_n))\in E_oG$

For the second statement f in addition has to satisfy

(c) $f(*,\ldots,*,s,t) = (e)$ $\qquad * =$ base point.

Since EG can be contracted by a homotopy which is natural in G we can find an extension for the first statement which is natural in G. If (G,e) is a NDR then $(e)\subset \Omega EG\subset L^oEG$ are closed cofibrations by [9;p.57], [52; Thm. 12], and Lemma (6.8). Let $X = X_1\times\ldots\times X_n$ and $*\in X$ its base point. Let $f' : X\times I^{k+1} \longrightarrow EG$ be the map just constructed. We inductively look for a map $f : X\times I^{k+1} \longrightarrow EG$ satisfying (a),(b),(c) and a homotopy $h_t : f \simeq f'$ which is given by induction on $X\times\partial I^{k+1}$. So we really want a map

$$H : X\times I^{k+1}\times I \longrightarrow EG$$

which is already given on $X\times\partial I^{k+1}\times I \cup X\times I^{k+1}\times 1 \cup *\times I^{k+1}\times 0$. Since EG is contractible we can extend $H|*\times\partial I^{k+2}$ to $*\times I^{k+2}$ and then extend $H|X\times I^{k+1}\times 0 \cup (X\times\partial I^{k+1} \cup *\times I^{k+1})\times I$ to the whole of $X\times I^{k+1}\times I$, which is possible because $(X\times I^{k+1}, X\times\partial I^{k+1} \cup *\times I^{k+1})$ is a NDR. ∎

Corollary 6.11: The maps $jG : G \longrightarrow \Omega BG$ define a natural transformation from the canonical functor $J : \mathfrak{Hom}_{\mathfrak{A}} \longrightarrow \mathfrak{Map}_{\mathfrak{A}}$ to the functor $\Omega\overline{B} : \mathfrak{Hom}_{\mathfrak{A}} \longrightarrow \mathfrak{Map}_{\mathfrak{A}}$.

__Definition 6.12__: A space X is called __numerably__ __contractible__ if it has a numerable __nullhomotopic__ __covering__, i.e. there is a covering $\mathfrak{U} = \{U_\alpha \mid \alpha \in A\}$ and maps $u_\alpha : X \longrightarrow I$ such that

(a) each $x \in X$ has a neighbourhood W such that $u_\alpha(W) = 0$ for all but finitely many $\alpha \in A$

(b) $\sum\limits_{\alpha \in A} u_\alpha(x) = 1$ for all $x \in X$

(c) $u_\alpha^{-1}(0,1] \subset U_\alpha$

(d) the inclusions $U_\alpha \longrightarrow X$ are nullhomotopic

__Example 6.13__: Any CW-complex is numerably contractible [13; Prop.6.7]

We now prove the main result of this section.

__Theorem 6.14__: Let G be a numerably contractible monoid such that (G,e) is a NDR and $\pi_o(G)$ is a group under the multiplication of G. Then $jG : G \longrightarrow \Omega BG$ is a based homotopy equivalence.

__Proof__: Let $\overline{G} = (G \cup I)/(e \sim 0)$ and extend the monoid structure of G to $\overline{G}$ by $t \cdot g = g \cdot t = e$ and $t \cdot u = tu$ for $g \in G$ and $t, u \in I$. Then $1 \in I$ is the unit of $\overline{G}$. The map $f : \overline{G} \longrightarrow G$ defined by $g \longmapsto g$, $g \in G$, and $t \longmapsto e$, $t \in I$, is a homomorphism and a based homotopy equivalence (A 4.3). By naturality, we have a commutative diagram of topological spaces

$$\begin{array}{ccc} \overline{G} & \xrightarrow{\;j\overline{G}\;} & \Omega B\overline{G} \\ \Big\downarrow f & & \Big\downarrow \Omega Bf \\ G & \xrightarrow{\;jG\;} & \Omega BG \end{array}$$

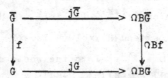

whose vertical maps are based homotopy equivalences (Use that $(X,*) \longrightarrow (Y,*)$ is a based homotopy equivalence provided $X \longrightarrow Y$ is a homotopy equivalence and $(X,*)$, $(Y,*)$ cofibred (see A 4.3)). Hence it suffices to prove the result for $\overline{G}$.

Our aim is to find a commutative diagram

$$\begin{array}{ccccc} B' & \xleftarrow{\;p'\;} & E' & \supset & F' \\ \downarrow & & \downarrow & & \downarrow \\ B\overline{G} & \xleftarrow{\;p\overline{G}\;} & E\overline{G} & \supset & \overline{G} \end{array}$$

in which p' is an h-fibration, i.e. a map having the weak covering homotopy property in the sense of Dold [13], and the vertical maps are homotopy equivalences. Inductively, we construct a sequence of spaces $E_o \subset E_1 \subset E_2 \subset \ldots$ and h-fibrations $q_i : E_i \longrightarrow B_i\overline{G}$ having the following properties

(a) $E_i\overline{G} \subset E_i$, and $E_i \subset E_{i+1}$ is an inclusion of pairs $(E_i, E_i\overline{G}) \subset (E_{i+1}, E_{i+1}\overline{G})$ such that

$$\begin{array}{ccc} E_i & \subset & E_{i+1} \\ {\scriptstyle q_i}\downarrow & & \downarrow{\scriptstyle q_{i+1}} \\ B_i\overline{G} & \subset & B_{i+1}\overline{G} \end{array} \qquad \text{and} \qquad \begin{array}{ccc} E_i\overline{G} & \subset & E_i \\ {\scriptstyle p_i\overline{G}}\searrow & & \swarrow{\scriptstyle q_i} \\ & B_i\overline{G} & \end{array}$$

commute

(b) (E_{i+1}, E_i) is a NDR

(c) there are deformation retractions $r_i : E_i \longrightarrow E_i\overline{G}$ and $d_i^b : q_i^{-1}(b) \longrightarrow (pG)^{-1}(b)$ for all $b \in B_i\overline{G}$ such that

$$\begin{array}{ccc} E_i & \subset & E_{i+1} \\ {\scriptstyle r_i}\downarrow & & \downarrow{\scriptstyle r_{i+1}} \\ E_i\overline{G} & \subset & E_{i+1}\overline{G} \end{array} \qquad \text{and} \qquad \begin{array}{ccc} q_j^{-1}(b) & \subset & q_i^{-1}(b) \\ {\scriptstyle d_j^b}\searrow & & \swarrow{\scriptstyle d_i^b} \\ & (pG)^{-1}(b) & \end{array} \quad b \in B_j\overline{G} \subset B_i\overline{G}$$

commute.

We start with $E_o = E_o\overline{G} = \overline{G}$ and suppose inductively that we have constructed E_{k-1}. Let $B = B_k\overline{G}$ and $A = B_{k-1}\overline{G} \subset B$. Then B is obtained from A by adjoining $\overline{G}^k \times I^k$ modulo the points $A' = D^k\overline{G} \times I^k \cup \overline{G}^k \times \partial I^k$ with

$D^k\overline{G} = \{(g_1,\ldots,g_k) \in \overline{G}^k \mid$ some g_i is the identity $1 \in I \subset \overline{G}\}$, and $E_k\overline{G}$ obtained from $E_{k-1}\overline{G}$ by adjoining $\overline{G}^{k+1} \times I^k$ modulo $\overline{G} \times A'$. Let Y be obtained from E_{k-1} by attaching $\overline{G}^{k+1} \times I^k$ by the map $\overline{G} \times A' \longrightarrow E_{k-1}\overline{G} \subset E_{k-1}$. Define a map $q : Y \longrightarrow B$ by $q|E_{k-1} = q_{k-1}$ and $q|E_k\overline{G} = p_k\overline{G}$. For $U \subset B$ denote $q^{-1}(U)$ by Y_U and $q|Y_U : Y_U \longrightarrow U$ by q_U.

If $Q = \{(g_1,\ldots,g_k) \in \overline{G}^k \mid$ some $g_i \in I \subset \overline{G}, g_i > \frac{3}{4}\}$, then $V' = Q \times I^k \cup \overline{G}^k \times ([0,\frac{1}{4}) \cup (\frac{3}{4},1])$ is a halo of A' in $\overline{G}^k \times I^k$ inducing a halo V of A. Since ∂I is a SDR of $[0,\frac{1}{4}) \cup (\frac{3}{4},1]$, the space A' is an SDR of V' and hence A an SDR of V. Let $\rho : V \longrightarrow A$ be the deformation retraction, then ρ is covered by a deformation retraction $\overline{\rho} : Y_V \longrightarrow Y_A$, i.e. we have a commutative diagram

$$\begin{array}{ccc} Y_V & \xrightarrow{\overline{\rho}} & Y_A \\ \downarrow{\scriptstyle q_V} & & \downarrow{\scriptstyle q_A} \\ V & \xrightarrow{\rho} & A \end{array}$$

Consider the diagram

$$\begin{array}{ccccc} Y_V & \xrightarrow{r} & Y' & \xrightarrow{\rho'} & Y_A \\ & {\scriptstyle q_V}\searrow & \downarrow{\scriptstyle q'} & & \downarrow{\scriptstyle q_A} \\ & & V & \xrightarrow{\rho} & A \end{array}$$

where Y' is the pullback and $r : Y_V \longrightarrow Y'$ is induced by $\overline{\rho}$ and q_V. Take E_k to be the double mapping cylinder of the maps $Y_V \subset Y$ and r, and take $q_k : E_k \longrightarrow B_k\overline{G}$ to be the map given by q on $Y \cup Y_V \times I$ and by q' and Y'. By [12;(17.8)] we find

(i) q_k is an h-fibration

(ii) Y is an SDR of E_k

(iii) Y_A is an SDR of $(E_k)_A$ over A

(iv) Y_{B-A} is an SDR of $(E_k)_{B-A}$ over $B-A$

provided we can show

(α) $q_A : Y_A \longrightarrow A$ and $q_{B-A} : Y_{B-A} \longrightarrow B-A$ are h-fibrations

(β) $V-A$ is numerably contractible

(γ) for all $b \in V-A$ the map $\overline{\rho}_b : Y_b \longrightarrow Y_{\rho(b)}$ is a homotopy equivalence.

The properties (i),...,(iv) imply (a),...,(c): The composite inclusion $E_{k-1} \subset Y \subset E_k$ satisfies (a). Since $(\overline{G}^{k+1} \times I^k, \overline{G} \times A')$ is a NDR so is (Y, E_{k-1}) and hence (E_k, E_{k-1}). By (ii) there is a deformation retraction $f : E_k \longrightarrow Y$. Define a deformation retraction $h : Y \longrightarrow E_k\overline{G}$ by

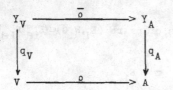

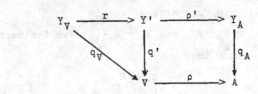

$h|E_{k-1} = r_{k-1}$ and $h|E_k\overline{G} = id$. Then $h \circ f = r_k$ is the required deformation retraction. Suppose $b \in B_{k-1}\overline{G} = A$, then there is a deformation retraction $u : q_k^{-1}(b) \longrightarrow q^{-1}(b)$ by (iii). Since $q^{-1}(b) = q_{k-1}^{-1}(b)$, we can put $d_k^b = d_{k-1}^b \circ u$. If $b \in B_k\overline{G} - B_{k-1}\overline{G} = B-A$, then (iv) provides a deformation retraction $d_k^b : q_k^{-1}(b) \longrightarrow g^{-1}(b) = (pG)^{-1}(b)$.

We now verify $(\alpha),(\beta),(\gamma)$. Since $q_A = q_{k-1}$ and since q_{B-A} is the projection $\overline{G}\times\overline{G}^k\times I^k \longrightarrow \overline{G}^k\times I^k$, both maps are h-fibrations. If $b \in V-A$ then $Y_b = \overline{G}\times\{b\}$. By construction, $\overline{\rho}(g,b)$ is represented by $(g \circ g'(b), \rho(b)) \in (pG)^{-1}(\rho b) \cong \overline{G}\times\{\rho(b)\}$ with $g'(b) \in \overline{G}$ depending on b. Since $\overline{G}$ is numerably contractible and $\pi_o(\overline{G})$ is a group, right translation is a homotopy equivalence [12;(12.7)]. Hence

$$\overline{\rho}_b : Y_b \simeq (pG)^{-1}(\rho b) \subset Y_{\rho(b)} = (E_{k-1})_{\rho(b)}$$

By induction hypothesis the inclusion $(pG)^{-1}(\rho b) \subset (E_{k-1})_{\rho(b)}$ is homotopy equivalence. It remains to check (β): We cover $V'-A'$ by the sets $V_o = \{(t_1,g_1,\ldots,t_k,g_k) \mid \text{some } g_i \in (\frac{3}{4},1) \subset \overline{G} \text{ or some } t_i \in (0,\frac{1}{4})\}$ and $V_1 = \{(t_1,g_1,\ldots,t_k,g_k) \mid \text{some } g_i \in (\frac{3}{4},1) \subset \overline{G} \text{ or some } t_i \in (\frac{3}{4},1)\}$. Then $(\overline{G}^k - D^k\overline{G})\times(\frac{1}{8},\frac{1}{8},\ldots,\frac{1}{8})$ is a SDR of V_o and $(\overline{G}^k - D^k\overline{G})\times(\frac{7}{8},\frac{7}{8},\ldots,\frac{7}{8})$ a SDR of V_1. Since $\overline{G}^k - D^k\overline{G}$ is homotopy equivalent to G^k and G is numerably contractible, so are V_o and V_1 (A 4.11). Define a map $v : V'-A' \longrightarrow I$ by

$$v(t_1,g_1,\ldots,t_k,g_k) = \prod_{i=1}^{k} \min\,[2\max(t_i - \tfrac{1}{4},0),1]$$

Then (V_o,V_1) is a numerable covering of $V'-A'$ with numeration $(1-v,v)$. Hence $V'-A'$ and therefore $V-A$ are numerably contractible.

Let $E = \varinjlim_n E_n$ and $q = \varinjlim_n q_n : E \longrightarrow B\overline{G}$. By (c), $E\overline{G} \subset E$ and $E_o\overline{G} \subset q^{-1}(e)$ are SDRs. Unfortunately, q might not be an h-fibration. To correct this, let TE be the telescope of the E_n and TB the telescope of the $B_n\overline{G}$, i.e.

$$TE = E_o\times[0,1] \cup E_1\times[1,2] \cup E_2\times[2,3] \cup \ldots$$

topologized as subspace of $E\times\mathbb{R}$. The maps q_i induce a map $u : TE \to TB$ and if we take $(e) \in B_o\overline{G}$ as basepoint, then $\overline{G} = u^{-1}(e)$. The composite maps $E_i \times [i,i+1] \longrightarrow E_i \subset E$ induce a map $kE : TE \longrightarrow E$ and similarly for TB giving rise to a commutative diagram

$$\begin{array}{ccc} \overline{G} & \xrightarrow{\ i\ } & q^{-1}(e) \\ \cap & & \cap \\ TE & \xrightarrow{\ kE\ } & E \\ {\scriptstyle u}\downarrow & & \downarrow{\scriptstyle q} \\ TB & \xrightarrow{\ kB\ } & B\overline{G} \end{array}$$

where i is the inclusion of $\overline{G}$ in $q^{-1}(e)$. The map is a homotopy equivalence by (c). By (A 4.4) the maps kE and kB are homotopy equivalences, too. Let $P = \bigcup_{i=0}^{\infty} E_i$ and $Q = \bigcup_{i=0}^{\infty} B_i\overline{G}$ be the disjoint unions. The maps $E_i \subset E_{i+1}$ and $B_i\overline{G} \subset B_{i+1}\overline{G}$ induce endomorphisms $f : P \longrightarrow P$ and $g : Q \longrightarrow Q$. It is easy to see that TE is homeomorphic to $P\times I/(x,1)\sim(fx,0)$ and similarly for TB. Consider P and Q embedded in $P\times I$ and $Q\times I$ at height $\frac{1}{2}$. Let $A = \{(x,0) \in Q\times I/\sim\}$ and $V = \{(x,t) \in Q\times I/\sim \mid t \in [0,\frac{1}{3}) \cup (\frac{2}{3},1]\}$. Then $TB-A = Q\times(0,1)$ and $A = \bigcup_{i=0}^{\infty} B_i\overline{G}$. Hence $u_A = \bigcup_{i=0}^{\infty} q_i : TE_A \longrightarrow A$ and $u_{TB-A} = (\bigcup_{i=0}^{\infty} q_i)\times id : TE_{TB-A} = P\times(0,1) \longrightarrow Q\times(0,1) = TB-A$ are h-fibrations. V-A is homotopy equivalent to Q and hence is numerably contractible. We have a canonical deformation retraction $\rho : V \longrightarrow A$ which is covered by the corresponding canonical deformation retraction $r : TE_V \longrightarrow TE_A$ which by (c) is a homotopy equivalence on each fibre. By [12;(17.8)] there is a commutative diagram

$$\begin{array}{ccc} \overline{G} & \overset{j}{\subset} & v^{-1}(e) \\ \cap & & \cap \\ TE & \subset & T \\ & {\scriptstyle u}\searrow \quad \swarrow{\scriptstyle v} & \\ & TB & \end{array}$$

whose horizontal maps are homotopy equivalences such that v is an h-fibration. We end up with a commutative diagram

$$\begin{array}{ccccccccc} v^{-1}(e) & \xleftarrow{\ j\ } & \overline{G} & \xrightarrow{\ i\ } & q^{-1}(e) & \xleftarrow{\ i\ } & \overline{G} & \xrightarrow{\ j\overline{G}\ } & \Omega B\overline{G} \\ \cap & & \cap & & \cap & & \cap & & \cap \\ T & \longleftarrow & TE & \longrightarrow & E & \longleftarrow & E\overline{G} & \xrightarrow{\ j\overline{G}\ } & LB\overline{G} \\ {\scriptstyle v}\downarrow & & \downarrow{\scriptstyle u} & & \downarrow{\scriptstyle q} & & \downarrow{\scriptstyle p\overline{G}} & & \downarrow{\scriptstyle \pi} \\ TB & = & TB & \longrightarrow & B\overline{G} & = & B\overline{G} & = & B\overline{G} \end{array}$$

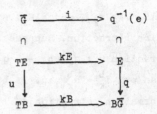

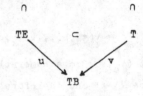

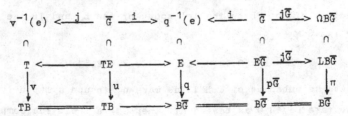

in which v and π are h-fibrations, π being the well-known path space fibration. We know of all horizontal maps with exception of $j\overline{G} : \overline{G} \longrightarrow \Omega B\overline{G}$ that they are homotopy equivalences, and since the inclusions of the base points are cofibrations, they are based homotopy equivalences. By the naturality of Puppe's h-fibration sequence [12; §14], the maps j,i and $j\overline{G}$ induce a homotopy equivalence $v^{-1}(e) \simeq \Omega B\overline{G}$. Hence $j\overline{G} : \overline{G} \simeq \Omega BG$. ▮

<u>Corollary 6.15</u>: Let G be a monoid which is Hausdorff such that (G,e) is a NDR and $\pi_o(G)$ is a group under the multiplication of G. Then jG is a weak homotopy equivalence.

<u>Proof</u>: Let $\mathfrak{Spl}$ denote the category of semisimplicial sets, $Sin : \mathfrak{Top} \longrightarrow \mathfrak{Spl}$ the singular functor and $R : \mathfrak{Spl} \longrightarrow \mathfrak{Top}$ the topological realization functor. Since we work with compactly generated spaces, $R \circ Sin$ preserves products so that $R \circ Sin(G)$ is a monoid. It is well-known that the back adjunction $\varepsilon : R \circ Sin \longrightarrow Id_{\mathfrak{Top}}$ induces isomorphisms of homotopy groups. By naturality, we have a commutative diagram

$$\begin{array}{ccc} RSin(G) & \xrightarrow{\quad jRSin(G) \quad} & \Omega BRSin(G) \\ \Big\downarrow \varepsilon(G) & & \Big\downarrow \Omega B\varepsilon(G) \\ G & \xrightarrow{\quad jG \quad} & \Omega B(G) \end{array}$$

Since $\pi_o RSin(G)$ is a group and RSin(G) a CW-complex, jRSin(G) is a homotopy equivalence by (6.14). Since Ω and B preserve weak homotopy equivalences, $\Omega B\varepsilon(G)$ is a weak homotopy equivalence. ▮

Let Y be a based space and $\Omega_M Y$ the Moore loop space on Y (cf.3.26). There is a canonical natural map $eY : B\Omega_M Y \longrightarrow Y$ given by

$$eY(t_1,x_1,t_2,x_2,\ldots,t_k,x_k) = \omega(\Sigma_{i=1}^{k}(1-t_1*t_2*\ldots*t_i)a_i)$$

where $x_i = (\omega_i,a_i) \in \Omega_M Y$ and $(\omega,\Sigma_i a_i) = x_1 \cdot x_2 \cdot \ldots \cdot x_k$. The composite

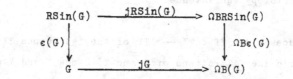

$$\Omega_M Y \xrightarrow{j\Omega_M Y} \Omega B\Omega_M Y \xrightarrow{\Omega eY} \Omega Y$$

carries an $\mathfrak{A}$-map structure, because $j\Omega_M Y$ and ΩeY do. It sends the Moore loop (ω, a) to the loop $\nu : I \longrightarrow Y$ given by $\nu(t) = \omega(ta)$. It is well-known that this is a homotopy equivalence [12;p.179].

Proposition 6.16: If Y has the homotopy type of a connected CW-complex the natural map $eY : B\Omega_M Y \longrightarrow Y$ is a homotopy equivalence.

Proof: Since $B\Omega_M Y$ and Y have the homotopy type of a CW-complex, it suffices to show that eY is a weak homotopy equivalence. By (6.15), $j\Omega_M Y : \Omega_M Y \longrightarrow \Omega B\Omega_M Y$ and hence ΩeY are weak homotopy equivalences. Hence eY is a weak homotopy equivalence, because Y is connected. ∎

The results (6.9 d) of p. 178 and (6.15) of p. 187 remain true if we drop the word "Hausdorff". For the proofs then use (A 4.8 b) and the following variant of (A 4.8 a):

If X and Y are properly filtered spaces and if $f : X \longrightarrow Y$ is a filtered map such that each f_n is a weak homotopy equivalence, then f is a weak homotopy equivalence.

Proof: f induces a map $Tf : TX \longrightarrow TY$ of the telescopes TX and TY associated with the filtrations of X and Y. Since X and Y are properly filtered, it suffices to show that Tf is a weak homotopy equivalence. Let $T_0X \subset T_1X \subset \ldots$ be the filtration of TX by partial telescopes. Then $\pi_i(TX) = \varinjlim \pi_i(T_nX)$. Since $(Tf|T_nX)_* : \pi_i(T_nX) \cong \pi_i(T_nY)$, the map Tf induces isomorphisms of homotopy groups.

We close this section with another elementary result on classifying spaces.

Lemma 6.17: Let G be a numerably contractible monoid such that (G,e) is a NDR. Then BG is numerably contractible.

Proof: Since $(G^n \times I^n, D^n G \times I^n \cup G^n \times \partial I^n)$ is a NDR and $G^n \times I^n$ is numerably contractible (A 4.11), BG is numerably contractible (A 4.12).

2. ACTIONS ON THE CLASSIFYING SPACE

We can extend the notion of a classifying space from monoids to W$\mathfrak{A}$-spaces by taking the composite

$$\bar{B}M : \mathfrak{Map}_{\mathfrak{A}} \longrightarrow \mathfrak{Hom}_{\mathfrak{A}}^{o} \longrightarrow \mathfrak{Top}^{o}$$

as classifying space functor. Here M is the functor of (4.49). Note that the unit e∈MX is a natural base point so that M can be interpreted as functor into $\mathfrak{Hom}_{\mathfrak{A}}^{o}$ instead of $\mathfrak{Hom}_{\mathfrak{A}}$. This definition makes the functor M important for our further investigations.

Lemma 6.18: (MX,e) is a NDR for any W$\mathfrak{A}$-space X.

Proof: MX is an iterated adjunction space filtered by the subspace M_nX of reduced cherry trees having at most n internal edges. The pairs $(M_{n+1}X, M_nX)$ and hence (MX, M_oX) are NDRs. Recall that

is the unit of MX. Since (M_oX, e) is a NDR, so is (MX,e).

Lemma 6.19: (a) (BMX,*) is a NDR for any W$\mathfrak{A}$-space X.
(b) BMX is numerably contractible if X is

Proof: (a) follows from (6.18) and (6.8) while (b) follows from (6.17),

(6.18) and (A 4.11) because $X \simeq MX$. $\blacksquare$

As an immediate application we find that for a well-pointed monoid (G,e) the spaces BMG and BG are homotopy equivalent. Indeed, the back adjunction $r : MG \longrightarrow G$ (see 4.49 ff) is a monoid homomorphism and a homotopy equivalence so that $Br : BMG \simeq BG$ by (6.9).

The functor $M : \mathfrak{Map}_{\mathfrak{A}} \longrightarrow \mathfrak{Hom}^{o}_{\mathfrak{A}}$ can be lifted to a functor $\overline{M} : \mathfrak{Mor}_{W\mathfrak{A}} \longrightarrow \mathfrak{Mor}^{o}_{\mathfrak{A}}$ such that

$$\begin{array}{ccc} \mathfrak{Mor}_{W\mathfrak{A}} & \xrightarrow{\overline{M}} & \mathfrak{Mor}^{o}_{\mathfrak{A}} \\ {\scriptstyle P}\downarrow & & \downarrow{\scriptstyle P'} \\ \mathfrak{Map}_{\mathfrak{A}} & \xrightarrow{M} & \mathfrak{Hom}^{o}_{\mathfrak{A}} \end{array}$$

commutes. P and P' denote the canonical projections. If $f : X \longrightarrow Y$ is a $W\mathfrak{A}$-homomorphism, then $\overline{M}f : MX \longrightarrow MY$ is the homomorphism determined on cherry trees by

$$\overline{M}f(A;x_1,\ldots,x_n) = (A;f(x_1),\ldots,f(x_n))$$

Note that $\overline{M}f$ is the unique homomorphism induced by the canonical composite of f and the universal reduced $\mathfrak{A}$-map $Y \longrightarrow MY$.

Let $\mathfrak{B}$ be a K-coloured PROP (for PROs the argument is analogous) and X a $(W\mathfrak{A} \otimes \mathfrak{B})$-space. By (2.19) we can consider X as $\mathfrak{B}$-space in the category of $W\mathfrak{A}$-spaces so that $b \in \mathfrak{B}(\underline{i},k)$ defines a $W\mathfrak{A}$-homomorphism $b : X_{\underline{i}} \longrightarrow X_k$ inducing a monoid homomorphism $\overline{M}b : M(X_{\underline{i}}) \longrightarrow M(X_k)$ and hence a based map $B\overline{M}b : BM(X_{\underline{i}}) \longrightarrow BMX_k$. By construction, $B\overline{M}b$ is continuous in $b \in \mathfrak{B}(\underline{i},k)$. Unfortunately, the maps $B\overline{M}b$ do not combine to a based action on $BMX = \{BM(X_k) \mid k \in K\} \in \mathfrak{Top}^{o}_{K}$, because $\overline{M}$ does not preserve products. Indeed, remark (3.26) shows that there is no product preserving functor $M : \mathfrak{Mor}_{W\mathfrak{A}} \longrightarrow \mathfrak{Mor}_{\mathfrak{A}}$ such that $MY \simeq Y$ for all $W\mathfrak{A}$-spaces Y.

Let $X = \{X_k,\alpha_k\}, k \in K$, be a collection of $W\mathfrak{A}$-spaces and $j : [p] \to K$

an object of $\mathfrak{S}_K$. By (II;§3) the product $W\mathfrak{A}$-space (X_j,α_j) is defined on objects $n\in W\mathfrak{A}$ by $\alpha_j(n) = (X_j)^n$ and on morphisms $b : m \longrightarrow n$ by the composite

$$(X_j)^m \cong (X_{j1})^m \times \ldots \times (X_{jp})^m \xrightarrow{\alpha_{j1}(b)\times\ldots\times\alpha_{jp}(b)} (X_{j1})^n \times \ldots \times (X_{jp})^n \cong (X_j)^n$$

The representing reduced cherry trees of $M(X_j)$ are of the form $(A;z_1,\ldots,z_n)$ with $z_r = (x_{r1},\ldots,x_{rp}) \in X_j$. The correspondence

$$(A;z_1,\ldots,z_n) \longmapsto [(A;x_{11},x_{21},\ldots,x_{n1}),\ldots,(A;x_{1p},x_{2p},\ldots,x_{np})]$$

defines a monoid homomorphism

$$h_j : M(X_j) \longrightarrow (MX)_j = MX_{j1} \times \ldots \times MX_{jp}$$

the unique homomorphism induced by the product of $W_r(\mathfrak{A}\otimes\mathfrak{L}_1)$-actions

$$(i_{\alpha_{j1}} \times\ldots\times i_{\alpha_{jp}}, \nu_{\alpha_{j1}} \times\ldots\times \nu_{\alpha_{jp}}) : (X_j,\alpha_j) \longrightarrow M(X_{j1},\alpha_{j1}) \times\ldots\times M(X_{jp},\alpha_{jp})$$

where $(i_{\alpha_k},\nu_{\alpha_k}) : (X_k,\alpha_k) \longrightarrow M(X_k,\alpha_k)$ is the universal reduced $\mathfrak{A}$-map of (4.49). The isomorphism subgroup S_j of $\mathfrak{S}_K(j,j)$ acts on $M(X_j)$ and $(MX)_j$ by permuting factors and h_j is S_j-equivariant.

<u>Lemma 6.20</u>: h_j embeds $M(X_j)$ as S_j-equivariant SDR in $(MX)_j$ such that $((MX)_j,M(X_j))$ is an S_j-NDR.

<u>Proof</u>: By definition, $h_j \circ i_{\alpha_j} = (i_{\alpha_{j1}} \times\ldots\times i_{\alpha_{jp}})$. Since i_{α_j} and $(i_{\alpha_{j1}} \times\ldots\times i_{\alpha_{jp}})$ are equivariant homotopy equivalences (see proof of 4.49), the map h_j is an equivariant homotopy equivalence. So we only have to show that it is an equivariant closed cofibration. A p-tuple of reduced cherry trees $(A_1,\ldots,A_p) \in (MX)_j$ lies in $M(X_j)$ iff the A_i have the same shapes if we neglect the edge colours and the same edge lengths up to the relation (4.49 a,c,d). Note that the shape uniquely determines the vertex labels because $\mathfrak{A}(n,1)$ has exactly one element and that (4.49 b) does not apply because $\mathfrak{A}$ is a PRO. Moreover, we may restrict our attention to reduced cherry trees having no vertex representing an identity so that (4.49 a) becomes redundant. We show

by induction on the product filtration of $(MX)_{\underline{j}}$ given by the filtration of M that there is an equivariant retraction

$$r : (MX)_{\underline{j}}\times I \longrightarrow (MX)_{\underline{j}}\times 0 \cup M(X_{\underline{j}})\times I$$

So suppose inductively that r has been given on $M_{q_1}(X_{j1})\times..\times M_{q_p}(X_{jp})\times I$ for $q_1+\ldots+q_p<n$. Let $\lambda_1,\ldots,\lambda_p$ be shapes with $q_1,\ldots,q_p$ internal edges, $q_1+\ldots+q_p = n$ and colours $\underline{j}(1),\ldots,\underline{j}(p)$. The space of all p-tuples of reduced cherry trees of shape $(\lambda_1,\ldots,\lambda_p)$ in $(MX)_{\underline{j}}$ is of the form $\prod_{i=1}^p Y_i\times I^{q_i}$ where Y_i is the space of cherries of the i-th tree. An element $(A_1,\ldots,A_p)\in\prod_{i=1}^p Y_i\times(I^{q_i}-\partial I^{q_i})$ lies in $M(X_{\underline{j}})$ iff the shapes $\lambda_1,\ldots,\lambda_p$ of $A_1,\ldots,A_p$ coincide disregarding edge colours and the lengths of corresponding edges in the A_i are the same, so that $q_1 = q_2 = \ldots = q_p = q$ and $(A_1,\ldots,A_p)\in \prod_{i=1}^p Y_i\times\Delta I^q$ where $\Delta I^q \subset I^{qp}$ is the diagonal. The group $S_{\underline{j}}$ acts on p-tuples of shapes $(\lambda_1,\ldots,\lambda_p)$ of cherry trees in $(MX)_{\underline{j}}$ by permuting factors. Let H be the subgroup leaving $(\lambda_1,\ldots,\lambda_p)$ fixed.

Case 1: At least two of the $\lambda_1,\ldots,\lambda_p$ are different neglecting edge colours. Then we need a H-equivariant retraction

$$(\prod Y_i\times I^{q_i})\times I \longrightarrow (\prod Y_i\times I^{q_i})\times 0 \cup (\prod Y_i\times\partial I^{q_i})\times I$$

Case 2: The λ_i coincide disregarding edge colours. Then we need a H-equivariant retraction

$$(\prod Y_i)\times I^{pq}\times I \to (\prod Y_i)\times I^{pq}\times 0 \cup (\prod Y_i)\times(\partial I^{pq} \cup \Delta I^q)\times I$$

Both retractions exist because $(I^q,\partial I^q)$ is a NDR and $(I^{pq},\partial I^{pq} \cup \Delta I^q)$ is a S_p-NDR. We extend to filtration n by making this process for a complete set of representatives of $S_{\underline{j}}$-orbits of shapes $(\lambda_1,\ldots,\lambda_p)$. ∎

<u>Proposition 6.21</u>: Let $\mathfrak{B}$ be a K-coloured PROP (or PRO) and X a $(W\mathfrak{A}\otimes\mathfrak{B})$-space. Then BMX admits a based $W\mathfrak{B}$-action.

<u>Proof</u>: Define a K-coloured PROP $\mathfrak{C}$ by taking as $\mathfrak{C}(\underline{i},k)$ the space of all pairs (b,f), $b\in\mathfrak{B}(\underline{i},k)$ and $f : (BMX)_{\underline{i}} \longrightarrow BMX_k$ a based map such that

the composite

$$BM(X_{\underline{i}}) \xrightarrow[Bn_{\underline{i}}]{} B(MX)_{\underline{i}} \cong (BMX)_{\underline{i}} \xrightarrow[f]{} BMX_k$$

is $B\overline{M}b$, with the subspace topology of $\mathfrak{B}(\underline{i},k)\times\mathfrak{Top}^{o}((BMX)_{\underline{i}},BMX_k)$. Composition in $\mathfrak{C}$ is given by the composition in $\mathfrak{B}$ and in $\mathfrak{Top}^{o}$. The projection $(b,f) \mapsto f$ defines a based action of $\mathfrak{C}$ on BMX and the projection $(b,f) \mapsto b$ a PROP-functor $P : \mathfrak{C} \to \mathfrak{B}$. By (6.9) and (6.20), $BM(X_{\underline{i}})$ is an $S_{\underline{i}}$-equivariant SDR of $(BMX)_{\underline{i}}$. Denote the retraction $(BMX)_{\underline{i}} \to BM(X_{\underline{i}})$ by r. Let $h_t : (BMX)_{\underline{i}} \to (BMX)_{\underline{i}}$ be the equivariant deformation with $h_o = id$ and $h_1 = Bn_{\underline{i}} \circ r$. Then $P : \mathfrak{C}(\underline{i},k) \to \mathfrak{B}(\underline{i},k)$ has an equivariant section $Q : \mathfrak{B}(\underline{i},k) \to \mathfrak{C}(\underline{i},k)$ defined by $Q(b) = (b, B\overline{M}b \circ r)$, and there is an equivariant deformation $(b,f) \mapsto (b, f\circ h_t)$ into this section. Hence, by the lifting theorem, there exists a PROP-functor $W\mathfrak{B} \to \mathfrak{C}$ making BMX into a based $W\mathfrak{B}$-space. ∎

3. n-FOLD AND INFINITE LOOP SPACES

Let $\mathfrak{D}_1$ be the first little-cube category of example (2.49) and $\tau : \mathfrak{D} \subset \mathfrak{D}_1$ the sub-PRO of example (2.53). The unique functor $\gamma : \mathfrak{D} \to \mathfrak{A}$ is a homotopy equivalence. Hence there is a PRO-functor $P : W\mathfrak{A} \to \mathfrak{D}$ such that $\gamma \circ P = \varepsilon(\mathfrak{A})$. Since the composite of augmentations $\varepsilon(\mathfrak{A}) \circ \varepsilon(W\mathfrak{A}) : W(W\mathfrak{A}) \to \mathfrak{A}$ is a homotopy equivalence, there is a functor $Q : W\mathfrak{A} \to W(W\mathfrak{A})$ such that $\varepsilon(\mathfrak{A}) \circ \varepsilon(W\mathfrak{A}) \circ Q = \varepsilon(\mathfrak{A})$. From the uniqueness part of (3.20) it follows that $\varepsilon(W\mathfrak{A}) \circ Q \simeq Id_{W\mathfrak{A}}$ through functors. Let $\mathfrak{B}$ be a K-coloured PROP (or PRO) and $i_k : W\mathfrak{A} \to W\mathfrak{A} \otimes \mathfrak{B}$ and $j_k : \mathfrak{D}_1 \to \mathfrak{D}_1 \otimes \mathfrak{B}$ the canonical inclusions (2.17). Then we have for each $k \in K$ PRO-functors

$$\kappa_k : Wi_k \circ Q : W\mathfrak{A} \to W(W\mathfrak{A} \otimes \mathfrak{B}) \qquad \rho_k : Wj_k \circ W\tau \circ WP \circ Q : W\mathfrak{A} \to W(\mathfrak{D}_1 \otimes \mathfrak{B})$$

Definition 6.22: Let $\mathfrak{B}$ be a K-coloured PROP and $X \in \mathfrak{Top}_K$. Two $W\mathfrak{B}$-actions α and β on X are called *equivalent*, if there exists a $\mathfrak{B}$-map

$(id_X,\mu) : (X,\alpha) \longrightarrow (X,\beta)$.

From (5.7) we obtain

Lemma 6.23: If $\alpha_t : W\mathfrak{B} \longrightarrow \mathfrak{T}_{OD}$ is a homotopy of $W\mathfrak{B}$-actions on X then α_0 and α_1 are equivalent.

Theorem 6.24: Let $\mathfrak{B}$ be a K-coloured PROP (or PRO) and $X=\{(X_k,\alpha_k)|k\in K\}$ a family of $W\mathfrak{A}$-spaces. Consider the statements

(a) $BMX = \{BM(X_k)|k\in K\}$ admits a based $W\mathfrak{B}$-action

(b) Up to equivalence of $W\mathfrak{A}$-actions, the $W\mathfrak{A}$-actions on the X_k come from a $W(W\mathfrak{A}\otimes\mathfrak{B})$-action on X via $\varkappa_k$.

(c) Up to equivalence of $W\mathfrak{A}$-actions, the $W\mathfrak{A}$-actions on the X_k come from a $W(\mathfrak{O}_1\otimes\mathfrak{B})$-action on X via ρ_k.

Then we have the following implications: (c) $\Rightarrow$ (b) $\Rightarrow$ (a). Moreover, if for all $k\in K$ the space X_k is numerably contractible, α_k induces a group structure on $\pi_0(X_k)$, and $\mathfrak{B}(0,k)$ has exactly one element, then (c) holds if BMX admits a (not necessarily based) $W\mathfrak{B}$-action.

Proof: (c) $\Rightarrow$ (b) because $j_k \circ \tau \circ P = ((\tau\circ P)\otimes Id)\circ i_k$. Hence $\rho_k = W((\tau\circ P)\otimes Id)\circ\varkappa_k$ and $W((\tau\circ P)\otimes Id)$ induces the required $W(W\mathfrak{A}\otimes\mathfrak{B})$-action on X.

(b) $\Rightarrow$ (a): By assumption, there is a $W(W\mathfrak{A}\otimes\mathfrak{B})$-structure η on X and there are $\mathfrak{A}$-maps $(id_{X_k},\mu_k) : (X_k,\alpha_k) \longrightarrow (X_k,\eta\circ\varkappa_k)$. By (4.49), there is a $(W\mathfrak{A}\otimes\mathfrak{B})$-space $(Y,\overline{\eta})$ and a homotopy equivalence $g : (X,\eta) \to (Y,\overline{\eta})$ carrying a $(W\mathfrak{A}\otimes\mathfrak{B})$-map structure. If $\beta_k = \overline{\eta}\circ\varepsilon(W\mathfrak{A}\otimes\mathfrak{B})\circ\varkappa_k$ then $g_k\circ f_k : (X_k,\alpha_k) \longrightarrow (Y_k,\beta_k)$ carries an $\mathfrak{A}$-map structure. Since $\varepsilon(W\mathfrak{A}\otimes\mathfrak{B})\circ\varkappa_k = i_k\circ\varepsilon(W\mathfrak{A})\circ Q \simeq i_k$, the $W\mathfrak{A}$-structure β_k on Y_k is equivalent to the $W\mathfrak{A}$-structure $\overline{\eta}\circ i_k$. Consequently, there are $\mathfrak{A}$-maps $(h_k,\xi_k) : (X_k,\alpha_k) \longrightarrow (Y_k,\overline{\eta}\circ i_k)$ which are homotopy equivalences. By (6.21), $BM(Y,\overline{\eta})$ admits a based $W\mathfrak{B}$-structure. Since $BMX \simeq BM(Y,\overline{\eta})$ by

(6.9) and both spaces are well-pointed, BMX admits a based $W\mathfrak{B}$-action by (5.20).

<u>Proof of the last part</u>: By (5.6 a) and (5.29) there is a based $\mathfrak{B}$-space (Z,η) and a based homotopy equivalence $f : BMX \to Z$, because BMX is connected and well-pointed. Since the loop space functor Ω preserves products and since $\mathfrak{D}_1$ acts on loop spaces, there is an action β of $\mathfrak{D}_1$ on ΩBMX and δ of $\mathfrak{D}_1 \otimes \mathfrak{B}$ on ΩZ making $\Omega f : \Omega BMX \to \Omega Z$ into a $\mathfrak{D}_1$-homomorphism and via $\tau \circ P$ into a $W\mathfrak{A}$-homomorphism. By (4.49),(6.10) and (6.14) there is a composite of $\mathfrak{A}$-maps

$$h_k : (X_k,\alpha_k) \to M(X_k,\alpha_k) \to (\Omega BM(X_k),\beta\circ\tau\circ P) \xrightarrow{\Omega f} (\Omega Z_k,\delta\circ j_k\circ\tau\circ P)$$

which is a homotopy equivalence. For any homotopy inverse $g : \Omega Z \to X$ of h there exists a $W(\mathfrak{D}\otimes\mathfrak{B})$-action λ on X making g into a $(\mathfrak{D}_1\otimes\mathfrak{B})$-map $(\Omega Z,\delta\circ\varepsilon(\mathfrak{D}_1\otimes\mathfrak{B})) \to (X,\lambda)$ by (4.20). Since $j_k\circ\tau\circ P\circ\varepsilon(W\mathfrak{A})=\varepsilon(\mathfrak{D}_1\otimes\mathfrak{B})\, W(j_k\circ\tau\circ P)$, the composite $g_k \circ h_k$ is an $\mathfrak{A}$-map from $(X_k,\alpha_k\circ\varepsilon(W\mathfrak{A})\circ Q)$ to $(X_k,\lambda\circ W(j_k\circ\tau\circ P)\circ Q)$. Hence, by (4.14), the two $W\mathfrak{A}$-structures on X_k are equivalent. Since $\varepsilon(W\mathfrak{A})\circ Q \simeq Id$, the actions $\alpha_k\circ\varepsilon(W\mathfrak{A})\circ Q$ and α_k on X_k are equivalent. So λ is the required $W(\mathfrak{D}_1\otimes\mathfrak{B})$-action. ∎

<u>Definition 6.25</u>: A map $f : X \to Y$ of based topological spaces is called n-<u>fold loop map</u>, $0\le n\le\infty$, if there exist based maps of based spaces $f_i : X_i \to Y_i$, $i=0,1,\dots,n$, and $h_i : X_{i-1} \to \Omega X_i$ and $k_i : Y_{i-1} \to \Omega Y_i$, $i=1,2,\dots,n$ such that $f=f_o$, each h_i and k_i is an unbased homotopy equivalence, and

$$\begin{array}{ccc} X_{i-1} & \xrightarrow{f_{i-1}} & Y_{i-1} \\ {\scriptstyle h_i}\downarrow & & \downarrow{\scriptstyle k_i} \\ \Omega X_i & \xrightarrow{\Omega f_i} & \Omega Y_i \end{array}$$

commutes up to a based homotopy. We call f an n-<u>fold based loop map</u> if each h_i and k_i is a based homotopy equivalence and a <u>strict</u> n-<u>fold loop map</u> if each h_i and k_i is a based homeomorphism and the diagrams commute. A based space X is called n-fold loop space [n-fold based loop space, strict n-fold loop space] if id_X is an n-fold loop map [n-fold based loop map, strict n-fold loop map].

Definition 6.26: Two [based] maps $f : X \longrightarrow Y$ and $g : X' \longrightarrow Y'$ are called [based] homotopy equivalent if there are [based] homotopy equivalences $h : X \simeq X'$ and $k : Y \simeq Y'$ such that

$$\begin{array}{ccc} X & \xrightarrow{\quad f \quad} & Y \\ {\scriptstyle h}\downarrow & & \downarrow{\scriptstyle k} \\ X' & \xrightarrow{\quad g \quad} & Y' \end{array}$$

commutes up to [based] homotopy.

In (2.49) we have constructed actions $\mu_n = \mu_n(X)$ of $\mathcal{Q}_n$, the n-th little cube PROP on $\Omega^n Y$, natural with respect to based maps $g : X \to Y$. Moreover, we have constructed inclusion PROP-functors $\iota_n^m : \mathcal{Q}_m \subset \mathcal{Q}_n$ for $n \geq m$ such that

$$\mu_n \iota_n^m = \mu_m$$

Hence, if $\iota_\infty^n : \mathcal{Q}_n \subset \mathcal{Q}_\infty$ is the inclusion into the direct limit, $\mu_\infty \iota_\infty^n = \mu_n$.

Recall that $\mathcal{Q}_n(r,1) = \{(x_1,y_1,x_2,y_2,\ldots,x_r,y_r) \in (I^n)^{2r} \mid$ the x_i are the lowest and y_i the upper vertices of r linearly embedded n-cubes in I^n with disjoint interior and axes parallel to those of $I^n\}$. Define PROP-functors $F_n : \mathcal{Q}_1 \longrightarrow \mathcal{Q}_n$ and $G_n : \mathcal{Q}_{n-1} \longrightarrow \mathcal{Q}_n$ by $F_n(x_1,y_1,\ldots,x_r,y_r) = (x_1',y_1',\ldots,x_r',y_r')$ and $G_n(v_1,w_1,\ldots,v_r,w_r) = (v_1',w_1',\ldots,v_r',w_r')$ with $x_i' = (x_i,0,\ldots,0)$, $y_i' = (y_i,1,\ldots,1)$, $v_i' = (0,v_{i1},\ldots,v_{in-1})$, and $w_i' = (1,w_{i1},\ldots,w_{in-1})$ if $v_i = (v_{i1},\ldots,v_{in-1})$ and $w_i = (w_{i1},\ldots,w_{in-1})$. Then F_n and G_n combine to a PROP-functor

$$\pi_n : \mathcal{Q}_1 \otimes \mathcal{Q}_{n-1} \longrightarrow \mathcal{Q}_n$$

determined on generators by $\pi_n(a \otimes id) = F_n(a)$ and $\pi_n(id \otimes b) = G_n(b)$.

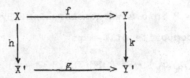

Since $\iota^n_{n+1}\cdot F_n = F_{n+1}$ and $\iota^n_{n+1}\cdot G_n = G_{n+1}\cdot\iota^{n-1}_n$, we have $\pi_{n+1}\cdot(\mathrm{Id}\otimes\iota^{n-1}_n) = \iota^n_{n+1}\cdot\pi_n$. Since for k-spaces finite products commute with indentifications, $\varinjlim_n [(\mathfrak{D}_1\otimes\mathfrak{D}_n)(r,1)] = (\mathfrak{D}_1\otimes\mathfrak{D}_\infty)(r,1)$, and the π_n induce a PROP-functor

$$\pi_\infty\colon \mathfrak{D}_1\otimes\mathfrak{D}_\infty \longrightarrow \mathfrak{D}_\infty$$

Let

$$\begin{array}{ccc} X_i & \xrightarrow{\ f_i\ } & Y_i \\ {\scriptstyle h_{i+1}}\downarrow & & \downarrow{\scriptstyle k_{i+1}} \\ \Omega X_{i+1} & \xrightarrow{\ \Omega f_{i+1}\ } & \Omega Y_{i+1} \end{array} \qquad i = 0,1,\ldots,n-1$$

be the data of an n-fold loop map, $0\le n\le\infty$. We identify the functors $\Omega^m\cdot\Omega$ with Ω^{m+1} by the exponential map (cf 2.49) Then

$$\Omega^{m-1}h_m\cdot\Omega^{m-2}h_{m-1}\cdot\ldots\cdot h_1\colon X_o \longrightarrow \Omega^m X_m$$

Let $f\colon X \longrightarrow Y$ be homotopy equivalent to the n-fold loop map f_o

$$\begin{array}{ccc} X & \xrightarrow{\ f\ } & Y \\ {\scriptstyle h}\downarrow & & \downarrow{\scriptstyle k} \\ X_o & \xrightarrow{\ g\ } & Y_o \end{array}$$

Denote $\Omega^{m-1}h_m\cdot\ldots\cdot h_1\cdot h$ by p_m and $\Omega^{m-1}k_m\cdot\ldots\cdot k_1\cdot k$ by q_m.

<u>Theorem 6.27:</u> If $f\colon X\longrightarrow Y$ is homotopy equivalent to an n-fold loop map, $0\le n\le\infty$, then X and Y admit $W\mathfrak{D}_n$-structures α and β and f a $\mathfrak{D}_n$-map structure $(f,\nu)\colon(X,\alpha)\longrightarrow(Y,\beta)$ such that (p_m,q_m) carries a $(\mathfrak{D}_m\otimes\mathfrak{L}_1)$-map structure

$$(f,\nu\cdot W(\iota^m_n\otimes\mathrm{Id})) \longrightarrow (\Omega^m f_m,(\mu_m\otimes\mathrm{Id})\cdot\varepsilon(\mathfrak{D}_m\otimes\mathfrak{L}_1)) \qquad m\le n$$

In particular, if X is an n-fold loop space, $0\le n\le\infty$, then X admits a $W\mathfrak{D}_n$-structure α such that $p_m\colon X\longrightarrow\Omega^m X_m$ carries a $\mathfrak{D}_m$-map structure $(X,\alpha\cdot\iota^m_n)\longrightarrow(\Omega^m X_m,\mu_m\cdot\varepsilon(\mathfrak{D}_m))$.

<u>Proof:</u> We proceed by induction on m. The homotopy commutative

square

$$\begin{array}{ccc} X & \xrightarrow{f} & Y \\ {\scriptstyle h}\downarrow & & \downarrow{\scriptstyle k} \\ X_o & \xrightarrow{f_o} & Y_o \end{array}$$

determines a $W(\mathfrak{L}_1 \otimes \mathfrak{L}_1) = W(\mathfrak{D}_o \otimes \mathfrak{L}_1 \otimes \mathfrak{L}_1)$-action such that (h,k) carries a $(\mathfrak{D}_o \otimes \mathfrak{L}_1)$-map structure from f to f_o. For the inductive step we are given a $W(\mathfrak{D}_{m-1} \otimes \mathfrak{L}_1)$-structure ν_{m-1} on f such that (p_{m-1}, q_{m-1}) admits a $(\mathfrak{D}_{m-1} \otimes \mathfrak{L}_1)$-map structure

$$(f, \nu_{m-1}) \longrightarrow (\Omega^{m-1} f_{m-1}, (\mu_{m-1} \otimes \mathrm{Id}) \cdot \varepsilon(\mathfrak{D}_{m-1} \otimes \mathfrak{L}_1))$$

We want to extend ν_{m-1} to a $W(\mathfrak{D}_m \otimes \mathfrak{L}_1)$-structure. Since

$$\begin{array}{ccc} X_{m-1} & \xrightarrow{f_{m-1}} & Y_{m-1} \\ {\scriptstyle h_m}\downarrow & & \downarrow{\scriptstyle k_m} \\ \Omega X_m & \xrightarrow{\Omega f_m} & \Omega Y_m \end{array}$$

commutes up to based homotopy, it is given by a based $W(\mathfrak{L}_1 \otimes \mathfrak{L}_1)$-action inducing a $\mathfrak{D}_{m-1} \otimes W(\mathfrak{L}_1 \otimes \mathfrak{L}_1)$-action.

$$\begin{array}{ccc} \Omega^{m-1} X_{m-1} & \xrightarrow{\Omega^{m-1} f_{m-1}} & \Omega^{m-1} Y_{m-1} \\ {\scriptstyle \Omega^{m-1} h_m}\downarrow & & \downarrow{\scriptstyle \Omega^{m-1} k_m} \\ \Omega^m X_m & \xrightarrow{\Omega^m f_m} & \Omega^m Y_m \end{array}$$

Now $\mathrm{id} \otimes \varepsilon(\mathfrak{L}_1 \otimes \mathfrak{L}_1) : \mathfrak{D}_{m-1} \otimes W(\mathfrak{L}_1 \otimes \mathfrak{L}_1) \longrightarrow \mathfrak{D}_{m-1} \otimes \mathfrak{L}_1 \otimes \mathfrak{L}_1$ is an equivariant homotopy equivalence. We apply (3.17) with $\mathfrak{B}$ generated by $W(\mathfrak{D}_{m-1} \otimes \mathfrak{L}_1 \otimes 0) \cup W(\mathfrak{D}_{m-1} \otimes \mathfrak{L}_1 \otimes 1) \cup \{\varphi_1, \varphi_2\}$ with φ_i being the image of $\mathrm{id}_0 \otimes \mathrm{id}_i \otimes j$, $j: 0 \longrightarrow 1$ in $\mathfrak{L}_1$, under the standard section $\mathfrak{D}_{m-1} \otimes \mathfrak{L}_1 \otimes \mathfrak{L}_1 \longrightarrow W(\mathfrak{D}_{m-1} \otimes \mathfrak{L}_1 \otimes \mathfrak{L}_1)$. The functor $H' : \mathfrak{B} \longrightarrow \mathfrak{D}_{m-1} \otimes W(\mathfrak{L}_1 \otimes \mathfrak{L}_1)$ is given on generators by $(\mathrm{Id} \otimes \eta) \cdot \varepsilon(\mathfrak{D}_{m-1} \otimes \mathfrak{L}_1 \otimes \mathfrak{L}_1)$ with $\eta : \mathfrak{L}_1 \otimes \mathfrak{L}_1 \longrightarrow W(\mathfrak{L}_1 \otimes \mathfrak{L}_1)$ the standard section. We obtain a $(\mathfrak{D}_{m-1} \otimes \mathfrak{L}_1)$-map

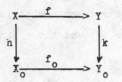

$(\Omega^{m-1}f_{m-1},(\mu_{m-1}\otimes Id)\cdot\varepsilon(\mathfrak{D}_{m-1}\otimes\mathfrak{L}_1)) \longrightarrow (\Omega^m f_m,(\mu_m\cdot\iota_m^{m-1}\otimes Id)\cdot\varepsilon(\mathfrak{D}_{m-1}\otimes\mathfrak{L}_1))$

with underlying maps $(\Omega^{m-1}h_m,\ \Omega^{m-1}k_m)$. Hence (p_m,q_m) carries a $(\mathfrak{D}_{m-1}\otimes\mathfrak{L}_1)$-map structure

$$(f,\nu_{m-1}) \longrightarrow (\Omega^m f_m,(\mu_m\cdot\iota_m^{m-1}\otimes Id)\cdot\varepsilon(\mathfrak{D}_{m-1}\otimes\mathfrak{L}_1))$$

Since (p_m,q_m) is a pair of homotopy equivalences, we can extend ν_{m-1} by (4.20) to a $(\mathfrak{D}_m\otimes\mathfrak{L}_1)$-map structure ν_m on f such that $\nu_m\cdot W(\iota_m^{m-1}\otimes Id) = \nu_{m-1}$ and the $(\mathfrak{D}_{m-1}\otimes\mathfrak{L}_1)$-map structure of (p_m,q_m) to a $(\mathfrak{D}_m\otimes\mathfrak{L}_1)$-map structure

$$(f,\nu_m) \longrightarrow (\Omega^m f_m,\ \mu_m\cdot\varepsilon(\mathfrak{D}_m\otimes\mathfrak{L}_1))$$

If n is finite, take $\nu=\nu_n$. If n is infinite, the ν_m induce an action ν of $W(\mathfrak{D}_\infty\otimes\mathfrak{L}_1)=\underrightarrow{\lim}\ W(\mathfrak{D}_n\otimes\mathfrak{L}_1)$ on f with the required properties. ∎

Corollary 6.28: Suppose $f : X \longrightarrow Y$ is homotopy equivalent to an n-fold loop map, $0\le n\le\infty$, as in (6.27). Then f admits a $\mathfrak{D}_n$-map structure $(f,\nu)\colon (X,\alpha) \longrightarrow (Y,\beta)$ which is a composite of $\mathfrak{D}_n$-maps $(f,\nu) = (u,\eta)\cdot(\Omega^n f_n,\ (\mu_n\otimes Id)\cdot\varepsilon(\mathfrak{D}_n\otimes\mathfrak{L}_1))\cdot(v,\xi)$ where $(u,\eta)\colon(X,\alpha)\longrightarrow(\Omega^n X_n,\mu_n\cdot\varepsilon(\mathfrak{D}_n))$ and $(v,\xi)\colon(\Omega^n Y_n\cdot\mu_n\cdot\varepsilon(\mathfrak{D}_n))\longrightarrow(Y,\beta)$ are homotopy equivalences of $W\mathfrak{D}_n$-spaces.

Proof: By (6.27) we are given a $W(\mathfrak{D}_n\otimes\mathfrak{L}_1\otimes\mathfrak{L}_1)$-structure whose restriction to $W(\mathfrak{D}_n\otimes\mathfrak{L}_1\otimes 0)$ is ν and to $W(\mathfrak{D}_n\otimes\mathfrak{L}_1\otimes 1)$ is $(\mu_n\otimes Id)\cdot\varepsilon(\mathfrak{D}_n\otimes\mathfrak{L}_1)$. The restrictions to $W(\mathfrak{D}_n\otimes i\otimes\mathfrak{L}_1)$, $i = 0,1$, give p_n and q_n structures of $\mathfrak{D}_n$-maps (p_n,η) and (q_n,ξ'). The two inclusions $W(\mathfrak{D}_n\otimes\mathfrak{L}_2)\to W(\mathfrak{D}_n\otimes\mathfrak{L}_1\otimes\mathfrak{L}_1)$ given by $F_i\colon \mathfrak{L}_2 \longrightarrow \mathfrak{L}_1\otimes\mathfrak{L}_1 = \mathfrak{L}_1 * \mathfrak{L}_1$, $F_i(0)=(0,0)$, $F_0(1)=(0,1)$, $F_1(1)=(1,0)$, $F_i(2)=(1,1)$, $i=0,1$, show that $(\Omega^n f_n,(\mu_n\otimes Id)\cdot\varepsilon(\mathfrak{D}_n\otimes\mathfrak{L}_1))\cdot(p_n,\eta)=(q_n,\xi')\cdot(f,\nu)$. Take $(u,\eta)=(p_n,\eta)$ and (v,ξ) any homotopy inverse of (q_n,ξ'). ∎

Originally we proved this corollary for n-fold loop spaces by substituting an n-fold loop space by a strict n-fold loop space, $n\le\infty$, and then using the results of (2.49). Indeed, by a refinement of a result of May [32; Thm.6] one can show (see [7] for a proof).

<u>Proposition 6.29:</u> An infinite loop space $(X_0, X_1, X_2, \dots)$ is homotopy equivalent to a strict infinite loop space provided the X_i are well-pointed. ∎

There is also a converse to Theorem 6.27. Let

$$\xi_k = W\iota_k^1 \cdot W(\tau \cdot Q) \cdot Q : W\mathfrak{A} \longrightarrow W(W\mathfrak{A}) \longrightarrow W\mathfrak{D}_1 \longrightarrow W\mathfrak{D}_k$$

<u>Theorem 6.30:</u> Let (X,α) and (Y,β) be numerably contractible $W\mathfrak{D}_n$-spaces, $0 \le n \le \infty$, such that α and β induce group structures on $\pi_0 X$ and $\pi_0 Y$ respectively and let $(f,\eta):(X,\alpha) \longrightarrow (Y,\beta)$ be a $\mathfrak{D}_n$-map. Then f is homotopy equivalent to an n-fold based loop map

$$\begin{array}{ccc} X_{i-1} & \xrightarrow{f_{i-1}} & Y_{i-1} \\ \Big\downarrow h_i & \text{I} & \Big\downarrow k_i \\ \Omega X_i & \xrightarrow{\Omega f_i} & \Omega Y_i \end{array} \quad i=1,2,\dots,n \qquad\qquad \begin{array}{ccc} X & \xrightarrow{f} & Y \\ \Big| h & \text{II} & \Big| k \\ X_0 & \xrightarrow{f_0} & Y_0 \end{array}$$

such that

(a) I and II commute strictly, $X_0 = \Omega BM(X, \alpha \cdot \xi_n)$, $Y_0 = \Omega BM(Y, \beta \cdot \xi_n)$

(b) each f_i admits a $\mathfrak{D}_{n-i}$-map structure $(f_i, \eta_i):(X_i, \alpha_i) \longrightarrow (Y_i, \beta_i)$

(c) h_i admits an $\mathfrak{A}$-map structure $(X_{i-1}, \alpha_{i-1} \cdot \xi_{n-i+1}) \longrightarrow (\Omega X_i, \mu_1 \cdot \tau \cdot P)$ and $h = jM(X, \alpha \cdot \xi_n)$. Similarly for k_i and k.

<u>Proof:</u> $\eta \cdot W(\pi_n \otimes \mathrm{Id}) : W(\mathfrak{D}_1 \otimes \mathfrak{D}_{n-1} \otimes \mathfrak{B}_1) \longrightarrow W(\mathfrak{D}_n \otimes \mathfrak{B}_1) \longrightarrow \mathfrak{Top}$ makes f to a $(\mathfrak{D}_1 \otimes \mathfrak{D}_{n-1})$-map. By (6.24), BMf admits a $\mathfrak{D}_{n-1}$-map structure $BMX \longrightarrow BMY$. Take $X_1 = BMX$, $Y_1 = BMY$ and $f_1 = BMf$. If we choose Mf such that

$$\begin{array}{ccc} X & \xrightarrow{f} & Y \\ i_X \cap & & \cap i_Y \\ MX & \xrightarrow[Mf]{} & MY \end{array}$$

commutes and put $h = jMX \cdot i_X$ and $k = jMY \cdot i_Y$, then (a) holds. Suppose inductively that X_i, Y_i are connected and numerably contractible for $1 \le i < m$ and that f_i, h_i, k_i with the required properties are found for $1 \le i < m$. Take $X_m = BM(X_{m-1}, \alpha_{m-1} \cdot \xi_{n-m+1})$ and Y_m analogous. Since f_{m-1} admits $\mathfrak{D}_{n-m+1}$-map structure and hence via $W(\pi_{n-m+1} \otimes Id)$ a $(\mathfrak{D}_1 \otimes \mathfrak{D}_{n-m})$-map structure, $f_m = BMf_{m-1}$ admits a $\mathfrak{D}_{n-m}$-map structure $(f_m, \eta_m): (X_m, \alpha_m) \to (Y_m, \beta_m)$. Define h_m and k_m by

$$\begin{array}{ccccc} & X_{m-1} & \xrightarrow{f_{m-1}} & Y_{m-1} & \\ & \cap & & \cap & \\ h_m \downarrow & MX_{m-1} & \xrightarrow{Mf_{m-1}} & MY_{m-1} & \downarrow k_m \\ & \downarrow j & & \downarrow j & \\ & \Omega BMX_{m-1} & \xrightarrow{\Omega BMf_{m-1}} & \Omega BMY_{m-1} & \end{array}$$

Then (a) and (c) are satisfied. By induction hypothesis jMX_{m-1} and jMY_{m-1} are based homotopy equivalences (6.14). Since the classifying space of a monoid is connected, X_m and Y_m are connected, and by (6.19) both are numerably contractible, so that induction can proceed. If $n = \infty$, we use $\pi_\infty : \mathfrak{D}_1 \otimes \mathfrak{D}_\infty \to \mathfrak{D}_\infty$ instead of π_n. ∎

4. HOMOTOPY-EVERYTHING H-SPACES, DYER-LASHOF OPERATIONS

In [8], we called an E-space (cf. 2.46) a homotopy-everything H-space motivated by the idea that an E-space satisfies all coherence conditions one can think of. This is not quite the case as we shall see in this section. We start with identifying E-map structures, i.e. actions of two-coloured PROPs $\mathfrak{C}$ admitting a PROP-functor $\mathfrak{C} \to \mathfrak{S} \otimes \mathfrak{L}_1$ and having contractible morphism spaces (for $\mathfrak{S}$ see II, §5).

<u>Theorem 6.31</u>: An E-map $(f, \xi) : (X, \alpha) \to (Y, \beta)$ is homotopy equivalent to an infinite based loop map such that (6.30 (a),(b),(c)) hold pro-

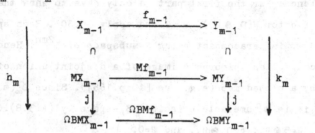

vided X and Y are numerably contractible and α and β induce group structures on $\pi_0 X$ and $\pi_0 Y$ respectively. Conversely, any map homotopy equivalent to an infinite loop map admits an E-map structure.

Proof: Let $\mathfrak{C}$ be a two-coloured PROP admitting a PROP-functor $P : \mathfrak{C} \longrightarrow \mathfrak{S} \otimes \mathfrak{L}_1$ and having contractible morphism spaces. Then P is a homotopy equivalence. For the first part we only have to show that there is a PROP-functor $W(\mathfrak{D}_\infty \otimes \mathfrak{L}_1) \longrightarrow \mathfrak{C}$ and apply (6.30). Each space $\mathfrak{D}_n(m,1)$, $0 \le n < \infty$, is paracompact being a subspace of I^{2mn}. Hence $\mathfrak{D}_\infty(m,1)$ is paracompact as epimorphic image of a disjoint union of paracompact spaces by a closed map (e.g. see [18;p.165]). Since $\mathfrak{D}_\infty(m,1)$ is also S_m-free it is a numerable principal S_n-space by (A 3.8). Apply (3.17) with $\mathfrak{D}=\mathfrak{C}$, $\mathfrak{C}=\mathfrak{S}\otimes\mathfrak{L}_1$, $\mathfrak{B}=\mathfrak{D}_\infty\otimes\mathfrak{L}_1$, and $\mathfrak{A}=\emptyset$.

Conversely, a map homotopy equivalent to an infinite loop map admits a $\mathfrak{D}_\infty$-map structure by (6.27). But $\mathfrak{D}_\infty$ is an E-category by (2.50) so that

$$(\rho \otimes \mathrm{Id}) \circ \varepsilon(\mathfrak{D}_\infty \otimes \mathfrak{L}_1) : W(\mathfrak{D}_\infty \otimes \mathfrak{L}_1) \longrightarrow \mathfrak{S} \otimes \mathfrak{L}_1$$

is the required PROP-functor which is a homotopy equivalence. Here $\rho : \mathfrak{D}_\infty \longrightarrow \mathfrak{S}$ is the unique PROP-functor. ∎

Remark: The important results (6.27), (6.30), (6.31) can, of course, be extended from maps to arbitrary diagrams by substituting $\mathfrak{L}_1$ by a suitable indexing category.

Putting some of our results together we obtain the following

Proposition 6.32: If $\mathfrak{B}$ is a monochrome PRO with contractible morphism spaces and X a $\mathfrak{B}$-space, then X is homotopy equivalent to a monoid.

Proof: The unique PRO-functor $P : \mathfrak{B} \longrightarrow \mathfrak{A}$ is a homotopy equivalence. Hence, by the lifting theorem (3.17) for PROs, there is a PRO-functor

$R : W\mathfrak{A} \longrightarrow \mathfrak{B}$ making X into a $W\mathfrak{A}$-space. But a $W\mathfrak{A}$-space is homotopy equivalent to an $\mathfrak{A}$-space, which is a monoid, by (4.49). ∎

This result becomes false if we replace $\mathfrak{A}$ by $\mathfrak{S}$. We cannot expect to replace an E-space by an $\mathfrak{S}$-space, or commutative monoid, because the k-invariants of a commutative monoid disappear [17;Satz 7.1], but there are E-spaces with non-trivial k-invariants. The essential difference between the two situations becomes clear if we go back to the split theories defined by PROs or PROPs (cf.(2.42),(2.44)). If $\mathfrak{B}$ is a PRO with contractible morphism spaces and Θ the associated split theory, then the unique theory functor $\Theta \longrightarrow \Theta_m$ is a homotopy equivalence. If $\mathfrak{B}$ is a PROP with contractible morphism spaces, then the unique theory functor $P : \Theta \longrightarrow \Theta_{cm}$ associated with $\mathfrak{B} \longrightarrow \mathfrak{S}$ need not be a homotopy equivalence: Let $\lambda_n \in \Theta_{cm}(n,1)$ be the operation

$$\lambda_n(x_1,\ldots,x_n) = x_1+x_2+\ldots x_n$$

then $\mathfrak{S}(n,1) = \{\lambda_n\}$. Recall that $\Theta(n,1)$ is obtained from $\mathfrak{B}$ by

$$\Theta(n,1) = \bigcup_k \mathfrak{B}(k,1) \times \mathfrak{S}(k,n)/\sim$$

with the relation $(b \circ \pi^*, \sigma) \sim (b, \sigma \circ \pi)$, $\sigma \in \mathfrak{S}(k,n)$, $\pi \in \mathfrak{S}_k \subset \mathfrak{S}(k,k)$ a permutation (see (2.37)). The functor P is given by

$$P(b,\sigma) = \lambda_k \circ \sigma^* \qquad b \in \mathfrak{B}(k,1)$$

Let $\sigma_k \in \mathfrak{S}(k,1)$ be the set map $\sigma_k : [k] \longrightarrow [1]$. Then $\lambda_k \circ \pi^* = \lambda_k$ and $\sigma_k \circ \pi = \sigma_k$ for $\pi \in S_k$. Hence $P^{-1}(\lambda_k \circ \sigma_k^*)$ is homeomorphic to the orbit space $\mathfrak{B}(k,1)/S_k$, which in general is not contractible. So the following is a more correct definition of a homotopy-everything H-space.

<u>Definition 6.33</u>: A topological space X is called a <u>homotopy-everything H-space</u> if it is a Θ-space for a theory Θ admitting a theory functor $\Theta \longrightarrow \Theta_{cm}$ which is a homotopy equivalence.

Theorem (4.58) shows that a <u>homotopy-everything H-space</u> X is homotopy equivalent to a commutative monoid if the space X and the morphism spaces $\mathfrak{B}(n,1)$ of its defining theory satisfy certain, not particularly restrictive, point set topological assumptions.

We now will show that Dyer-Lashof operations are connected with obstructions to the existence of homotopy-everything H-structures. Let $\mathfrak{B}$ be a monochrome E-category. Fix an element $m_2 \in \mathfrak{B}(2,1)$ and define $m_p \in \mathfrak{B}(p,1)$, $p \geq 2$, inductively by $m_p = m_2(m_{p-1} \oplus id)$. Let G be a discrete group and EG any contractible numerable principal G-space. To stay inside the setting of Dyer and Lashof we take EG to be the realization of the simplicial complex determined by the partially ordered set $\{(g,n) \mid g \in G,\ n \text{ a non-negative integer}\}$ with the ordering $(g,n) \leq (g',n')$ if $g=g'$ and $n=n'$ or if $n<n'$ (i.e. EG has the elements (g,n) as vertices and a p-simplex with vertices $(g_o,n_o),\ldots,(g_p,n_p)$ iff $(g_o,n_o) \lneq (g_1,n_o) \lneq \ldots \lneq (g_p,n_p)$). There is a S_p-equivariant map

$$\theta_p : ES_p \longrightarrow \mathfrak{B}(p,1) \qquad p \geq 2$$

such that $\theta_p(e,0)=m_p$, where $e \in S_p$ is the unit. (We could take θ_p to be the composite

$$ES_p \longrightarrow \mathfrak{D}_\infty(p,1) \xrightarrow[\text{stand. section}]{} W\mathfrak{D}_\infty(p,1) \xrightarrow[\text{lift}]{} \mathfrak{B}(p,1)$$

with a suitable equivariant homotopy equivalence $ES_p \simeq \mathfrak{D}_\infty(p,1)$.

Let π be the cyclic group of order p with generator T and let W be the complex

$$\ldots \xrightarrow{\partial} \mathbb{Z}[\pi] \xrightarrow{\partial} \mathbb{Z}[\pi] \xrightarrow{\partial} \mathbb{Z}[\pi] \xrightarrow{\partial} \ldots \xrightarrow{\partial} \mathbb{Z}[\pi] \xrightarrow{\partial} \mathbb{Z}[\pi] \xrightarrow{\varepsilon} \mathbb{Z}$$

where $\mathbb{Z}[\pi]$ is the group ring of π. Then W_i has a single π-generator e_i, $i \geq 0$, and we define

$$\begin{aligned} \partial e_{2i+1} &= (T-1)e_{2i} && i \geq 0 \\ \partial e_{2i+2} &= (1+T+\ldots+T^{p-1})e_{2i+1} \\ \varepsilon(e_o) &= 1 \end{aligned}$$

The Eilenberg-Zilber map defines a chain map

$$F : W \otimes_{\mathbb{Z}[\pi]} C_*(X)^p \longrightarrow C_*(E\pi \times_\pi X^p)$$

for any space X with the obvious $\mathbb{Z}[\pi]$-action on $C_*(X)^p$ and π-action on X^p. Moreover, we can choose F such that

$$F(e_o \otimes x_1 \otimes \ldots \otimes x_p) = ((e,0), x_1, \ldots, x_p)$$

If X admits a $\mathfrak{B}$-space structure α then α, θ_p, and the inclusion $E\pi \subset ES_p$ define a map

$$\varphi_p : E\pi \times_\pi X^p \longrightarrow X$$

The Dyer-Lashof operations on $H_*(X;\mathbb{Z}_p)$ are then defined by

$$Q_i^{(p)} : H_j(X;\mathbb{Z}_p) \longrightarrow H_{pj+i}(X;\mathbb{Z}_p)$$
$$x \longmapsto \varphi_p \cdot F(e_i \otimes_\pi x^p)$$

where we write $e_i \otimes_\pi x^p$ for the homology class in $H_*(W \otimes_\pi C_*(X)^p; \mathbb{Z}_p)$ represented by this cycle. We list a few elementary properties

(6.34) (a) $Q_i^{(p)}$ is a homomorphism

(b) $Q_o^{(p)}(x) = x^p$, the multiplication on $H_*(X;\mathbb{Z}_p)$ is induced by m_2.

(c) If ∂_p is the homology Bockstein operator of the sequence $0 \to \mathbb{Z}_p \to \mathbb{Z}_{p^2} \to \mathbb{Z}_p \to 0$ then $Q_{2i-1}^{(p)} = \partial_p Q_{2i}^{(p)}$

(d) $Q_{2i}^{(p)} = 0 : H_j(X;\mathbb{Z}_p) \to H_{jp+2i}(X;\mathbb{Z}_p)$ unless $2i = (2k-j)(p-1)$

For a proof see [19]. In view of (d) one usually puts

$$Q^i_{(p)} = c_{p,i} \cdot Q^{(p)}_{(2i-j)(p-1)} : H_j(X;\mathbb{Z}_p) \longrightarrow H_{j+2i(p-1)}(X;\mathbb{Z}_p) \qquad p \neq 2$$

$$Q^i_{(2)} = Q^{(2)}_{i-j} : H_j(X;\mathbb{Z}_2) \longrightarrow H_{i+j}(X;\mathbb{Z}_2)$$

For further properties of $Q^i_{(p)}$ we refer to the fundamental paper of Dyer and Lashof [19] and the recent papers of May, for example [33].

We call the map φ_p π-<u>commutative</u> if it is independent of the $E\pi$-

coordinate. Then φ_p factors as

$$\varphi_p : E\pi\times_\pi X^p \longrightarrow X^p/\pi \xrightarrow{\bar{\varphi}} X$$

and the Dyer-Lashof operations are trivial with exception of $Q_o^{(p)}(x)=x^p$.

Let $\Delta X \subset X^p$ be the diagonal copy of X in X^p. Then ΔX is precisely the set of fixed points of π. By restriction, φ_p induces a map

$$\delta_p : B\pi\times X \longrightarrow X$$

with $B\pi = E\pi/\pi$ the classifying space of π.

Lemma 6.35: Suppose X^p is paracompact and $(X^p,\Delta X)$ is a π-NDR. Then φ_p is homotopic to a π-commutative map iff δ_p is homotopic to a map which is independent of the $B\pi$-coordinate.

Proof: Obviously, if φ_p is homotopic to a π-commutative map, then, by restriction, δ_p is homotopic to a map which is independent of the $B\pi$-coordinate. Conversely, given a homotopy $h_t : \delta_p \simeq f : B\pi\times X \longrightarrow X$ with f independent of the $B\pi$-coordinate. Since $(X^p,\Delta X)$ is a π-NDR, $(E\pi\times_\pi X^p, B\pi\times\Delta X)$ is a NDR so that φ_p is homotopic to an extension f' of f. We show that f' is homotopic to a π-commutative map. The projection $E\pi\times(X^p-\Delta X) \longrightarrow (X^p-\Delta X)$ is an ordinary homotopy equivalence of numerable principal π-spaces (A 3.8). Hence $q : E\pi\times_\pi(X^p-\Delta X) \to (X^p-\Delta X)/\pi$ is a homotopy equivalence (A 3.4). Moreover, since q is a fibre bundle map, there is a section s of q and a homotopy $l_t : id \simeq s\circ q$ such that $q\circ l_t = q$ for all t [13; Thm. 6.1]. Since $(X^p,\Delta X)$ is a π-NDR, there is an equivariant map $u : X^p \longrightarrow I$ and an equivariant homotopy $r_t : X^p \longrightarrow X$ such that $\Delta X = u^{-1}(0)$, $r_o(x) = x$ for all $x\in X^p$, $r_t(y) = y$ for all $y\in\Delta X$ and all $t\in I$, and $r_1(x)\in\Delta X$ for $x\in U = u^{-1}[0,1)$. Define $f_t, k_t : E\pi\times_\pi X^p \longrightarrow X$ by $k_t(e,x) = f'(e,r_t(x))$ and

$$f_t(e,x) = \begin{cases} k_t\circ l_{\max(2u(x)t-1,0)}(e,x) & (e,x)\in E\pi\times_\pi(X^p-\Delta X) \\ k_t & (e,x)\in E\pi\times_\pi\Delta X \end{cases}$$

Then $f_o = k_o = f'$ and f_1 is independent of the $E\pi$-coordinate because

$k_1(e,x)$ is independent of the $E\pi$-coordinate for $x\in U$, and $l_t(e,x)\in E\pi\times_\pi U$ if $x\in U$, and because l_1 is independent of the $E\pi$-coordinate. ∎

Now suppose X is a homotopy-everything H-space with PROP $\mathfrak{C}$ and associated split theory Θ. Since the unique theory functor $\Theta \longrightarrow \Theta_{cm}$ is a homotopy equivalence, the spaces $\mathfrak{C}(k,1)/S_k$ are contractible. Since δ_p factors as

$$\delta_p : B\pi\times X \longrightarrow BS_p\times X \longrightarrow \mathfrak{C}(p,1)\times_{S_p} X \longrightarrow \mathfrak{C}(p,1)\times_{S_p} X^p \longrightarrow X$$

and $\mathfrak{C}(p,1)\times_{S_p} X = (\mathfrak{C}(p,1)/S_p)\times X$, it is independent of the $B\pi$-coordinate so that all Dyer-Lashof operations with exception of $Q_o^{(p)}$ are trivial for all primes p.

5. EXAMPLES OF INFINITE LOOP SPACES

In this section we describe a method of imposing E-structures on some well-known H-spaces. Since our examples will satisfy the assumptions of (6.31), we obtain a number of infinite loop spaces.

Consider the category $\mathfrak{Q}\mathfrak{J}$ of real inner-product spaces of countable (algebraic) dimension and linear isometric maps between them. Then each object of $\mathfrak{Q}\mathfrak{J}$ is isomorphic to $\mathbb{R}^\infty$ with orthonormal base $\{e_1,e_2,e_3,...\}$ or one of its subspaces $\mathbb{R}^n$ with base $\{e_1,...,e_n\}$. We topologize $A\in\mathrm{ob}\mathfrak{Q}\mathfrak{J}$ by giving its finite dimensional subspaces the metric topology and A itself the direct limit topology of the diagram of its finite dimensional subspaces. The morphism sets $\mathfrak{Q}\mathfrak{J}(A,B)$ obtain the k-function space topology (Appendix I).

Lemma 6.36: $\mathfrak{Q}\mathfrak{J}(V,\mathbb{R}^\infty)$ is contractible for all $V\in\mathrm{ob}\mathfrak{Q}\mathfrak{J}$.

Proof: Let $i_1,i_2 : V \longrightarrow V\oplus V$ be the inclusions onto the first respectively second summand. If $\{v_i\}$ is an orthonormal basis of V then

$$f_t(v_i) = \frac{1}{2t^2-2t+1}[(1-t)(v_i,0)+t(0,v_i)]$$

is a homotopy through isometries from i_1 to i_2. By applying the Gram-Schmidt orthogonalization process to

$$g_t(e_n) = (1-t)e_n+te_{2n}$$

we obtain a homotopy through isometries from $id_{\mathbb{R}^\infty}$ to $g : \mathbb{R}^\infty \longrightarrow \mathbb{R}^\infty$ given by $g(e_n) = e_{2n}$. Finally, let $h : \mathbb{R}^\infty \longrightarrow \mathbb{R}^\infty \oplus \mathbb{R}^\infty$ be the isometry

$$h(e_{2n}) = (e_n,0) \qquad h(e_{2n-1}) = (0,e_n),$$

let $k : V \longrightarrow \mathbb{R}^\infty$ be a fixed isometry, and $i \in \mathfrak{Q}\mathfrak{J}(V,\mathbb{R}^\infty)$ arbitrary. Then

$$i = h^{-1}\circ h\circ i \underset{g_t}{\simeq} h^{-1}\circ h\circ g\circ i = h^{-1}\circ i_1\circ i = h^{-1}\circ(i\oplus k)\circ i_1 \underset{f_t}{\simeq} h^{-1}\circ(i\oplus k)\circ i_2 = h^{-1}\circ(k\oplus k)\circ i_2$$

is continuous in i and contracts $\mathfrak{Q}\mathfrak{J}(V,\mathbb{R}^\infty)$ to the point $h^{-1}\circ(k\oplus k)\circ i_2$. ∎

We next consider bifunctors $\odot : \mathfrak{Q}\mathfrak{J} \times \mathfrak{Q}\mathfrak{J} \longrightarrow \mathfrak{Q}\mathfrak{J}$ which make $\mathfrak{Q}\mathfrak{J}$ into a symmetric monoidal category in the sense of Eilenberg-Kelly [20].

<u>Definition 6.37</u>: A <u>symmetric-monoidal category</u> $\mathfrak{C}$ consists of the following data:

(i) a category $\mathfrak{C}$

(ii) a functor $\odot : \mathfrak{C} \times \mathfrak{C} \longrightarrow \mathfrak{C}$

(iii) an object I of $\mathfrak{C}$

(iv) natural isomorphisms
$$\begin{aligned} r &= r_A : A \odot I \longrightarrow A \\ l &= l_A : I \odot A \longrightarrow A \\ a &= a_{ABC} : (A \odot B) \odot C \longrightarrow A \odot (B \odot C) \\ c &= c_{AB} : A \odot B \longrightarrow B \odot A \end{aligned}$$

These data satisfy the following axioms:

(a) $l_I = r_I : I \odot I \longrightarrow I$

(b) $c_{AB} \circ c_{BA} = id : B \odot A \longrightarrow B \odot A$

(c) The following diagrams commute

$$(A \odot B) \odot I \xrightarrow{\ a\ } A \odot (B \odot I)$$
$$r \searrow \quad \swarrow id \odot r$$
$$A \odot B$$

$$(I \odot A) \odot B \xrightarrow{\ a\ } I \odot (A \odot B)$$
$$l \odot id \searrow \quad \swarrow l$$
$$A \odot B$$

$$(A \odot I) \odot B \xrightarrow{\ a\ } A \odot (I \odot B)$$
$$r \odot id \searrow \quad \swarrow id \odot l$$
$$A \odot B$$

$$((A \odot B) \odot C) \odot D \xrightarrow{\ a \odot id\ } (A \odot (B \odot C)) \odot D \xrightarrow{\ a\ } A \odot ((B \odot C) \odot D)$$
$$\downarrow a \qquad\qquad\qquad\qquad\qquad\qquad\qquad\qquad \downarrow id \odot a$$
$$(A \odot B) \odot (C \odot D) \xrightarrow{\qquad\qquad a\qquad\qquad} A \odot (B \odot (C \odot D))$$

$$(A \odot B) \odot C \xrightarrow{\ a\ } A \odot (B \odot C) \xrightarrow{\ c\ } (B \odot C) \odot A$$
$$c \odot id \downarrow \qquad\qquad\qquad\qquad\qquad\qquad\qquad \downarrow a$$
$$(B \odot A) \odot C \xrightarrow{\quad a\quad} B \odot (A \odot C) \xrightarrow{\ id \odot c\ } B \odot (C \odot A)$$

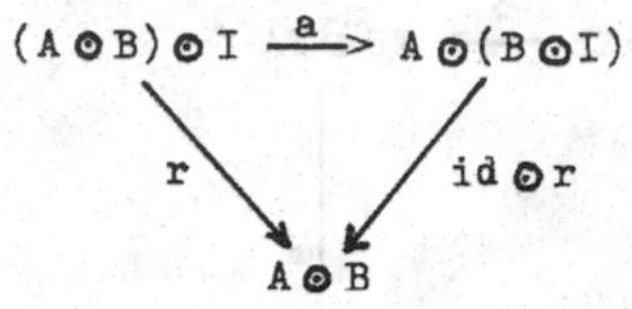

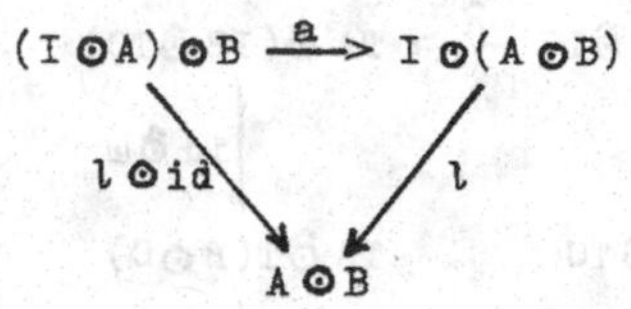

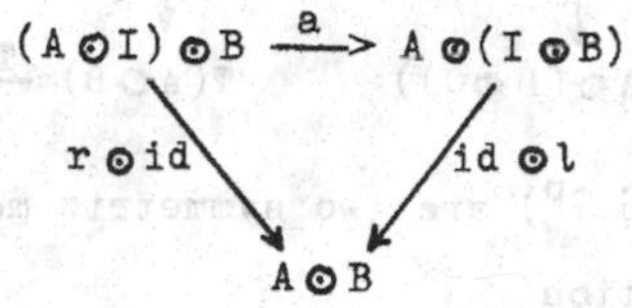

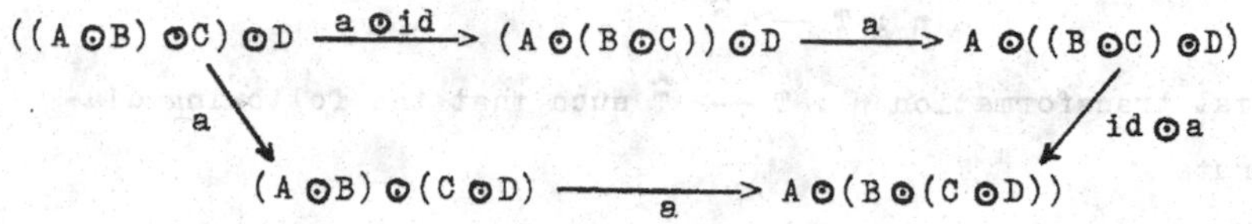

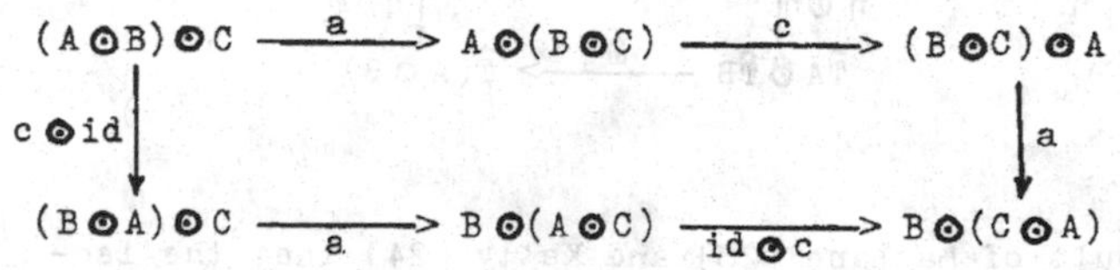

Given symmetric categories $\mathfrak{C} = (\mathfrak{C}, \odot, I, r, l, a, c)$ and $\hat{\mathfrak{C}} = (\hat{\mathfrak{C}}, \hat{\odot}, \hat{I}, \hat{r}, \hat{l}, \hat{a}, \hat{c})$ a
<u>symmetric</u> <u>monoidal</u> <u>functor</u>

$$T = (T, \omega, \omega^o) : \mathfrak{C} \longrightarrow \hat{\mathfrak{C}}$$

consists of

(i)   a functor $T : \mathfrak{C} \longrightarrow \hat{\mathfrak{C}}$

(ii) a natural transformation $\omega = \omega_{AB} : TA \,\hat{\odot}\, TB \longrightarrow T(A \odot B)$

(iii) a morphism $\omega^o : \hat{I} \longrightarrow TI$

such that the following diagrams commute

$$TI \,\hat{\odot}\, TA \xrightarrow{\ \omega\ } T(I \odot A)$$
$$\omega^o \hat{\odot} id \downarrow \qquad\qquad \downarrow Tl$$
$$\hat{I} \,\hat{\odot}\, TA \xrightarrow{\ \hat{l}\ } TA$$

$$TA \,\hat{\odot}\, TI \xrightarrow{\ \omega\ } T(A \odot I)$$
$$id \,\hat{\odot}\, \omega^o \downarrow \qquad\qquad \downarrow Tr$$
$$TA \,\hat{\odot}\, \hat{I} \xrightarrow{\ \hat{r}\ } TA$$

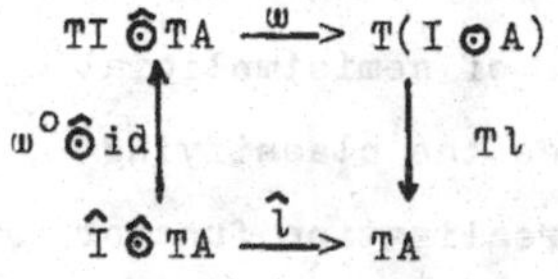

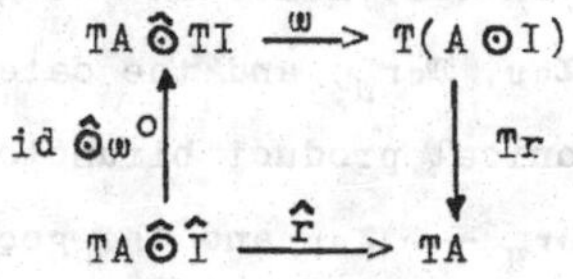

$$\begin{array}{ccc}
(TA\,\hat{\odot}\,TB)\,\hat{\odot}\,TC & \xrightarrow{\hat{a}} & TA\,\hat{\odot}\,(TB\,\hat{\odot}\,TC) \\
{\scriptstyle \omega\,\hat{\odot}\,id}\downarrow & & \downarrow{\scriptstyle id\,\hat{\odot}\,\omega} \\
T(A\odot B)\,\hat{\odot}\,TC & & TA\,\hat{\odot}\,T(B\odot C) \\
{\scriptstyle \omega}\downarrow & & \downarrow{\scriptstyle \omega} \\
T((A\odot B)\odot C) & \xrightarrow{Ta} & T(A\odot(B\odot C))
\end{array}
\qquad
\begin{array}{ccc}
TA\,\hat{\odot}\,TB & \xrightarrow{\hat{c}} & TB\,\hat{\odot}\,TA \\
{\scriptstyle \omega}\downarrow & & \downarrow{\scriptstyle \omega} \\
T(A\odot B) & \xrightarrow{Tc} & T(B\odot A)
\end{array}$$

If $T = (T,\omega,\omega^{o})$ and $\hat{T} = (\hat{T},\hat{\omega},\hat{\omega}^{o})$ are two symmetric monoidal functors then a <u>monoidal</u> <u>transformation</u>

$$\eta : T \longrightarrow \hat{T}$$

is a natural transformation $\eta : T \longrightarrow \hat{T}$ such that the following diagrams commute

$$\begin{array}{ccc}
 & \overset{\omega^{o}}{\nearrow} & TI \\
\hat{I} & & \downarrow{\scriptstyle \eta} \\
 & \underset{\hat{\omega}^{o}}{\searrow} & \hat{T}I
\end{array}
\qquad
\begin{array}{ccc}
TA\,\hat{\odot}\,TB & \xrightarrow{\omega} & T(A\odot B) \\
{\scriptstyle \eta\,\hat{\odot}\,\eta}\downarrow & & \downarrow{\scriptstyle \eta} \\
\hat{T}A\,\hat{\odot}\,\hat{T}B & \xrightarrow{\hat{\omega}} & \hat{T}(A\odot B)
\end{array}$$

It is now a result of MacLane [28] and Kelly [24] that the isomorphisms r, l, a, and c are coherent. Roughly speaking, this means that all diagrams obtained from them, their inverses, and constructions involving id and $\odot$ are commutative.

For us the interesting example of a symmetric monoidal category is $\mathfrak{Q}\mathfrak{Z}$ with the direct sum functor $\oplus$ and the canonical isomorphisms r, l, a, c. Recall that the inner product of $A\oplus B$ is given by

$$\langle (a,b),(a',b')\rangle_{A\oplus B} = \langle a,a'\rangle_{A} + \langle b,b'\rangle_{B}$$

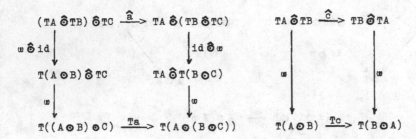

Other examples of symmetric monoidal categories of importance for us are the categories $\mathfrak{Top}$, $\mathfrak{Mor}_{\mathfrak{A}}$, and the category $\mathfrak{Spl}$ of semisimplicial sets with their canonical product bifunctors. Since the classifying space functor $B : \mathfrak{Mor}_{\mathfrak{A}} \longrightarrow \mathfrak{Top}$ and the geometric realization functor $R : \mathfrak{Spl} \longrightarrow \mathfrak{Top}$ preserve products (we work with k-spaces), they are in a canonical way symmetric monoidal functors.

The following result explains the importance of symmetric monoidal functors for the construction of examples of infinite loop spaces.

<u>Theorem 6.38:</u> Any symmetric monoidal structure $(\mathfrak{B},\odot, I,r,l,a,c)$ with $\odot$ continuous on $\mathfrak{B}$ determines an E-category $\mathfrak{E}$, any continuous symmetric monoidal functor $\mathfrak{B} \longrightarrow \mathfrak{Top}$ induces an $\mathfrak{E}$-structure on $T\mathbb{R}^{\infty}$, and any monoidal transformation $\eta : T \longrightarrow \hat{T}$ induces an $\mathfrak{E}$-space homomorphism $T\mathbb{R}^{\infty} \longrightarrow \hat{T}\mathbb{R}^{\infty}$.

<u>Proof:</u> Let $\overset{n}{\odot}\mathbb{R}^{\infty} = \mathbb{R}^{\infty}\odot\ldots\odot\mathbb{R}^{\infty}$, n times, with a fixed choice of bracketing. Put $\mathfrak{E}(n,1) = \mathfrak{B}(\overset{n}{\odot}\mathbb{R}^{\infty},\mathbb{R}^{\infty})$. The isomorphisms c extend uniquely to an action of S_n on $\overset{n}{\odot}\mathbb{R}^{\infty}$, denoted by $(\xi,b)\longmapsto b\cdot\xi^*, \xi\in S_n$. The other morphism spaces of $\mathfrak{E}$ are given by

$$\mathfrak{E}(n,r) = \bigcup_{n_1+\ldots+n_r=n} \mathfrak{E}(n_1,1)\times\ldots\times\mathfrak{E}(n_r,1)\times S_n/\sim$$

with $(b_1\cdot\pi_1^*,\ldots,b_r\cdot\pi_r^*,\xi)\sim(b_1,\ldots,b_r,\xi\cdot(\pi_1\oplus\ldots\oplus\pi_r)),\pi_i\in S_{n_i}$. The elements ξ represent the set operations. Hence composition with $\xi^* = (id,\ldots,id,\xi)$ on the left is fixed by (2.43) and determined in general by

$$a\cdot(b_1,\ldots,b_r,\xi)=a\cdot(b_1\odot\ldots\odot b_r)\cdot\xi^* \qquad a\in\mathfrak{E}(r,1)$$

with the composition in $\mathfrak{B}$ on the right.

Let $\eta:(T,\omega,\omega^o)\longrightarrow(\hat{T},\hat{\omega},\hat{\omega}^o)$ be a monoidal transformation of symmetric monoidal functors. Define an E-action $\alpha:\mathfrak{E}\longrightarrow\mathfrak{Top}$ on $T\mathbb{R}^{\infty}$ by

$$\alpha(b_1,\ldots,b_r,\xi):(T\mathbb{R}^{\infty})^n \xrightarrow{\xi^*} (T\mathbb{R}^{\infty})^n \xrightarrow{(\omega^{n_1},\ldots,\omega^{n_r})} T(\overset{n_1}{\odot}\mathbb{R}^{\infty})\times\ldots\times T(\overset{n_r}{\odot}\mathbb{R}^{\infty}) \xrightarrow{(Tb_1,\ldots,Tb_r)} (T\mathbb{R}^{\infty})^r$$

for $(b_1,\ldots,b_r,\xi)\in\mathfrak{E}(n,r)$, $b_i\in\mathfrak{E}(n_i,1)$. Here $\omega^n:(T\mathbb{R}^{\infty})^n\longrightarrow T(\overset{n}{\odot}\mathbb{R}^{\infty})$

is a suitable composite of ω or ω^o. The coherence conditions ensure that α is a multiplicative functor. Similary define an E-Structure on $\hat{T}\mathbb{R}^\infty$. Then $\eta: T\mathbb{R}^\infty \longrightarrow \hat{T}\mathbb{R}^\infty$ is obviously an $\mathfrak{E}$-space homomorphism. ∎

Remark: (a) To obtain $\mathfrak{E}$-spaces is suffices to construct a symmetric monoidal functor $\mathfrak{V} \longrightarrow \mathfrak{Spl}$, because composition with the geometric realization gives a symmetric monoidal functor $\mathfrak{V} \longrightarrow \mathfrak{Top}$.
(b) If a symmetric monoidal functor $T : \mathfrak{V} \longrightarrow \mathfrak{Top}$ happens to be monoid valued, then we can follow it by the classifying space functor B to obtain another symmetric monoidal functor.

We now list a number of infinite loop spaces and infinite loop maps. For this we construct symmetric monoidal functors (T,ω,ω^o). As monoidal structure on $\mathfrak{V}$ we take $\oplus$. We define T and ω for finite dimensional inner product spaces and extend them to all of $\mathfrak{V}$ by taking direct limits over the diagrams of finite dimensional subspaces. Since $A\oplus-$ and $X\times-$, $A\in\mathfrak{V}$, $X\in\mathfrak{Top}$ or $\mathfrak{Spl}$ preserve direct limits, this suffices.

(6.39) Examples: The unit I of $\oplus$ in $\mathfrak{V}$ is $\mathbb{R}^o$.

(a) $TA = O(A)$, the orthogonal group of A. As $\omega: TA \times TB \longrightarrow T(A\oplus B)$ take the Whitney sum: $\omega(f,g) = f\oplus g: A\oplus B \longrightarrow A\oplus B$, $f\in O(a)$, $g\in O(B)$. Since $O(\mathbb{R}^o)$ consist of one point, $\omega^o: * \longrightarrow O(\mathbb{R}^o)$ is uniquely determined. Then $O(\mathbb{R}^\infty)$ is the stable orthogonal group O. Since O is numerably contractible and $\pi_o(O)$ is the cyclic group $\mathbb{Z}_2$ under Whitney sum, O is an infinite loop space with multiplication given by Whitney sum. In matrix form it reads

$$O(\mathbb{R}^n) \times O(\mathbb{R}^m) \longrightarrow O(\mathbb{R}^{n+m})$$
$$M \;,\; N \longmapsto \begin{pmatrix} M & 0 \\ 0 & N \end{pmatrix}$$

(b) $TA = U(A\otimes\mathbb{C})$, the unitary group of $A\otimes\mathbb{C}$ ($\mathbb{C}$ denotes the complex numbers). Again ω is Whitneysum and ω^o the unique map $* \longrightarrow T(\mathbb{R}^o)$.

Then $T\mathbb{R}^{\infty}$ is the stable unitary group U. Since U is connected and numerably contractible, it is an infinite loop space with multiplication given by Whitney sum.

(c) $TA = Sp(A \otimes \mathbb{H})$, the symplectic group of $A \otimes \mathbb{H}$ ($\mathbb{H}$ denotes the quaternions). Again ω is Whitney sum and ω^0 the unique map $* \longrightarrow T(\mathbb{R}^0)$. Then $T\mathbb{R}^{\infty}$ is the stable symplectic group Sp. Since Sp is connected and numerably contractible, it is an infinite loop space with respect to the Whitney sum-E-structure.

(d) $TA = SO(A)$, the special orthogonal group of A. Take ω to be the Whitney sum. The stable group $SO = T\mathbb{R}^{\infty}$ is connected and numerably contractible. Hence SO is an infinite loop space.

(e) $TA = F(A)$, the space of based homotopy equivalences of the sphere S^A, which is the one-point compactification $A \cup \infty$ of A, with ∞ as base point. The Whitney sum ω takes here the form of the smash product since $S^A \wedge S^B \cong S^{A \oplus B}$. The Whitney sum multiplication makes $\pi_0(F)$ into the cyclic group $\mathbb{Z}_2$. Since F is numerably contractible, it is an infinite loop space.

(f) TA = space of homeomorphisms of A. Take ω to be Whitney sum. Then $T\mathbb{R}^{\infty}$ is an E-space with $\pi_0(T\mathbb{R}^{\infty}) = \mathbb{Z}_2$, but we do not know whether or not $T\mathbb{R}^{\infty}$ is numerably contractible. So we follow T by the composite

$$\mathfrak{Top} \xrightarrow{\ Sin\ } \mathfrak{Spl} \xrightarrow{\ R\ } \mathfrak{Top}$$

where Sin associates with $X \in \mathfrak{Top}$ its singular complex. Since both preserve products, they are symmetric monoidal functors. Then $R \circ Sin\ (T\mathbb{R}^{\infty})$ is the stable group Top. It is an infinite loop space, because it is a CW-complex.

(g) SU and the orientation preserving versions of (e) and (f) are infinite loop spaces under Whitney sum.

(h) We can do to examples (a),...,(e),(g) what we have done in (f): We follow T by the symmetric monoidal functor $R \circ Sin$. The resulting stable groups $R\cdot\mathrm{Sin}\, T\mathbb{R}^{\infty}$ are all infinite loop spaces. The back adjunction $R\cdot\mathrm{Sin} \longrightarrow \mathrm{Id}_{\mathcal{I}op}$ is a monoidal transformation so that $R\cdot\mathrm{Sin}\, T\mathbb{R}^{\infty} \longrightarrow T\mathbb{R}^{\infty}$ is a homomorphism of E-spaces and hence homotopy equivalent to an infinite loop map. Moreover, in all of our examples with exception of (f) it is a homotopy equivalence so that $R\cdot\mathrm{Sin}\, T\mathbb{R}^{\infty}$ and $T\mathbb{R}^{\infty}$ are homotopy equivalent by infinite loop maps.

(k) In all examples (a),...,(g) the spaces $T(A)$ are monoids under composition. Moreover, they are well pointed (in (f) take $R\cdot\mathrm{Sin}\, T(A)$) and numerably contractible. Hence $BT(A)$, A finite dimensional, B the classifying space functor, defines another symmetric monoidal functor T' making $T'(\mathbb{R}^{\infty})$ into an infinite loop space under Whitney sum

(l) We identify $\mathbb{C}$ with $\mathbb{R}^2$ and $\mathbb{H}$ with $\mathbb{C}^2 = \mathbb{R}^4$. Also identify A and $A \oplus \mathbb{R}^1$. Then the canonical inclusions $\mathbb{R}^1 \subset \mathbb{C} = \mathbb{R}^2 \subset \mathbb{H} = \mathbb{C}^2$ define monoidal transformations

$$\begin{array}{ccccc} O(A) & \longrightarrow & U(A\otimes\mathbb{C}) & \longrightarrow & Sp(A\otimes\mathbb{H}) \\ & & \downarrow & & \downarrow \\ & & O(A\otimes\mathbb{R}^2) & \longrightarrow & O(A\otimes\mathbb{R}^4) \end{array}$$

making the diagram commute. Since $O(\mathbb{R}^{\infty}\oplus\mathbb{R}^n) \cong O(\mathbb{R}^{\infty}) = O$, the monoidal transformations make the canonical inclusion maps

$$O \subset U \subset S_p \subset O$$

to infinite loop maps under Whitney sum.

(m) Since any orthogonal transformation of A is a homeomorphism and any homeomorphism induces a based homotopy equivalence of S^A, we have inclusions

$$O(A) \subset \text{homeomorphisms of } A \subset F(A),$$

which define monoidal transformations. Passing to the topological realization of the singular complexes we find that the canonical maps

$$O \subset Top \subset F$$

are infinite loop maps under Whitney sum

(n) Since all inclusion maps listed are monoid homomorphisms, we may again pass to the classifying spaces and find that the canonical maps

$$BO \longrightarrow BU \longrightarrow BSp \longrightarrow BO \longrightarrow Top \longrightarrow F$$

are infinite loop maps.

(o) Let $(T_1,\omega_1,\omega_1^o),(T_2,\omega_2,\omega_2^o):\mathfrak{B} \longrightarrow \mathfrak{Top}$ be group valued symmetric monoidal functors and $\eta:T_1 \longrightarrow T_2$ a monoidal transformation which is a homomorphism. Define $T_3A = T_2A/T_1A$ the factor set, A finite dimensional. Then ω_1 and ω_2 induce a natural transformation ω_3: $T_3A \times T_3B \longrightarrow T_3(A \bullet B)$ and ω_1^o, ω_2^o a map $* \longrightarrow T_3(I)$. We obtain a symmetric monoidal functor T_3 and $T_3(\mathbb{R}^\infty)$ is an E-space As application we obtain that the coset spaces Top/O, Top/Sp, Top/U, O/Sp, O/U, Sp/U, Sp/O, and U/O are infinite loop spaces under Whitney sum.

Since the projections $T_2A \longrightarrow T_3A = T_2A/T_1A$ induce a monoidal transformation the various canonical maps Top $\longrightarrow$ Top/O, etc. are infinite loop maps.

(p) Suppose $(T_1,\omega_1,\omega_1^o)$, $(T_2,\omega_2,\omega_2^o)$: $\mathfrak{B} \longrightarrow \mathfrak{Top}$ are monoid valued and $\eta:(T_1,\omega_1,\omega_1^o) \longrightarrow (T_2,\omega_2,\omega_2^o)$ is a monoidal transformation and homomorphism. Then define T_3A, A finite dimensional, to be the homotopy theoretic fibre of the map

$$BT_1A \longrightarrow BT_2A$$

induced by η. There is a canonical map ω_3: $T_3A \times T_3B \longrightarrow T_3(A \bullet B)$ making

$$\begin{array}{ccc} T_3A \times T_3B & \xrightarrow{\omega_3} & T_3(A \bullet B) \\ \downarrow & & \downarrow \\ BT_1A \times BT_1B & \longrightarrow & BT_1(A \bullet B) \\ \downarrow & & \downarrow \\ BT_2A \times BT_2B & \longrightarrow & BT_2(A \bullet B) \end{array}$$

commute, because the homotopy theoretical fibre construction preserves products. The maps ω_1^o and ω_2^o induce a unique map $\omega_3^o : * \longrightarrow T_3(\mathbb{R}^o)$. Hence $T_3\mathbb{R}^\infty$ is an E-space. As application, we obtain that $F/Top = \underrightarrow{\lim}$ (homotopy theoretic fibre of $BTop(n) \longrightarrow BF(n)$) is an infinite loop space under Whitney sum and the canonical map

$$F/Top \longrightarrow BTop$$

is an infinite loop map.

There is another symmetric monoidal structure on $\mathfrak{L}\mathfrak{J}$ given by the tensor product. Unfortunately, we have no examples to apply it to, the reason being that it is very difficult to arrange a commutative diagram

$$\begin{array}{ccc} T(A) \times T(B) & \xrightarrow{\omega} & T(A \otimes B) \\ \downarrow & & \downarrow \\ T(A') \times T(B') & \xrightarrow{\omega} & T(A' \otimes B') \end{array}$$

if $A \subsetneq A'$ and $B \subsetneq B'$. If one for example tries $T(A) = O(A)$ with $\omega(f,g) = f \otimes g$, then the diagram does not commute because $(f \oplus id) \otimes (g \oplus id) \neq (f \otimes g) \oplus id$. If one wants to show that O, U, BO, and BU are infinite loop spaces under the tensor product structure, one should use the theory of G. Segal [45] instead of trying to define a tensor product E-structure. A detailed treatment can be found in [7].

We have seen that Theorem 6.38 enables us to show that most of the stable groups which are of interest in the topology of manifolds are infinite loop spaces under Whitney sum structure. Our machine fails if we want to impose an E-structure on PL. The reason is that we kept too close to the linear group. Contrary to what one might think, the action of the general linear group is not linear. Let $\sigma: \Delta^k \longrightarrow GL(n,R)$ be a singular simplex. It determines a map $\Delta^k \times \mathbb{R}^n \longrightarrow \Delta^k \times \mathbb{R}^n$, which is piecewise linear iff σ is a constant map. We therefore fail to find enough homotopies to make the program work. The remedy is to use a PL machine instead, and forget isometries. We do not want to go into detail here and refer to [7].

VII. Chapter

HOMOTOPY COLIMITS

To illustrate that our theory has more applications than just loop spaces we show that it gives rise to a more or less satisfactory definition of homotopy colimits. We only sketch our results; a more detailed treatment including homotopy limits will appear in [56].

1. HOMOTOPY DIAGRAMS

Let $\mathfrak{C}$ be an arbitrary small category such that each pair $(\mathfrak{C}(A,A),\{id_A\})$ is a NDR. We consider $\mathfrak{C}$ as an ob $\mathfrak{C}$-coloured PRO (cf. example 2.48), i.e. we have only 1-ary operations.

__Definition 7.1__: A __homotopy-$\mathfrak{C}$-diagram__, or __$h\mathfrak{C}$-diagram__ for short, is a $W\mathfrak{C}$-space, i.e. a continuous functor $W\mathfrak{C} \longrightarrow \mathfrak{Top}$.

__Example__: Let $\mathfrak{C}$ be the category given by the commutative diagram

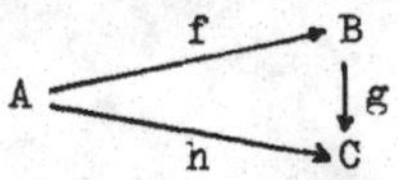

Then $W\mathfrak{C}(A,B)$ and $W\mathfrak{C}(B,C)$ consist of a single point, while $W\mathfrak{C}(A,C)$ is a unit interval, and a $W\mathfrak{C}$-space is a homotopy commutative diagram

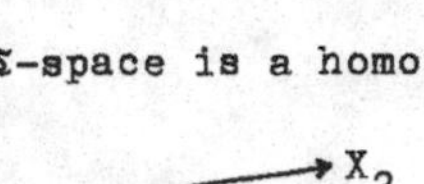

As maps between h$\mathfrak{C}$-diagrams we use $\mathfrak{C}$-maps. Since $\mathfrak{C}$ has only 1-ary operations, $W(\mathfrak{C} \otimes \mathfrak{L}_1) = W(\mathfrak{C} \times \mathfrak{L}_1)$, (see 2.22). We have the category $\mathfrak{Map}_{\mathfrak{C}}$ of h$\mathfrak{C}$-diagrams and $\mathfrak{C}$-maps.

2. HOMOTOPY COLIMITS

By (4.51), there is a functor

$$M : \mathfrak{Map}_{\mathfrak{C}} \longrightarrow \mathfrak{Hom}_{\mathfrak{C}}$$

which is left adjoint to the obvious functor $J : \mathfrak{Hom}_{\mathfrak{C}} \longrightarrow \mathfrak{Map}_{\mathfrak{C}}$. By definition, $\mathfrak{Hom}_{\mathfrak{C}}$ is the category of $\mathfrak{C}$-diagrams in the usual sense, i.e. continuous functors $\mathfrak{C} \longrightarrow \mathfrak{Top}$, and of homotopy classes of homomorphisms. Let

$$K : \mathfrak{Top}_h \longrightarrow \mathfrak{Hom}_{\mathfrak{C}}$$

be the functor assigning to each space X of the homotopy category the constant $\mathfrak{C}$-diagram on X (i.e. each morphism of $\mathfrak{C}$ is mapped to id_X). It is well-known that K is a right adjoint of the functor

$$\underrightarrow{\lim}_h : \mathfrak{Hom}_{\mathfrak{C}} \longrightarrow \mathfrak{Top}_h$$

induced by the colimit functor $\underrightarrow{\lim} : \mathfrak{Mor}_{\mathfrak{C}} \longrightarrow \mathfrak{Top}$.

<u>Definition 7.2</u>: The <u>homotopy</u> <u>colimit</u> <u>functor</u> h-$\underrightarrow{\lim}$: $\mathfrak{Map}_{\mathfrak{C}} \longrightarrow \mathfrak{Top}_{\mathfrak{C}}$ is defined to be the composite functor $\underrightarrow{\lim}_h \circ M$.

<u>Theorem 7.3</u>: The homotopy colimit functor is left adjoint to the obvious functor $\mathfrak{Top}_h \longrightarrow \mathfrak{Map}_{\mathfrak{C}}$ assigning to $X \in \mathfrak{Top}_h$ the constant h$\mathfrak{C}$-diagram on X.

This result justifies the notation "homotopy colimit", because, as mentioned above, it has the same universal property as the usual colimit functor, which is the left adjoint of the constant diagram functor $\mathfrak{Top} \longrightarrow \mathfrak{Mor}_{\mathfrak{C}}$.

Examples:

(1) If $\mathfrak{C}$ is the infinite linear category

$$0 \longrightarrow 1 \longrightarrow 2 \longrightarrow 3 \longrightarrow \cdots$$

then a sequence of spaces and maps

$$X_o \longrightarrow X_1 \longrightarrow X_2 \longrightarrow X_3 \longrightarrow \cdots$$

determines a h$\mathfrak{C}$-diagram whose homotopy colimit contains the Milnor telescope (cf. proof of (A 4.10)) of this sequence as a SDR.

(2) If $\mathfrak{G}$ is a topological category with exactly one object. Then its morphism space is a topological monoid G. Let D be the constant h$\mathfrak{G}$-diagram on a space with exactly one point. Then, by (VI §1), MD is homeomorphic to EG and h-$\underrightarrow{\lim}$ D to BG the total space and base space of Milgram's classifying space construction for G.

(3) Let $\mathfrak{C}$ be the category

$$\bullet \longleftarrow \bullet \longrightarrow \bullet$$

Then a h$\mathfrak{C}$-diagram is a diagram

$$D:\ B \xleftarrow{\ f\ } A \xrightarrow{\ g\ } C$$

of topological spaces and h-$\underrightarrow{\lim}$ D is the double mapping cylinder Z(f,g). Hence h-$\underrightarrow{\lim}$ D is the mapping cone of f if C = * and the (unreduced) suspension ΣA if B = * = C.

The based case is obtained by a slight modification. We adjoin a 0-ary operation to each object of C with the obvious definition of composition. Its action is the inclusion of the base point. The based homotopy colimit is then $\underrightarrow{\lim}$ M"D with M" of (V, §6).

(4) Let $\mathfrak{C}$ be the category

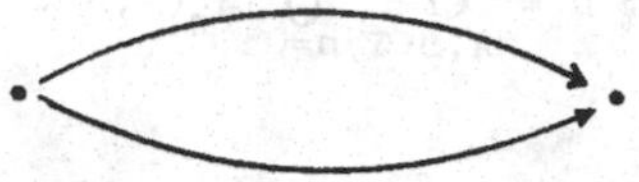

and

D :

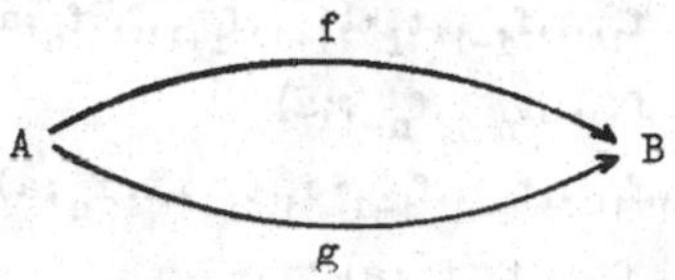

a h$\mathfrak{C}$-diagram. Then h-$\underrightarrow{\lim}$ D is the mapping torus of f and g.

The examples show that homotopy colimits crop up in many places in homotopy theory.

3. SPECTRAL SEQUENCES FOR HOMOTOPY COLIMITS

Since $\mathfrak{C}$ has only 1-ary operations, it is easy to give a direct definition of h-lim D for a h$\mathfrak{C}$-diagram D. The representing trees of MD are linear and vertical, and may be specified by giving in order, going up the tree the vertex labels and edge lengths and finally the cherry as

$$(g_o, t_1, g_1, t_2, \ldots, g_k; x)$$

$g_o \circ \ldots \circ g_k : A \longrightarrow B$ is defined in $\mathfrak{C}$, $t_i \in I$, and $x \in D(A)$. The $\mathfrak{C}$-action on MD is given

$$g(g_o, t_1, \ldots, g_k; x) = (g \circ g_o, t_1, \ldots, g_k; x)$$

This $\mathfrak{C}$-action has to be factored out to obtain h-$\underrightarrow{\lim}$ D = $\underrightarrow{\lim}$ MD. Define $\mathfrak{C}_n(A,B) = \{(f_1, f_2, \ldots, f_n) \in (\text{mor } \mathfrak{C})^n \mid f_1 \circ \ldots \circ f_n : A \longrightarrow B \text{ is defined in } \mathfrak{C}\}$ with the subspace topology of $(\text{mor } \mathfrak{C})^n$. If n=0, define

$$\mathfrak{C}_o(A,B) = \begin{cases} \{id_A\} & \text{if } B=A \\ \emptyset & \text{otherwise} \end{cases}$$

Then

$$h\text{-}\underrightarrow{\lim}\, D = \bigcup_{A,B\in\mathfrak{C}} \bigcup_{n=0}^{\infty} \mathfrak{C}_n(A,B)\times I^n\times D(A)/\sim$$

with the relations

(7.4)

$$(t_1,f_1,t_2,f_2,\dots,t_n,f_n;a)=\begin{cases}(t_1,f_1,\dots,f_{i-1},t_i*t_{i+1},f_{i+1},\dots,f_n;a) & \text{if } f_i=id,\ i<n\\ (t_1,f_1,\dots,t_{n-1},f_{n-1};a) & \text{if } f_n=id\\ (t_1,f_1,\dots,t_{i-1},f_{i-1}\circ f_i,t_{i+1},\dots,f_n;a) & \text{if } t_i=0,\ i>1\\ (t_2,f_2,\dots,t_n,f_n;a) & \text{if } t_1=0\\ (t_1,f_1,\dots,f_{i-1};D(f_i,t_{i+1},\dots,f_n)(a)) & \text{if } t_i=1\end{cases}$$

(recall that $t_1*t_2 = t_1+t_2-t_1t_2$). The filtration on MD hence induces a filtration FD of $h\text{-}\underrightarrow{\lim}\, D$ by the images F_pD of $\bigcup_{A,B\in\mathfrak{C}} \bigcup_{n=0}^{p} \mathfrak{C}_n(A,B)\times I^n\times D(A)$ in h-lim D.

Let k_* be an arbitrary homology and k^* an arbitrary cohomology theory. Since a $h\mathfrak{C}$-diagram $D : W\mathfrak{C} \longrightarrow \mathfrak{Top}$ is a $\mathfrak{C}$-diagram up to coherent homotopies and k_* and k^* are homotopy functors, the functors

$$k_q \circ D,\ k^q \circ D : W\mathfrak{C} \longrightarrow \mathfrak{Ab} = \text{abelian groups}$$

factor through $\epsilon : W\mathfrak{C} \longrightarrow \mathfrak{C}$, the augmentation. In other words, the composites

$$k_q \circ D \circ \eta,\ k^q \circ D \circ \eta : \mathfrak{C} \longrightarrow W\mathfrak{C} \longrightarrow \mathfrak{Top} \longrightarrow \mathfrak{Ab}$$

where η is the standard section (III, 3.5) are functors, although η is not.

For a proof of the following result we refer the reader to [56].

Theorem 7.5: Let $\mathfrak{C}$ be a small category with discrete morphism spaces, k_* a homology and k^* a cohomology theory, both additive. Then $E^2_{p,q}D \cong \underrightarrow{\lim}^{(p)}(k_qD\eta)$ in the spectral sequence $\{E^rD\}$ derived from the k_* exact couple of the filtration of $h\text{-}\underrightarrow{\lim}\, D$ and $E_2^{p,q}D \cong \underleftarrow{\lim}^{(p)}(k^qD\eta)$ in the spectral sequence $\{E_rD\}$ derived from the k^* exact couple of the filtration of $h\text{-}\underrightarrow{\lim}\, D$. Here $\underrightarrow{\lim}^{(p)}$ and $\underleftarrow{\lim}^{(p)}$ denote the p-th left derived of $\underrightarrow{\lim}$ and the p-th right derived of $\underleftarrow{\lim}$. ▮

This spectral sequence generalizes some well-known results:
Let $\mathfrak{C}$ be the infinite linear category of §2, Example 1, and

$$D: \qquad X_0 \xrightarrow{f_0} X_1 \xrightarrow{f_1} X_2 \xrightarrow{f_2} \cdots$$

a sequence of spaces. Let k* be an arbitrary additive cohomology theory. Then the spectral sequence $\{E_r D\}$ converges and collapses, giving rise to a short exact sequence

$$0 \longrightarrow \underleftarrow{\lim}^{(1)} k^{q-1} D \longrightarrow k^q(h\text{-}\underrightarrow{\lim}\, D) \longrightarrow \underleftarrow{\lim}\, k^q D \longrightarrow 0$$

If the f_i are cofibrations, then $h\text{-}\underrightarrow{\lim}\, D$ is homotopy equivalent to $\underrightarrow{\lim}\, D$, and we obtain Milnor's $\underleftarrow{\lim}^{(1)}$-Lemma [40].

Let $\mathfrak{C}$ be the category $\cdot \longleftarrow \cdot \longrightarrow \cdot$ and

$$D: \qquad B \xleftarrow{f} A \xrightarrow{g} C$$

a $h\mathfrak{C}$-diagram. The again the k* spectral sequence converges and collapses giving rise to an exact sequence

$$\cdots \longrightarrow k^{q-1}A \longrightarrow k^q(Z(f,g)) \longrightarrow k^q B \oplus k^q C \longrightarrow k^q A \longrightarrow k^{q+1}(Z(f,g)) \longrightarrow \cdots$$

If one of the maps, f say, is a cofibration, then the double mapping cylinder is of the homotopy type of $B \cup_g C$ and we obtain the Mayer-Vietoris sequence.

Analogous results hold for homology theories k_*.

4. HOMOTOPY COLIMITS OF COVERINGS

G. Segal [44] associated with each covering of a topological space its homotopy colimit and used this construction for a classification theorem for very general types of bundles. The essential step in this study was to show that the homotopy colimit of a numerable covering is naturally homotopy equivalent to the original space (recall that according to Dold [13] a covering $\mathfrak{U} = (U_\lambda | \lambda \in \Lambda)$ of X is numerable if there exists a locally finite partition of unity on X, $(v_\mu : X \longrightarrow I | \mu \in M)$

such that the covering $\mathfrak{B} = (v_\mu^{-1}(0,1]|\mu\in M)$ is a refinement of $\mathfrak{U}$). Using Segal's result, tom Dieck [11] proved the following theorem.

Theorem 7.6: Let $\mathfrak{U}=(U_\alpha|\alpha\in A)$ and $\mathfrak{B}=(V_\alpha|\alpha\in A)$ be numerable coverings of X and Y. For any non-empty subset $\sigma \subset A$ put $U_\sigma = \bigcap_{\alpha\in\sigma} U_\alpha$. Let $f:X \to Y$ be a map which carries each $U_\sigma, \sigma \subset A$ finite, into V_σ by a homotopy equivalence. Then f is a homotopy equivalence.

This result has a number of interesting consequences which we shall not discuss here. In the remainder of this section we give a detailed proof of Segal's result and show that the theorem is then an immediate consequence of our theory. As always before, we work in the category of k-spaces, but Segal's result is true for arbitrary topological spaces (by a similar type of argument using that the partition of unity is locally finite).

Segal's homotopy colimit of a covering $\mathfrak{U} = (U_\alpha|\alpha\in A)$ of X is defined as

$$B\mathfrak{U} = \bigcup_{\sigma_0\subset\ldots\subset\sigma_n} U_{\sigma_n}\times\Delta^n/\sim \qquad , \sigma_n \subset A \text{ finite, non-empty}$$

with the relations

$$(\sigma_0,u_1,\sigma_1,\ldots,u_n,\sigma_n;x) = (\sigma_0,u_1,\sigma_1,\ldots,\hat{u}_i,\hat{\sigma}_i,\ldots,u_n,\sigma_n;x) \text{ if } \sigma_{i-1}=\sigma_i \text{ or } u_{i-1}=u_i$$

where $u_0=0\le u_1\le\ldots\le u_n\le 1=u_{n+1}\in\Delta^n$, $x\in U_{\sigma_n}$ and $\wedge$ means "delete".

Our version of the homotopy colimit $h\mathfrak{U}$ of the covering $\mathfrak{U}$ is the homotopy colimit of the commutative diagram of spaces $U_\sigma, \sigma \subset A$ finite non-empty, and inclusions $U_\sigma \subset U_\tau$ whenever $\tau\subset\sigma$. Hence

$$h\mathfrak{U} = \bigcup_{\sigma_0\subset\ldots\subset\sigma_n} U_{\sigma_n}\times I^n/\sim$$

with

$$(\sigma_o,t_1,\sigma_1,\dots,t_n,\sigma_n;x) = \begin{cases} (\sigma_o,t_1,\dots,\sigma_{i-1},t_i*t_{i+1},\sigma_{i+1},\dots,t_n,\sigma_n;x) & \text{if } \sigma_{i-1}=\sigma_i,\ i<n \\ (\sigma_o,t_1,\dots,t_{n-1},\sigma_{n-1};x) & \text{if } \sigma_{n-1}=\sigma_n \\ (\sigma_o,t_1,\dots,\widehat{\sigma}_{i-1},\widehat{t}_i,\dots,\sigma_n;x) & \text{if } t_i=0 \\ (\sigma_o,t_1,\dots,\sigma_{i-1};x) & \text{if } t_i=1 \end{cases}$$

Note that the pair (σ_{i-1},σ_i) stand for the unique inclusion $U_{\sigma_i}\subset U_{\sigma_{i-1}}$ in our diagram.

The map $(\sigma_o,t_1,\sigma_1,\dots,t_n,\sigma_n;x) \longmapsto (\sigma_o,u_1,\sigma_1,\dots,u_n,\sigma_n;x)$ with $u_i=t_1*t_2*\dots*t_i$ is a natural homeomorphism

$$hU \cong BU$$

(cf. 6.6). We now simplify the simplicial structure of BU. In terms of barycentric coordinates, the relations for BU read

$$(b_o,\sigma_o,b_1,\sigma_1,\dots,b_n,\sigma_n;x) = \begin{cases} (b_o,\sigma_o,\dots,\widehat{b}_i,\widehat{\sigma}_i,\dots,\sigma_n;x) & \text{if } b_i=0 \\ (b_o,\sigma_o,\dots,\sigma_{i-1},b_i+b_{i+1},\sigma_{i+1},\dots,\sigma_n;x) & \text{if } \sigma_i=\sigma_{i+1} \end{cases}$$

$b_i\geq 0$, $\Sigma b_i=1$.

Give the indexing set A of the covering a well-ordering and define

$$MU = \bigcup_{\alpha_o<\dots<\alpha_n} (U_{\alpha_o}\cap\dots\cap U_{\alpha_n})\times\Delta^n/\sim \qquad \alpha_i\in A$$

with

$$(b_o,\alpha_o,b_1,\dots,b_n,\alpha_n;x) = (b_o,\alpha_o,\dots,\widehat{b}_i,\widehat{\alpha}_i,\dots,\alpha_n;x) \quad \text{if } b_i=0$$

It is readily seen that BU is just the barycentric subdivision of MU, so that BU and MU are naturally homeomorphic.

<u>Proposition 7.7</u> (Segal): The canonical map

$$\pi : MU \longrightarrow X, \qquad (b_o,\alpha_o,\dots,b_n,\alpha_n;x) \longmapsto x$$

has a section which embeds X as a SDR in MU.

<u>Proof</u>: We may assume that there is a locally finite partition of unity on X, $\{\lambda_\alpha : X \longrightarrow I \mid \alpha\in A\}$ with the indexing set of the cover U.

For $x \in X$, there is a finite number of indices $\alpha_o < \alpha_1 < \ldots < \alpha_n$ such that $\lambda_{\alpha_i}(x) \neq 0$. To obtain the section $s : X \longrightarrow MU$ of π map

$$x \longmapsto (\lambda_{\alpha_o}(x), \alpha_o, \ldots, \lambda_{\alpha_n}(x), \alpha_n; x)$$

If $y = (b_o, \sigma_o, \ldots, b_m, \sigma_m; , x) \in MU$ and $s\pi(y) = (v_o, \alpha_o, \ldots, v_n, \alpha_n; x)$, then y and $s\pi(y)$ are points in the simplex $x \times \Delta^r$, spanned by $(\gamma_o, \ldots, \gamma_r)$, the ordered collection of elements in $(\sigma_o, \ldots, \sigma_m) \cup (\alpha_o, \ldots, \alpha_n)$. Hence we can deform MU linearly into the section. It remains to check continuity. Let MA be the space associated with the A-indexed covering $\{V_\alpha = *\}$ of a single point. There is a canonical map $\rho : MU \longrightarrow MA$, given by $(b_o, \alpha_o, \ldots, b_n, \alpha_n; x) \longmapsto (b_o, \alpha_o, \ldots, b_n, \alpha_n; *)$. Then

$$(\pi, \rho) : MU \longrightarrow X \times MA$$

is injective. To show that it is an inclusion we have to prove that a function $f : C \longrightarrow MU$ from a compact Hausdorff space C to MU is continuous, provided $(\pi, \rho) \circ f$ is continuous (because we work with k-spaces). Since C is compact and MA is a simplicial complex, fC is contained in a finite subcomplex. Hence it suffices to show that (π, ρ) is an inclusion for finite coverings U. In this case, MA can be thought of as a subcomplex of the standard m-Simplex whose vertices are indexed by the elements of A. Let

$$p : \bigcup_{\alpha_o < \ldots < \alpha_n} (U_{\alpha_o} \cap \ldots \cap U_{\alpha_n}) \times \Delta^n \longrightarrow MU$$

be the identification map and $V \subset MU$ closed. Denote $p^{-1}V \cap (U_{\alpha_o} \cap \ldots \cap U_{\alpha_n}) \times \Delta^n$ by $V(\alpha_o, \ldots, \alpha_n)$. If $\overline{V(\alpha_o, \ldots, \alpha_n)}$ is the closure of $V(\alpha_o, \ldots, \alpha_n)$ in $X \times \Delta^n$, then $V(\alpha_o, \ldots, \alpha_n) = \overline{V(\alpha_o, \ldots, \alpha_n)} \cap (U_{\alpha_o} \cap \ldots \cap U_{\alpha_n}) \times \Delta^n$. The union of the $\overline{V(\alpha_o, \ldots, \alpha_n)}$ is closed in $X \times \Delta^m$ because it is finite. Hence MU is a subspace of $X \times \Delta^m$ and therefore of $X \times MA$. We obtain a diagram

$$\begin{array}{ccccc} & & MU & & \\ & \overset{s}{\nearrow} & \downarrow{\scriptstyle (\pi,\rho)} & \overset{\pi}{\searrow} & \\ X & & & & X \\ & \underset{(id,\varphi)}{\searrow} & X \times MA & \underset{proj}{\nearrow} & \end{array}$$

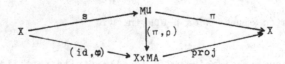

with $\varphi = \rho \circ s$. The section s and the deformation H of $M\mathfrak{U}$ into the section are continuous if φ and the deformation $(\pi, \rho) \circ H$ are continuous. If C is a compact Hausdorff space and $r : C \longrightarrow X$ is continuous, then $\varphi \circ r(C)$ lies in a finite subcomplex of MA and the composition of $\varphi \circ r$ with the barycentric coordinate functions is just the collection of maps $\{\lambda_\alpha \circ r \mid \alpha \in A\}$. Consequently $\varphi \circ r$ and hence φ are continuous. Similarly, one can prove the continuity of H by restriction to a finite subcomplex of MA. ∎

<u>Proof of Theorem 7.6</u>: Let $\mathfrak{D}$ be the dual of the category of finite non-empty subsets of A and inclusions. The coverings $\mathfrak{U}$ of X and $\mathfrak{V}$ of Y give rise $\mathfrak{D}$-diagrams with vertices $U_{\alpha_o} \cap \ldots \cap U_{\alpha_n}$ respectively $V_{\alpha_o} \cap \ldots \cap V_{\alpha_n}$ for the finite subset $\{\alpha_o, \ldots, \alpha_n\} \subset A$ and inclusions as morphisms. The map $f : X \longrightarrow Y$ is then a homomorphism of $\mathfrak{D}$-diagrams whose underlying maps are homotopy equivalences. Hence we have a commutative diagram

$$\begin{array}{ccc} h\mathfrak{U} & \xrightarrow{hf} & h\mathfrak{V} \\ {\scriptstyle \pi_{\mathfrak{U}}}\downarrow & & \downarrow{\scriptstyle \pi_{\mathfrak{V}}} \\ X & \xrightarrow[f]{} & Y \end{array}$$

in which hf is a homotopy equivalence by (4.21) and $\pi_{\mathfrak{U}}$ and $\pi_{\mathfrak{V}}$ are homotopy equivalences by (7.7). ∎

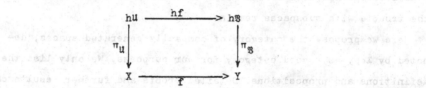

Appendix

1. COMPACTLY GENERATED SPACES

The category $\mathfrak{T}$ of topological spaces and continuous maps is inconvenient for the study of algebraic structures on spaces mainly for two reasons: The exponential law does not hold in general, and a product of two identifications need not be an identification. Steenrod [49] proposed the category $\mathfrak{CG}$ of compactly generated Hausdorff spaces as a convenient locus for dealing with these questions. Its objects are Hausdorff spaces X such that $U \subset X$ is open provided $U \cap C$ is open in C for each compact subspace C of X. This category has two defects namely it does not automatically contain all the subspaces and quotient spaces of its spaces and hence not the usual colimits. The second defect evaporates if one drops the Hausdorff condition. However, the trouble with subspaces remains.

Here we propose the category of compactly generated spaces, denoted by $\mathfrak{Top}$, as a good category for our purposes. We only list the definitions and propositions. Detailed proofs and further results can be found in [55].

<u>Definition 1.1</u>: A k-<u>space</u> is a topological space X such that $U \subset X$ is open whenever $r^{-1}(U)$ is open in C for every map $r : C \longrightarrow X$ from a compact Hausdorff space C into X. The category of k-spaces and continuous maps is denoted by $\mathfrak{Top}$.

Every topological space X has a finer topology making it into a k-space kX : Add to its open sets all subsets U satisfying the condition of Definition 1.1. Obviously, kX = X if X is a k-space.

__Proposition 1.2__: The correspondence $X \longrightarrow kX$ defines a functor $k : \mathfrak{T} \longrightarrow \mathfrak{T}op$, which is right adjoint to the inclusion $\mathfrak{T}op \subset \mathfrak{T}$. Hence k preserves limits.

__Proposition 1.3__: Let D be a small diagram of k-spaces and $\varinjlim D$, $\varprojlim D$ its colimit and limit in $\mathfrak{T}$. Then $\varinjlim D$ is a k-space and $\varinjlim D$ and $k(\varprojlim D)$ are the colimit and limit of D in $\mathfrak{T}op$.

__Corollary 1.4__: Quotient spaces of k-spaces are k-spaces.

Unfortunately, limits of k-spaces have to be retopologized. This applies in particular to products. Let $X \times Y$ denote the cartesian product and $X \times_k Y = k(X \times Y)$ the retopologized product of X and Y.

__Proposition 1.5__: Let X and Y be k-spaces and suppose each point of Y has a base of compact neighbourhoods. Then $X \times Y = X \times_k Y$.

Hence the notion of homotopy of continuous mappings does not change and the functor k preserves homotopy.

As an immediate consequence of the definitions one has that the maps of compact Hausdorff spaces into X factor through $id : kX \longrightarrow X$. This implies

__Proposition 1.6__: The identity map $kX \longrightarrow X$ induces isomorphisms for homotopy groups and singular homology and cohomology groups.

As mentioned before, subspaces of k-spaces need not be k-spaces. But we have

Proposition 1.7: Let X be a k-space and A a subspace of X.

(a) If A is open or closed, then A is a k-space.

(b) Let Z be a k-space. A function $f : Z \longrightarrow kA$ is continuous iff the composite $Z \xrightarrow{f} kA \subset X$ is continuous.

Part (b) shows that $kA \subset X$ has the universal property for k-spaces which characterises the relative topology of A in X in the category of all topological spaces.

For topological spaces X and Y let CO(X,Y) be the space of all continuous maps from X to Y with the compact-open topology and $C(X,Y) = k(CO(X,Y))$. If X and Y are k-spaces, we also denote C(X,Y) by $\mathfrak{Top}(X,Y)$.

Proposition 1.8: If Y is a k-space, then the evaluation map $e_{Y,Z} : C(Y,Z) \times_k Y \longrightarrow Z$, defined by $e_{Y,Z}(f,y) = f(y)$, is continuous.

Proposition 1.9 (Exponential law): Let X and Y be k-spaces. Then the correspondence $f \longmapsto e_{Y,Z} \circ (f \times id_Y)$ determines a natural homeomorphism

$$C(X,C(Y,Z)) \cong C(X \times_k Y,Z)$$

This result has a number of consequences.

Proposition 1.10: Let X be a k-space

(a) The functor $\mathfrak{Top}(X,-) : \mathfrak{Top} \longrightarrow \mathfrak{Top}$ preserves limits. In particular

$$\mathfrak{Top}(X,Y \times_k Z) \cong \mathfrak{Top}(X,Y) \times_k \mathfrak{Top}(X,Z)$$

(b) The functor $- \times_k X : \mathfrak{Top} \longrightarrow \mathfrak{Top}$ preserves colimits.

(c) The functor $\mathfrak{Top}(-,X) : \mathfrak{Top} \longrightarrow \mathfrak{Top}$ transfers colimits to limits.

Corollary 1.11: If $p : X \longrightarrow X'$ and $q : Y \longrightarrow Y'$ are identification maps of k-spaces, then $p \times_k q : X \times_k Y \longrightarrow X' \times_k Y'$ is an identification map.

Proposition 1.12: If X and Y are k-spaces, then composition of maps induces a continuous map

$$C(Y,Z) \times_k C(X,Y) \longrightarrow C(X,Z)$$

Our last statement shows that $\mathfrak{Top}$ is sufficiently large:

Proposition 1.13: $\mathfrak{Top}$ contains the category $\mathfrak{CG}$ of compactly generated Hausdorff spaces.

2. EQUIVARIANT COFIBRATIONS

In this chapter we need not restrict ourselves to k-spaces. The results hold in the category of all topological spaces as well as in the category of k-spaces.

Let G be a topological group. An equivariant map $i : A \longrightarrow X$ is called an equivariant cofibration or G-cofibration, if for all equivariant maps $f : X \longrightarrow Z$ and $H : A \times I \longrightarrow Z$ (with trivial G-action on I) such that $H(a,0) = f \circ i(a)$ for $a \in A$ there exists an equivariant map $F : X \times I \longrightarrow Z$ such that $F \circ (i \times id) = H$ and $F(x,0) = f(x)$ for $x \in X$.

As in the non-equivariant case (e.g. see [12; (1.17)]) one can show that a G-cofibration has to be an inclusion. So we also say that (X,A) is G-cofibered or that (X,A) has the G-HEP (homotopy extension property).

Again as in the non-equivariant case ([52; Thm. 2 and Lemma 4]) one shows

Proposition 2.1: Let X be a G-space and $A \subset X$ an invariant subspace. Then the following statements are equivalent

(a) (X,A) is G-cofibred

(b) $X \times 0 \cup A \times I$ is an equivariant retract of $X \times I$

(c) There exists an equivariant map $u : X \longrightarrow I$ such that $A \subset u^{-1}(0)$ and an equivariant homotopy $H : X \times I \longrightarrow X$ such that

$$H(x,0) = x \qquad x \in X$$
$$H(a,t) = a \qquad a \in A,\ t \in I$$
$$H(x,t) \in A \qquad \text{for } t > u(x)$$

If in addition A is an equivariant SDR of X, we may assume that $u(x) < 1$ for all $x \in X$. ∎

Let A be an invariant subspace of a G-space X. We say that (X,A) is an equivariant NDR (neighbourhood deformation retract) or G-NDR if there is a G-map $u : X \longrightarrow I$ and a G-homotopy $H : X \times I \longrightarrow X$ such that $A = u^{-1}(0)$ and

$$H(x,0) = x \qquad x \in X$$
$$H(a,t) = a \qquad a \in A,\ t \in I$$
$$H(x,1) \in A \qquad \text{for } u(x) < 1$$

If (X,A) is G-cofibred and A closed in X, then the conditions on u and H in (2.1 c) imply that $H(x,u(x)) \in A$ whenever $u(x) < 1$ and hence $u^{-1}(0) = A$. Hence (X,A) is a G-NDR. Conversely, the proof of [51; Thm. 2] also works for the equivariant case and shows that (X,A) is G-cofibred if (X,A) is a G-NDR. Hence we have

Proposition 2.2: Let A be a closed invariant subspace of a G-space X. Then (X,A) is G-cofibred iff (X,A) is a G-NDR. ∎

As a direct consequence we have

Lemma 2.3: If $i : A \subset X$ is a G-cofibration, then

(a) $i/G : A/G \subset X/G$ is a cofibration

(b) If $H \subset G$ is a subgroup, then $i : A \subset X$ is a H-cofibration.

Proof: (a) follows directly from the definition and (b) from (2.1 b). ∎

As a consequence of (2.1 c), the proof for the non-equivariant case [52;Thm. 6] of the following result carries over.

Proposition 2.4: Let G,H be topological groups and suppose (X,A) is G-cofibred, (Y,B) is H-cofibred, and A is closed in X. Then the product pair

$$(X,A) \times (Y,B) = (X\times Y,\ A\times Y \cup X\times B)$$

is (G×H)-cofibred. If in addition A [or B] is an equivariant SDR of X [or Y], then A×Y ∪ X×Y is a (G×H)-equivariant SDR of X×Y. ∎

Corollary 2.5: Let (X,A) and (Y,B) be G-cofibred and A closed in X, then (X×Y, A×Y ∪ X×B) is G-cofibred under the diagonal action.

Proof: Use (2.4) and (2.3 b). ∎

Recall from (2.43 d) that the action of the symmetric group from the left on X^n is our cases given by

$$\pi(x_1,\dots,x_n) = (x_{\pi^{-1}1},\dots,x_{\pi^{-1}n})$$

If X is a G-space, then X^n admits an action of the wreath product $G \wr S_n$ (our definition of the wreath product differs slightly from the usual one).

Let G be an arbitrary topological group. Define the wreath product

$$G \wr S_n = \{(f,\pi) \mid f : [n] \longrightarrow G,\ \pi \in S_n\}$$

with the topology of $G^n \times S_n$. The continuous multiplication is given by

$$(f_1,\pi_1)\cdot(f_2,\pi_2) = (h,\pi_1\cdot\pi_2)$$

with $h(i) = f_1(i)\cdot f_2(\pi_1^{-1}(i))$.

If X is a G-space, we have an action of $G \wr S_n$ on X^n given by

$$(f,\pi)\cdot(x_1,\ldots,x_n) = (f(1)\cdot x_{\pi^{-1}1},\ldots,f(n)\cdot x_{\pi^{-1}n})$$

The non-equivariant proof of (2.4) generalizes to give

Proposition 2.6: Suppose (X,A) is a G-NDR. Let Y_r be the subspace of all points in X^n having at least r coordinates in A. Then (X^n,Y_r) is a $G \wr S_n$-NDR. If, in addition, A is a G-equivariant SDR of X, then Y_r is a $G \wr S_n$-equivariant SDR of X^n.

Proof: Let $u : X \longrightarrow I$ be the G-map and $H : X\times I \longrightarrow X$ the G-homotopy of (2.1 c) for (X,A). Since A is closed, $H(x,u(x)) \in A$ whenever $u(x)<1$. Let M and M_i be the set of all subsets of cardinality r of $[n]=\{1,2,..,n\}$ respectively $[n]-\{i\}$. Then

$$v(x_1,\ldots,x_n) = \min\,(u(x_{i_1})+\ldots+u(x_{i_r})\,|\,\{i_1,\ldots i_r\}\in M)$$

$$F(x_1,\ldots,x_n,t) = (y_1,\ldots,y_n)$$

with $y_i = H_i(x_i,\min(t,u(x_{i_1})+\ldots+u(x_{i_r})\,|\,\{i_1,\ldots,i_r\}\in M_i))$ are $G\wr S_n$-equivariant maps satisfying (2.1 c) for (X^n,Y_r). ∎

Utilizing an idea of Lillig [27] we generalize Proposition 2.3. Let A be an arbitrary subspace of a G-space X. The subgroup $St(A) = \{g\in G\,|\,ga\in A \text{ for all } a\in A\}$ is called the stabilizer of A.

Theorem 2.7: Let G be a topological group. Let A be an invariant subspace of the G-space X and suppose $A = A_1 \cup A_2 \cup \ldots \cup A_n$ with A_i closed in X. Suppose G also acts on the set $[n] = \{1,2,\ldots,n\}$ such that $g\cdot A_i = A_{gi}$ for all $g\in G$ and $i \in [n]$. The G-action on [n] induces a G-action on the set of all subsets of [n]. For a subset $\sigma \subset [n]$ let $A_\sigma = \bigcap_{i\in\sigma} A_i$ and let $H_\sigma = St(\sigma)$. Then (X,A) is a G-NDR if each pair (X,A_σ) is a H_σ-NDR for all sunsets $\sigma \neq \emptyset$ of [n].

Proof: Let P_k be the set of all subsets of [n] with cardinality k. If

l is the cardinality of P_k define

$$X_k = \bigcup_{\sigma\in P_k} A_\sigma \qquad\qquad Y_k = X \times \Delta^{l-1}/\sim \qquad\qquad k>0$$

with $(x,t) \sim (x,t')$ for $x \in X_{k+1} \subset X$ and $t,t' \in \Delta^{l-1}$. Then X_k is a closed invariant subspace of X. Starting with X_n, we inductively show that (X,X_k) is a G-NDR for $k>0$, which will prove the theorem.

Suppose we know that (X,X_{k+1}) is a G-NDR for some k, $1\le k<n$. Choose a representative σ in each G-orbit $\mathfrak{O}$ of P_k and for each $\alpha\in\mathfrak{O}$ a $g_\alpha\in G$ such that $g_\alpha\alpha = \sigma$ and $g_\sigma = \mathrm{id}$. Since (X,A_σ) is an H_σ-NDR there is an H_σ-equivariant map $v_\sigma : X \longrightarrow I$ with $v_\sigma^{-1}(0) = A_\sigma$ and an H_σ-equivariant retraction

$$r_\sigma : X\times I \longrightarrow A_\sigma\times I \cup X\times 0$$

We define a map $u : P_k\times X \longrightarrow I$ such that $u^{-1}(0) \cap \{\alpha\}\times X = \{\alpha\}\times A_\alpha$ and $u(\alpha,x) = u(g\alpha,gx)$ for all $g\in G$ by putting $u(\alpha,x) = v_\sigma(g_\alpha x)$ if α is in the orbit represented by σ. If $\beta = g\alpha$, then $g_\beta g g_\alpha^{-1}\sigma = \sigma$, so that $g_\beta g g_\alpha^{-1} \in H_\sigma$. Hence

$$u(\alpha,x) = v_\sigma(g_\alpha x) = v_\sigma((g_\beta g g_\alpha^{-1})g_\alpha x) = u(g\alpha,gx)$$

Index the barycentric coordinates of Δ^{l-1} by the elements of P_k. The G-action on P_k then determines a G-action on Δ^{l-1}, and the diagonal G-action on $X\times\Delta^{l-1}$ defines a G-action on Y_k. Define a map $j_k : X \longrightarrow Y_k$ by

$$j_k(x) = (x, t_{\sigma_1}(x),\ldots,t_{\sigma_l}(x)) \qquad \{\sigma_1,\ldots,\sigma_l\} = P_k$$

where

$$t_\sigma(x) = \frac{(\prod u(\alpha,x)\,|\,\alpha\in P_k-\{\sigma\})}{\sum\limits_{\sigma\in P_k} (\prod u(\alpha,x)\,|\,\alpha\in P_k-\{\sigma\})}$$

is the barycentric coordinate indexed by $\sigma\in P_k$. One easily checks that j_k is continuous and equivariant.

If α is in the orbit represented by σ, define

$$r_\alpha : X\times I \longrightarrow A_\alpha\times I \cup X\times 0$$

by $r_\alpha(x,y) = g_\alpha^{-1} r_\sigma(g_\alpha x)$. Then r_α is a H_α-equivariant retraction, because $H_\alpha = g_\alpha^{-1} H_\sigma g_\alpha$.

The symmetric group S_l acts on Δ^{l-1} by permuting the barycentric coordinates. Obviously the inclusion of the 0-skeleton $\Delta_0^{l-1} \subset \Delta^{l-1}$ is an S_l-cofibration and hence a G-cofibration. By (2.4), the pair $(X\times\Delta^{l-1}, X_{k+1}\times\Delta^{l-1} \cup X\times\Delta_0^{l-1})$ is a G-NDR. Hence there is a G-retraction

$$p : X\times\Delta^{l-1}\times I \longrightarrow X\times\Delta^{l-1}\times 0 \cup X\times\Delta_0^{l-1}\times I \cup X_{k+1}\times\Delta^{l-1}\times I$$

Define an equivariant map

$$f : X\times\Delta^{l-1}\times 0 \cup X\times\Delta_0^{l-1}\times I \cup X_{k+1}\times\Delta^{l-1}\times I \longrightarrow X\times 0 \cup X_k\times I$$

by $f(x,u,0) = (x,0)$ $\qquad x\in X,\ u\in\Delta^{l-1}$

$f(x,\alpha,t) = r_\alpha(x,t)$ $\qquad x\in X,\ \alpha\in\Delta_0^{l-1},\ t\in I$

$f(x,u,t) = (x,t)$ $\qquad x\in X_{k+1},\ u\in\Delta^{l-1},\ t\in I$

Then $f \circ p : X\times\Delta^{l-1}\times I \longrightarrow X\times 0 \cup X_k\times I$ factors through the identification $X\times\Delta^{l-1}\times I \longrightarrow Y_k\times I$ inducing a G-equivariant map $q : Y_k\times I \longrightarrow X\times 0 \cup X_k\times I$. Then

$$q \circ (j_k\times id) : X\times I \longrightarrow X\times 0 \cup X_k\times I$$

is a G-equivariant retraction. Hence (X,X_k) is a G-NDR. ∎

As a consequence of (2.7) we have

<u>Proposition 2.8</u>: Let $\Delta_k X\subset X^k$ denote the diagonal and $\Delta' X^k \subset X^k$ the fat diagonal, i.e. the subspace of all points $(x_1,\ldots,x_k)$ for which two coordinates agree. Suppose $(X^k,\Delta_k X)$ is an S_k-NDR for all $k\le n$. Then $(X^n,\Delta' X^n)$ is an S_n-NDR.

<u>Proof</u>: Let M be the set of partitions of $[n]$ into less than n subsets. For an arbitrary partition P of $[n]$ let $A(P)=\{(x_1,\ldots,x_n)\in X^n \mid x_i = x_j$ if i and j lie in the same element of P$\}$. Then

$$\Delta' X^n = \bigcup_{P\in M} A(P)$$

Any intersection $A(P_1)\cap\ldots\cap A(P_r)$ is another space $A(Q)$ and $St(A(Q))=$

$= St(Q) = St(\{P_1,\ldots,P_r\})$ under the obvious action of S_n on M. Suppose Q has k_r elements of cardinality r. Then A(Q) is homeomorphic to $(\Delta_1X)^{k_1}\times\ldots\times(\Delta_nX)^{k_n}$ with $\Delta_1X = X$, and St(A(Q)) mapped to $S_1 \wr S_{k_1}\times\ldots\times S_n \wr S_{k_n}$ under this homeomorphism ($S_i \wr S_{k}$ is the trivial group for k=0). By (2.4) and (2.5), the pair $(X^n,(\Delta_1X)^{k_1}\times\ldots\times(\Delta_nX)^{k_n})$ is an $(S_1 \wr S_{k_1}\times\ldots\times S_n \wr S_{k_n})$-NDR. By the following lemma, $(X^n,A(Q))$ is a St(A(Q))-NDR, so that $(X^n,\Delta' X^n)$ is an S_n-NDR by (2.7). ∎

Lemma 2.9: Let A and B be arbitrary subspaces of a G-space X. Let $\varphi : G_1 \cong G_2$ be an isomorphism of subgroups of G and assume that (X,A) is G_1-cofibred. Suppose there is a homeomorphism $f : X \longrightarrow X$ such that f(A) = B and $f(gx) = \varphi(g)f(x)$ for $x \in X$ and $g \in G_1$. Then (X,B) is G_2-cofibred.

Proof: If $r : X\times I \longrightarrow X\times 0 \cup A\times I$ is a G_1-retraction then

$$X\times I \xrightarrow{f^{-1}\times id} X\times I \xrightarrow{r} X\times 0 \cup A\times I \xrightarrow{f\times id} X\times 0 \cup B\times I$$

is a G_2-retraction. ∎

We also need a strange generalization of (2.8). Let G be a finite discrete group and X be a G-space. Let A be an invariant subspace of X. Put

$$D = \{(x_1,\ldots,x_n)\in X^n \mid gx_i = x_j \text{ for some } g\in G,\ i\neq j, \text{ or some } x_i \in A\}.$$

Proposition 2.10: Suppose (X^k,Δ_kX) is an $(S_k\times G)$-NDR for all $k\leq n$, the pair (X,A) is a G-NDR, and G acts freely on X-A. Then (X^n,D) is a $G \wr S_n$-NDR.

We prove this result in steps: Suppose $k\leq n$ and $X-A\neq\emptyset$.

Step 1: Let $Y_k = \{(x_1,\ldots x_k)\in X^k \mid$ all x_i are in the same G-orbit$\}\cup A^k$. Then (X^k,Y_k) is a $G \wr S_k$-NDR.

Proof: The space Y_k is the union of the spaces

$$Y(g_2,g_3,\dots,g_k) = \{(x,g_2x,g_3x,\dots,g_kx)\,|\,x\in X\}\cup A^k \qquad g_i\in G,$$

and $Y(g_2,g_3,\dots,g_k)\cap Y(g_2',g_3',\dots,g_k') = A^n$ if $(g_2,g_3,\dots,g_k)\neq(g_2',g_3',\dots,g_k')$, because G acts freely on $X-A$. By (2.6), the pair (X^k,A) is a $G\wr S_k$-NDR. Hence, by (2.7), we only have to show that each pair $(X^k,Y(g_2,\dots,g_k))$ is a H-NDR, where $H = St(Y(g_2,\dots,g_k))$. Define a homeomorphism $f : X^k \longrightarrow X^k$ with $f(\Delta_kX\cup A^k) = Y(g_2,\dots,g_k)$ by

$$f(x_1,\dots,x_k) = (x_1,g_2x_2,\dots,g_kx_k),$$

and an isomorphism $\varphi : G\times S_k \cong H$ by

$$\varphi(g,\pi) = (h,\pi)\in H\subset G\wr S_k$$

with $h(i) = g_i\cdot g\cdot g_{\pi^{-1}(i)}^{-1}$ with $g_1 = id$. Then $f(\Delta_kX)=X(g_2,\dots,g_k) = \{(x,g_2x,\dots,g_kx)\,|\,x\in X\}$, $f(A^k) = A^k$, and $f(\Delta_kX\cap A^k)=X(g_2,\dots,g_k)\cap A^k$. Since (X^k,A^k) is a $G\wr S_k$-NDR by (2.6) and (X^k,Δ_kX) is a $(G\times S_k)$-NDR, and since $(X^k,\Delta X_k\cap A^k) = (X^k,\Delta_kA)$ is a $G\wr S_k$-NDR because (X,A) is a G-NDR, (2.7) and (2.9) imply that $(X^k,Y(g_2,\dots,g_k))$ is an H-NDR.

Step 2: (X^n,D) is a $G\wr S_n$-NDR

We proceed as in the proof of (2.8). Let M be the set of all partitions of [n] into n-1 subsets and let $V = \{(x_1,\dots,x_n)\in X^n\,|\,\text{some } x_i\in A\}$. Then

$$D = \bigcup_{P\in M} (Y(P)\cup V)$$

where $Y(P) = \{(x_1,\dots,x_n)\in X^n\,|\,x_i = gx_j$ for some $g\in G$ if i and j are in the same element of P$\}$. An intersection of spaces $Y(P)\cup V$ is just another space $Y(P)\cup V$. By (2.7) the result holds if each $(X^n,Y(P)\cup V)$ is a $St(Y(P))$-NDR. Suppose P has k_r elements of cardinality r. Then

$$Y(P)\cup V \cong (Y_1)^{k_1}\times\dots\times(Y_n)^{k_n}\cup V$$

and $St(Y(P))$ corresponds to $(G\wr S_1)\wr S_{k_1}\times\dots\times(G\wr S_n)\wr S_{k_n}$ under this homeomorphism. By (2.6), the pair (X^n,V) is a $G\wr S_n$-NDR and $(X^n,(Y_1)^{k_1}\times\dots\times(Y_n)^{k_n})$ a $[(G\wr S_1)\wr S_{k_1}\times\dots\times(G\wr S_n)\wr S_{k_n}]$-NDR. The intersection $(Y_1)^{k_1}\times\dots\times(Y_n)^{k_n}\cap V$ can be written as union of A^n and products of spaces Y_r and A^k. Since the family of spaces consisting of A^n and products of spaces Y^r and A^k is closed under intersection, $(Y_1)^{k_1} \times \ldots \times (Y_n)^{k_n} \cap V$ is a $[(G \wr S_1) \wr S_{k_1} \times \ldots \times (G \wr S_n) \wr S_{k_n}]$-NDR by (2.6) and (2.7). Hence, by (2.7) and (2.8) the pair $(X^n, Y(P) \cup V)$ is a St(Y(P))-NDR. ∎

3. NUMERABLE PRINCIPAL G-SPACES

In this section we work in the category of all topological spaces. The results hold in the category of k-spaces, too.

Let G be an arbitrary topological group. We call a space X a <u>numerable principal G-space</u> if X is a free G-space and the projection $X \longrightarrow X/G$ is a numerable principal G-bundle in the sense of Dold [13].

<u>Lemma 3.1:</u> A space X is a numerable principal G-space iff there is a numerable cover $\mathfrak{U} = \{U_\alpha | \alpha \in A\}$ of X (see VII,§4) by G-invariant subspaces with equivariant numeration (i.e. equivariant partition of unity subordinate to $\mathfrak{U}$) such that there are equivariant maps $r_\alpha : U_\alpha \longrightarrow G$ for all $U_\alpha \in \mathfrak{U}$.

<u>Proof:</u> ⇒ Let $p : X \longrightarrow X/G$ be the projection and $\mathfrak{B} = \{V_\alpha | \alpha \in A\}$ a numerable cover of X/G over which p is locally trivial. Define $\mathfrak{U} = \{p^{-1}(V_\alpha)\}$, a numeration $\{f_\alpha\}$ by $f_\alpha = \nu_\alpha \circ p$ where $\{\nu_\alpha\}$ is a numeration of $\mathfrak{B}$, and $r_\alpha : p^{-1}(V_\alpha) \cong G \times V_\alpha \xrightarrow{\text{proj}} G$.

⇐ Let $\mathfrak{B} = \{p(U_\alpha), \alpha \in A\}$. Then $\mathfrak{B}$ is a cover of X/G. The f_α induce maps $v_\alpha : X/G \longrightarrow I$ defining a numeration of $\mathfrak{B}$. It remains to show that p is locally trivial over $\mathfrak{B}$. Now $p^{-1}pU_\alpha = U_\alpha$ since U_α is G-equivariant. So we have to find a G-homeomorphism

$$\begin{array}{ccc} G \times p(U_\alpha) & \xleftarrow{\;h\;} & U_\alpha \\ & {\scriptstyle proj}\searrow \quad \swarrow{\scriptstyle p} & \\ & pU_\alpha & \end{array}$$

making the triangle commute. Define $h(x) = (r_\alpha(x), p(x))$. The inverse k of h is given by $(g, p(x)) \longmapsto g \cdot r_\alpha(x)^{-1} \cdot x$, which is a well defined function. It remains to check the continuity of k. Let $Y = r_\alpha^{-1}(id) \subset U_\alpha$. Then k factors as

$$G \times pU_\alpha \xrightarrow{\;id \times u\;} G \times Y \xrightarrow{\;v\;} U_\alpha$$

with $u(p(x)) = r_\alpha(x)^{-1} \cdot x$ and $v(g,y) = g \cdot y$. Evidently v is continuous, and it remains to show the continuity of u. But u is induced by the map $U_\alpha \longrightarrow Y$ given by $x \longmapsto r_\alpha(x)^{-1}$, which is continuous. ∎

This result implies

Lemma 3.2: (a) If G is a discrete group, then X is a numerable principal G-space iff X has an open cover $\mathfrak{U} = \{U_\alpha | \alpha \in A\}$, with a subordinate partition of unity $\{f_\alpha : X \longrightarrow I | \alpha \in A\}$ such that for all $g \in G$ different from the identity, $gU_\alpha \cap U_\alpha = \emptyset$, and gU_α is some $U_\beta \in \mathfrak{U}$, and $f_\beta(gx) = f_\alpha(x)$, $x \in X$.

(b) If G is a finite discrete group, then X is a numerable principal G-space iff X has an open cover $\mathfrak{U} = \{U_\alpha | \alpha \in A\}$ with a subordinate partition of unity, such that $gU_\alpha \cap U_\alpha = \emptyset$ for all $g \in G$ different from the identity.

Proof: Part (a) is an immediate consequence of Lemma 3.1. Now suppose we have a cover $\mathfrak{U}$ of X as described in (b). Enlarge it to an open cover $\mathfrak{B}$ of X by taking all subsets $g \cdot U_\alpha, g \in G$. If $\{f_\alpha : X \longrightarrow I\}$ is the partition of unity subordinate to $\mathfrak{U}$, enlarge it to a partition of unity subordinate to $\mathfrak{B}$ by associating the map

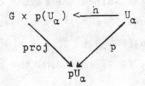

$$x \longmapsto \frac{1}{|G|} f_\alpha(g^{-1}x) \qquad |G| = \text{order of } G$$

with the element g U_α of $\mathfrak{B}$. Then $\mathfrak{B}$ and its partition of unity satisfy (a). ∎

<u>Lemma 3.3</u>: If $f : X \longrightarrow Y$ is a G-map and Y a numerable principal G-space, then so is X. In particular, any invariant subspace of a numerable principal G-space is a numerable principal G-space.

<u>Proof</u>: Let $\mathfrak{B} = \{V_\alpha | \alpha \in A\}$ be an invariant numerable cover of Y with numeration $\{f_\alpha\}$ and equivariant maps $r_\alpha : V_\alpha \longrightarrow G$ associated with Y. Then $\{f^{-1}(V_\alpha), f_\alpha \circ f, r_\alpha \circ f\}$ makes X into a numerable principal G-space. ∎

<u>Lemma 3.4</u>: Let $r : X \longrightarrow Y$ be a G-map from a G-space X to a numerable principal G-space Y. Then r is an equivariant homotopy equivalence iff it is an ordinary homotopy equivalence.

<u>Proof</u>: By Lemma 3.3, both X and Y are numerable principal G-spaces. Consider

$$\begin{array}{ccc} EG\times X & \xrightarrow{\ id\times r\ } & EG\times Y \\ \downarrow{\scriptstyle proj} & & \downarrow{\scriptstyle proj} \\ X & \xrightarrow{\ r\ } & Y \end{array}$$

By [10; Lemma 2, Bemerkung], $id\times r$ is an equivariant homotopy equivalence, and by [10; Lemma 4], the projections are equivariant homotopy equivalences. ∎

We now prove the homotopy extension lifting property (HELP) of a homotopy equivalence.

<u>Theorem 3.5</u>: Given a diagram of G-spaces and G-maps

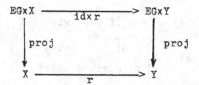

$$\begin{array}{ccc} Y & \xrightarrow{\;p\;} & Z \\ {\scriptstyle f_A}\uparrow & & \uparrow{\scriptstyle h} \\ A & \subset & X \end{array}$$

and a G-homotopy $H_A : h|A \simeq p \circ f_A$. Assume that (X,A) is G-cofibred. Then there is a G-map $f : X \longrightarrow Y$ extending f_A and a G-homotopy $H : h \simeq p \circ f$ extending H_A provided

(a) p is an equivariant homotopy equivalence

OR (b) p is an ordinary homotopy equivalence and X-A is a numerable principal G-space.

Proof: Replace p by the equivariantly homotopy equivalent G-fibration $q : E \longrightarrow Z$, where $E = \{(\omega,y) \in Z^I \times Y | \omega(1) = p(y)\}$ and $q(\omega,y) = \omega(0)$. The G-action on E is given by $g(\omega,y) = (g\cdot\omega, g\cdot y)$, where $(g\cdot\omega)(t) = g\cdot\omega(t)$. Let $r : F \longrightarrow X$ be the G-fibration over X induced by h, i.e. $F = \{(x,\omega,y) \in X \times Z^I \times Y | \omega(0) = h(x), \omega(1) = p(y)\}$ and $r(x,\omega,y) = x$.

$$\begin{array}{ccccccc} & & F & \longrightarrow & E & & Y \\ & {\scriptstyle k}\nearrow & \downarrow{\scriptstyle r} & & \downarrow{\scriptstyle q} & \swarrow{\scriptstyle p} & \\ A & \subset & X & \xrightarrow{\;h\;} & Z & & \end{array}$$

Define $k : A \longrightarrow F$ by $k(a) = (a, \omega_a, f_A(a))$ with

$$\omega_a(t) = \begin{cases} h(a) & 0 \le t \le \frac{1}{2} \\ H_A(a, 2t-1) & \frac{1}{2} \le t \le 1 \end{cases}$$

Then k is an equivariant section of r over A. The theorem is proved if we can extend k to an equivariant section of r over X.

Since (X,A) is G-cofibred, there is an equivariant map $u : X \longrightarrow I$ and a G-homotopy $K : X \times I \longrightarrow X$ such that $A \subset u^{-1}(0)$, $K(x,0) = x$, $K(a,t) = a$ for all $a \in A$ and $t \in I$, and $K(x,1) \in A$ for $x \in u^{-1}[0,1)$. Put $U = u^{-1}[0,1)$. Extend k to an equivariant section of r over U by $k(x) = [x, \omega_x, f_A(K(x,1))]$ with

$$\omega_x(t) = \begin{cases} h(K(x,2t)) & 0 \le t \le \frac{1}{2} \\ H_A(K(x,1), 2t-1) & \frac{1}{2} \le t \le 1 \end{cases} \qquad x \in U$$

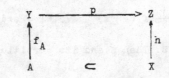

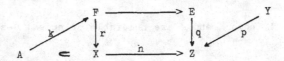

We claim now and prove later:

Let $r_{X-A} : F_{X-A} \longrightarrow X-A$ be the restriction of r to $F_{X-A} = r^{-1}(X-A)$. Then r_{X-A} has an equivariant section s', and there is a G-homotopy $L : F_{X-A} \times I \longrightarrow F_{X-A}$ from the identity to $s' \cdot r_{X-A}$ such that $r_{X-A}(L(e,t)) = r_{X-A}(e)$ for all $e \in F_{X-A}$ and $t \in I$.

The required equivariant sections s of r over X is then given by

$$s(x) = \begin{cases} s'(x) & x \in X-U \\ L(k(x), \max[2u(x)-1,0] & x \in U-A \\ k(x) & x \in A \end{cases}$$

We now prove the claim: q is a G-fibration and a homotpy equivalence. By [13; Cor.6.2], there is a section $\bar{q}$ of q and a homotopy $Q : id_E \simeq \bar{q} \cdot q$ such that $q \cdot Q(e,t) = q(e)$. Both $\bar{q}$ and Q are equivariant if p is an equivariant homotopy equivalence. Define a section $\bar{r} : X \longrightarrow F$ of r and a homotopy $R : F \times I \longrightarrow F$ from id_F to $\bar{r} \cdot r$ by $\bar{r}(x) = (x, \bar{q} \cdot h(x))$ and $R(x,e,t) = (x, Q(e,t))$, $e \in E$. Then $r \cdot R(x,e,t) = r(x,e)$ for all $(x,e) \in F$ and $t \in I$. Both $\bar{r}$ and R are equivariant if $\bar{q}$ and Q are, i.e. if p is an equivariant homotopy equivalence, and provide the section and homotopy of the claim. If p is an ordinary homotopy equivalence, then $r_{X-A} : F_{X-A} \longrightarrow X-A$ is a G-fibration and a homotopy equivalence because of the existence of $\bar{r}$ and R. Since X-A is a numerable principal G-space, r_{X-A} is an equivariant homotopy equivalence (Lemma 3.4).By the equivariant version of [13; Cor. 6.2] the equivariant section and the homotopy of the claim exist. ∎

Proposition 3.6: Let $p : (X,A) \longrightarrow (Y,B)$ be an equivariant map of pairs of G-spaces such that $p_A = p|A : A \longrightarrow B$ is an equivariant homotopy equivalence and $p : X \longrightarrow Y$ is an ordinary homotopy equivalence. Suppose X-A and Y-B are numerable principal G-spaces and (X,A),(Y,B) are G-cofibred. Then any equivariant homotopy inverse q_B of p_A can be extended to an equivariant homotopy inverse q of p and any equivariant homotopy $H_B : id_B \simeq p_A \cdot q_B$ to an equivariant homotopy $H : id_Y \simeq p \cdot q$.

Proof: Let $i : B \subset Y$ be the inclusion. By part (b) of the previous theorem, there is an equivariant extension $q : Y \longrightarrow X$ of q_B and $H : Y \times I \longrightarrow Y$ of $i \circ H_B$ such that $H : id_Y \simeq p \circ q$. Hence $(p,p_A)\circ(q,q_B) \simeq id$ equivariantly as maps of pairs. Analogously, we can find an extension $\overline{p} : X \longrightarrow Y$ of p_A such that $(q,q_B)\circ(\overline{p},p_A) \simeq id$ equivariantly as maps of pairs. Hence

$$(q,q_B)\circ(p,p_A) \simeq (q,q_B)\circ(p,p_A)\circ(q,q_B)\circ(\overline{p},p_A) \simeq (q,q_B)\circ(\overline{p},p_A) \simeq id$$

equivariantly as maps of pairs. ▮

Corollary 3.7: Let X be a G-space and A an invariant subspace. Suppose that (X,A) is G-cofibred, that $A \subset X$ is a homotopy equivalence, and X-A is a numerably principal G-space. Then A is an equivariant SDR of X.

Proof: Apply the previous proposition to the inclusion $(A,A) \subset (X,A)$ with $q_B = id_A$ and H_B the constant homotopy. ▮

Lemma 3.8: If X is a paracompact G-space, $A \subset X$ a subspace such that G acts freely on X-A, and $u : X \longrightarrow I$ a map with $A = u^{-1}(0)$, then X-A is a numerable principal G-space provided G is a compact Lie group.

Proof: X-A is an F_σ, i.e. a countable union of closed subspaces of X, and hence paracompact [36]. Since X-A is normal, the projection $p : X\text{-}A \longrightarrow (X\text{-}A)/G$ is a principal fibre bundle (e.g. see [5; p.88]). Since p is closed, (X-A)/G is paracompact [18; p.165]. ▮

4. FILTERED SPACES AND ITERATED ADJUNCTION SPACES

As in the previous sections the results of this section hold in the category of all topological spaces as well as in the category of k-spaces.

Let G be an arbitrary topological group. A *filtration* of a G-space X is an increasing sequence of invariant subspaces

$$\emptyset = X_{-1} \subset X_0 \subset X_1 \subset \dots$$

with X as colimit (direct limit). Given such a sequence, we call X a *filtered G-space*. If each (X_n, X_{n-1}) is G-cofibred, we call X *properly filtered*. A *filtered G-map* is a G-map $f : X \longrightarrow Y$ of filtered G-spaces such that $f(X_n) \subset Y_n$. We denote $f|X_n : X_n \longrightarrow Y_n$ by f_n. A filtered space is called an *iterated adjunction G-space* if X_n is obtained from X_{n-1} by adjoining a G-space A_n relative to an invariant subspace B_n by an equivariant map. We say, X_n is obtained from X_{n-1} by adjoining (or attaching) (A_n, B_n). If each (A_n, B_n) is G-cofibred, we call X a *proper iterated adjunction G-space*.

We list a few elementary properties

Lemma 4.1: (a) A proper iterated adjunction G-space is properly filtered.

(b) If X is a properly filtered G-space, then each (X, X_n) is G-cofibred.

(c) If Y is obtained from X by attaching a NDR (A,B), then Y is Hausdorff if X and A are.

(d) If X is a properly filtered space and each X_n is Hausdorff, then so is X.

Proof:

$$\begin{array}{ccc} B_n & \subset & A_n \\ \downarrow & & \downarrow \\ X_{n-1} & \subset & X_n \end{array}$$

is a push-out diagram in the category of G-spaces. Hence $X_{n-1} \subset X_n$ is a G-cofibration if $A_n \subset B_n$ is a G-cofibration. If (A_n, B_n) is a NDR, then there is a map $u : A_n \longrightarrow I$ with $u^{-1}(0) = B_n$. Using this one readily checks that X_n is Hausdorff if A_n and X_{n-1} are.

Now suppose X is a properly filtered G-space. We construct a retraction $r : X\times I \longrightarrow X\times 0 \cup X_n\times I$. Since $X\times I = \varinjlim (X_k\times I)$ it suffices to construct compatible G-retractions $r_k : X_k\times I \longrightarrow X_k\times 0 \cup X_n\times I$ for $k\geq n$ (Use [12; Satz 1.16 and Satz 1.19] to show that the subspace $X\times 0 \cup X_n\times I$ is the colimit of the subspaces $X_k\times 0 \cup X_n\times I$). The retractions r_k are obtained inductively by

$$r_k = (id \cup r_{k-1})\circ r : X_k\times I \longrightarrow X_k\times 0 \cup X_{k-1}\times I \longrightarrow X_k\times 0 \cup X_n\times I$$

where r is the G-retraction of (X_k, X_{k-1}). For the continuity of r_k use [12; Satz 1.1.9] observing that $X_k\times 0 \cup X_n\times I$ is a retract of $X_k\times I$, because composites of cofibrations are cofibrations.

For a proof of (d) see [49; Thm. 9.4]. ∎

It is well-known that a filtered map $f : X \longrightarrow Y$ of properly filtered spaces is a homotopy equivalence provided each $f_n : X_n \longrightarrow Y_n$ is a homotopy equivalence. Usually one proves this using the Milnor telescope construction [9; IV,§5]. We give an alternative proof in the category of G-spaces, the intermediate results of which we will need for other purposes.

<u>Lemma 4.2</u>: Given a commutative diagram of G-spaces

$$\begin{array}{ccc} A & \xrightarrow{\ f\ } & A' \\ \downarrow{\scriptstyle i} & & \downarrow{\scriptstyle j} \\ B & \xrightarrow{\ g\ } & B' \end{array}$$

where i and j are G-cofibrations and f and g are G-homotopy equivalences. Let $\bar{f}$ be a G-homotopy inverse of f and $h_t : id_{A'} \simeq f\circ\bar{f}$ a G-homotopy. Then there is a G-homotopy inverse $\bar{g}$ of g extending $\bar{f}$ and a G-homotopy $k_t : id_{B'} \simeq g\circ\bar{g}$ extending h_t, i.e. $(k_t, h_t) : (id_{B'}, id_{A'}) \simeq (g\circ\bar{g}, f\circ\bar{f})$ in the category of pairs of G-spaces.

<u>Proof</u>: Apply Theorem 3.5 to the diagram

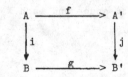

$$\begin{array}{ccc} B & \xrightarrow{\ g\ } & B' \\ {\scriptstyle i\cdot\overline{f}}\uparrow & & \uparrow{\scriptstyle id} \\ A' & \xrightarrow[\subset]{\ j\ } & B' \end{array}$$

and the homotopy $j\cdot h_t : id_{B'}|A' = j \simeq j\cdot f\cdot\overline{f} = g\cdot i\cdot\overline{f}$ to obtain the required extensions. ∎

<u>Corollary 4.3:</u> Given the assumptions of Lemma 4.2, the G-map $(g,f) : (B,A) \longrightarrow (B',A')$ is a G-homotopy equivalence in the category of pairs of G-spaces.

<u>Proof</u>: Let $(\overline{g},\overline{f}) : (B',A') \longrightarrow (B,A)$ and $(k_t,h_t) : (id_{B'},id_{A'}) \simeq (g\cdot\overline{g}, f\cdot\overline{f})$ be the map and G-homotopy of pairs of (4.2). Applying the lemma once again to the pair $(\overline{g},\overline{f})$, we obtain a G-map of pairs $(g',f'):(B,A)\to(B',A')$ and a G-homotopy of pairs $(id_B,id_A) \simeq (\overline{g}\cdot g', \overline{f}\cdot f')$. By general non-sense, (g,f) is a G-homotopy equivalence of pairs (cf. the proof of (3.6)). ∎

<u>Theorem 4.4</u>: Let $f : X \longrightarrow Y$ be a filtered G-map of property filtered G-spaces such that each $f_n : X_n \longrightarrow Y_n$ is a G-homotopy equivalence. Then f is a G-homotopy equivalence.

<u>Proof</u>: Using Lemma 4.3, we inductively construct homotopy inverse $g_n : Y_n \longrightarrow X_n$ extending g_{n-1} and homotopies $H_n(t) : id_{Y_n} \simeq f_n\cdot g_n$ extending $H_{n-1}(t)$. Taking the colimit we obtain a G-map $g : Y \longrightarrow X$ and a G-homotopy $H(t) : f\cdot g \simeq id_Y$. Apply the same procedure to the filtered map g to obtain a G-map $h : X \longrightarrow Y$ and a G-homotopy $K : id_X \simeq g\cdot h$. As in (3.6), we find that f is a G-homotopy equivalence. ∎

<u>Theorem 4.5</u>: Let X be a properly filtered G-space such that each X_n is an equivariant SDR of X_{n+1}. Then each X_n is an equivariant SDR of X.

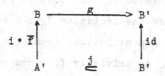

Proof: By the equivariant version of [14; 3.7] it suffices to show that (X,X_n) is G-cofibred and the inclusion $X_n \subset X$ is a G-homotopy equivalence. The first requirement follows from (4.1) and the second from (4.4) if we consider X_n as trivially filtered by itself. ∎

We now investigate conditions which make maps of iterated adjunction spaces to filtered homotopy equivalences.

Proposition 4.6: Let Y and Y' be G-spaces obtained from X and X' by adjoining (A,B) and (A',B') by maps $f : B \longrightarrow X$ and $g : B' \longrightarrow X'$ respectively. Suppose (A,B) and (A',B') are G-cofibred. Given a commutative diagram of G-maps

$$\begin{array}{ccccc} X & \xleftarrow{\;f\;} & B & \overset{\subset}{\underset{i}{}} & A \\ \downarrow{\scriptstyle h} & & \downarrow{\scriptstyle l} & & \downarrow{\scriptstyle k} \\ X' & \xleftarrow[\;g\;]{} & B' & \overset{\subset}{\underset{j}{}} & A' \end{array}$$

whose vertical maps are G-homotopy equivalences. Then the induced map $r : (Y,X) \longrightarrow (Y',X')$ is a G-homotopy equivalence of pairs.

Proof: Let Z and Z' be the double mapping cylinders of the horizontal sequences. E.g. Z is obtained by identifying the mapping cylinders Z_f and Z_i along their common subspace B. Fix a homotopy inverse $\bar{l}$ of l and homotopies $\bar{l} \cdot l \simeq id_B$, $l \cdot \bar{l} \simeq id_{B'}$. The pairs (h,l) and (k,l) induce G-homotopy equivalences $Z_f \longrightarrow Z_g$ and $Z_i \longrightarrow Z_j$ and a map $r' : Z \longrightarrow Z'$. Since the inclusions $Z_f \supset B \subset Z_i$ and $Z_g \supset B' \subset Z_j$ are G-cofibrations, the map $\bar{l}$ and the homotopies can be extended by (4.2) to G-maps of triads $p,q : (Z,Z_f,Z_i) \longrightarrow (Z',Z_g,Z_j)$ and G-homotopies of triads $p \cdot r' \simeq id$ and $r' \cdot q \simeq id$. Hence, by general nonsense, $r' : (Z,Z_f,Z_i) \longrightarrow (Z',Z_f,Z_j)$ is a G-homotopy equivalence of triads. Since i and j are G-cofibrations, the natural projections $(Z,Z_f) \rightarrow (Y,X)$ and $(Z',Z_g) \longrightarrow (Y',X')$ are G-homotopy equivalences of pairs. Since

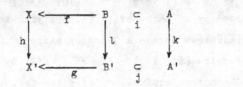

$r : (Y,X) \longrightarrow (Y',X')$ is induced by r', it is a G-homotopy equivalence of pairs. ▮

We also need a strange generalization of this result.

<u>Proposition 4.7</u>: Let Y and Y' be spaces obtained from X and X' by adjoining (A,B) and (A',B') by maps $f : B \longrightarrow X$ and $g : B' \longrightarrow X'$ respectively (no G-action). Let (K,L) and (K',L') be G-cofibred pairs such that $K-L$ and $K'-L'$ are numerable principal G-spaces and $(K/G,L/G) = (A,B)$ and $(K'/G,L'/G) = (A',B')$. Given a commutative diagram of maps

$$\begin{array}{ccccccc} X & \xleftarrow{\ f\ } & B & \xleftarrow{\ proj\ } & L & \subset & K \\ \downarrow{\scriptstyle h} & & & & \downarrow{\scriptstyle q'} & & \downarrow{\scriptstyle q} \\ X' & \xleftarrow{\ g\ } & B' & \xleftarrow{\ proj\ } & L' & \subset & K' \end{array}$$

with q a G-map and h, q', q ordinary homotopy equivalences. Then the induced map $r : (Y,X) \longrightarrow (Y',X')$ is a homotopy equivalence of pairs.

<u>Proof</u>: Consider X and X' as trivial G-spaces. Let Z and Z' be the G-spaces obtained from X and X' by adjoining (K,L) respectively (K',L'). Then q and h induce a G-map $p : (Z,X) \longrightarrow (Z',X')$, which is an ordinary homotopy equivalence of pairs. By (4.1) the pairs (Z,X) and (Z',X') are G-cofibred. Since $Z-X$ and $Z'-X'$ are numerable principal G-spaces and $p : X \longrightarrow X'$ is a G-homotopy equivalence, p is a G-homotopy equivalence of pairs by (3.6). Passing to the orbit spaces we find that $r : (Y,X) \longrightarrow (Y',X')$ is a homotopy equivalence of pairs. ▮

We have a similar result for weak homotopy equivalences.

<u>Proposition 4.8</u>: (a) If X and Y are filtered T_1-spaces and $f : X \to Y$ a filtered map such that each f_n is a weak homotopy equivalence, then f is a weak homotopy equivalence

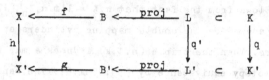

(b) Let Y and Y' be spaces obtained from X and X' by adjoining (A,B) and (A',B') by maps $f : B \to X$ and $g : B' \to X'$ respectively. Suppose (A,B) and (A',B') are cofibred. Given a diagram of maps

$$\begin{array}{ccccc} X & \xleftarrow{\;f\;} & B & \overset{i}{\subset} & A \\ \Big\downarrow h & & \Big\downarrow l & & \Big\downarrow k \\ X' & \xleftarrow{\;g\;} & B' & \overset{j}{\subset} & A' \end{array}$$

whose vertical maps are weak homotopy equivalences. Then the induced map $r : Y \to Y'$ is a weak homotopy equivalence.

Proof: Part (a) follows from the fact that $\pi_i X = \lim \pi_i X_n$ [17;(2.14)]. For part (b) let Z and Z' be the double mapping cylinders of the horizontal sequences. Then the triple (h,l,k) induces a map $r' : Z \to Z'$, which is a weak homotopy equivalence by [35; Thm.6]. The canonical projections $Z \to Y$ and $Z' \to Y'$ are homotopy equivalences because i and j are cofibrations. Hence r is a weak homotopy equivalence. ∎

Proposition 4.9: Let $f : X \to Y$ be a filtered G-map of filtered G-spaces. Assume that the maps $X_{n-1} \subset X_n$, $Y_{n-1} \subset Y_n$, and $f : X_n \to Y_n$ are closed G-cofibrations. Then f is a closed G-cofibration if $X_n \cap Y_{n-1} = X_{n-1}$.

Proof: We construct inductively a G-retraction $Y \times I \to X \times I \cup Y \times 0$. In the inductive step we have a G-retraction $Y_{n-1} \times I \to Y_{n-1} \times 0 \cup X_{n-1} \times I$ so that we need a G-retraction

$$q : Y_n \times I \to (Y_{n-1} \cup X_n) \times I \cup Y_n \times 0$$

Since $X_n \cap Y_{n-1} = X_{n-1}$ and (Y_n, X_{n-1}) is a G-NDR, the pair $(Y_n, Y_{n-1} \cup X_n)$ is a G-NDR by (2.7). Hence the required G-retraction exists. ∎

Proposition 4.10: Let X be a properly filtered space such that each X_{n-1} is contractible in X_n. Then X is contractible.

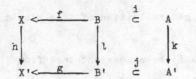

<u>Proof</u>: Inductively, define spaces Y_n, inclusions $X_n \subset Y_n$, and retractions $q_n : Y_n \longrightarrow X_n$. Put $X_o = Y_o$ and $Y_n = X_n \cup CY_{n-1}/\sim$, where CY_{n-1} is the (unreduced) cone on Y_{n-1} and $X_{n-1} \subset X_n$ is identified with $X_{n-1} \subset Y_{n-1} \subset CY_{n-1}$. The retraction $q_n : Y_n \longrightarrow X_n$ is given by

$$CY_{n-1} \xrightarrow{Cq_{n-1}} CX_{n-1} \xrightarrow{h} X_n$$

where h is the contracting homotopy. Let Y be the colimit of the Y_n. Then the inclusions $X_n \subset Y_n$ and the retractions $Y_n \longrightarrow X_n$ define filtered maps $i : X \subset Y$ and $q : Y \longrightarrow X$ such that $q \bullet i = id_X$. We show that Y is contractible.

First note that $j_{n-1} : Y_{n-1} \subset Y_n$ and the inclusion of the cone point $\{y_n\} \subset CY_{n-1} \subset Y_n$ are cofibrations and that Y_{n-1} is contractible in Y_n to the cone point. For any sequence Σ of spaces and maps

$$A_o \xrightarrow{p_o} A_1 \xrightarrow{p_1} A_2 \xrightarrow{p_2} A_3 \longrightarrow \ldots$$

define $T_n\Sigma$, inclusions $A_n \subset T_n\Sigma \subset T_{n+1}\Sigma$, and retractions $r_n: T_n\Sigma \to A_n$ inductively. Put $T_o\Sigma = A_o$ and $T_n\Sigma = T_{n-1}\Sigma \cup Z_{p_{n-1}}/\sim$, where $A_{n-1} \subset T_{n-1}\Sigma$ is identified with $A_{n-1} \subset Z_{p_{n-1}}$, the mapping cylinder of p_{n-1}. The retraction r_n is the composite

$$T_n\Sigma \longrightarrow Z_{p_{n-1}} \longrightarrow A_n$$

whose first map is induced by r_{n-1} and the second is the standard retraction of the mapping cylinder. The colimit $T\Sigma$ of the $T_n\Sigma$ is called the telescope of Σ.

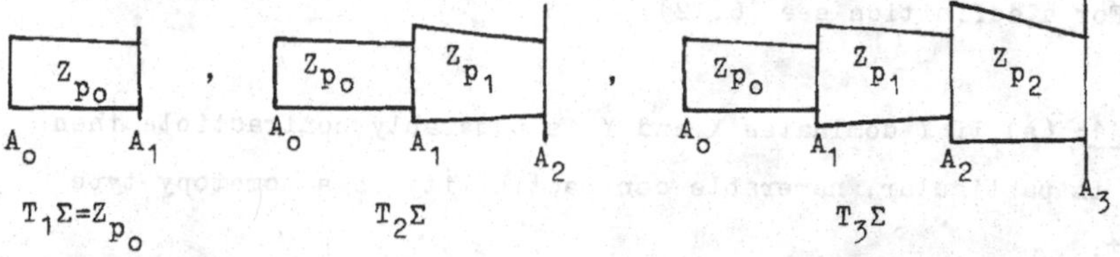

The $T_n\Sigma$ define a proper filtration of $T\Sigma$. The r_n are homotopy equivalences and they induce a map r from $T\Sigma$ to the colimit A of Σ. If Σ defines a proper filtration of A, i.e. if each p_i is a cofiltration, then $r : T\Sigma \longrightarrow A$ is a homotopy equivalence by (4.4).

Consider the sequences

$$\Sigma_1 : Y_o \overset{j_o}{\subset} Y_1 \overset{j_1}{\subset} Y_2 \overset{j_2}{\subset} \dots$$

$$\Sigma_2 : Y_o \xrightarrow{c_o} Y_1 \xrightarrow{c_1} Y_2 \xrightarrow{c_2}$$

where c_i is the constant map to the cone point $\{y_{i+1}\}$. We have shown that $j_k \simeq c_k$. It is well-known (and can easily be deduced from (3.5) and (4.2)) that if $f \simeq g : A \longrightarrow B$ there is a homotopy equivalence of pairs $(Z_f, A) \simeq (Z_g, A)$. Hence there is a filtered map $h : T\Sigma_1 \to T\Sigma_2$ such that each h_n is a homotopy equivalence, whence $Y \simeq T\Sigma_1 \simeq T\Sigma_2$. Filter $T\Sigma_2$ differently: Put $Q_o = Y_o = T_o\Sigma_2$, $Q_1 = CY_o$ with the obvious inclusion of the cone point y_1. Inductively, let $Q_n = Q_{n-1} \cup CY_{n-1}/\sim$ with the cone point y_{n-1} in Q_{n-1} identified with $y_{n-1} \in Y_{n-1} \subset CY_{n-1}$. Again we have the inclusion $\{y_n\} \subset Q_n$ of the cone point.

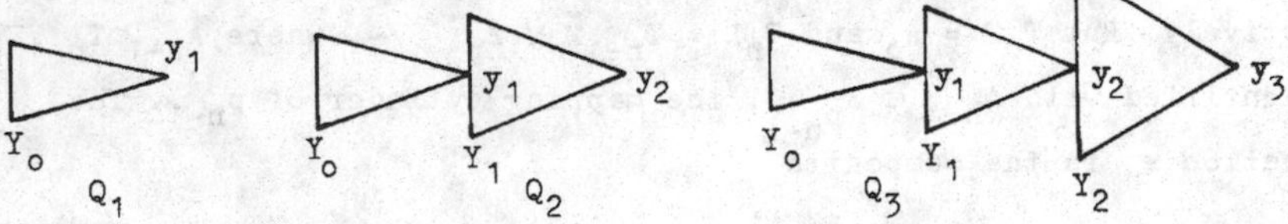

Since $\{y_n\} \subset Y_n \subset CY_n$ are cofibrations, the Q_i define a proper filtration of $T\Sigma_2$. The Q_i are obviously contractible. Hence, by (4.4), $T\Sigma_2 \simeq$ point, hence Y and therefore X are contractible. ∎

We close this chapter with some results on numerably contractible spaces. For a definition see (6.12).

<u>Lemma 4.11</u>: (a) If Y dominates X and Y is numerably contractible then so is X. In particular, numerable contractibility is a homotopy type invariant.

(b) A finite product of numerably contractible spaces is numerably contractible.

(c) Let X be a properly filtered space such that each X_n is numerably contractible. Then so is X.

For proofs and further references see [43]. ∎

<u>Proposition 4.12</u>: Let X be a proper iterated adjunction space such that the spaces A_n which are attached are numerably contractible. Then X is numerably contractible.

<u>Proof</u>: By (4.1) and (4.11) it suffices to show that each X_n is numerably contractible. Suppose inductively that X_{n-1} is numerably contractible. The subspaces $U = X_{n-1} \cup B_n \times [0,1)$ and $V = A_n \cup B \times (0,1]$ of the double mapping cylinder Z of $X_{n-1} \longleftarrow B_n \subset A_n$ form a numerable covering of Z. By (4.11 a), both U and V and hence Z are numerably contractible. Since $Z \simeq X_n$, also X_n is numerably contractible. ∎

REFERENCES

1. J.F. Adams: The sphere considered as an H-space mod p (Quart. J. Math. Oxford Ser.(2),12 (1961), 52-60)

2. J. Beck: On H-spaces and infinite loop spaces (Category Theory, Homology Theory, Their Applications III, Lecture Notes in Math.99 (1969), 139-153, Springer Verlag)

3. J. Beck: Classifying spaces for homotopy-everything H-spaces (H-spaces, Neuchatel (Suisse), Août 1970, Lecture Notes in Math.196 (1971), 54-62, Springer Verlag)

4. J. Bénabou: Structures algébriques dans les catégories (Thèse, Fac.Sci., U. de Paris, 1966)

5. G.E. Bredon: Introduction to compact transformation groups (Academic Press, New York 1972)

6. J.M. Boardman: Homotopy structures and the language of trees (Proceedings of the Summer Institute on Algebraic Topology, University of Wisconsin (1970), 37-58, Amer. Math. Soc.)

7. J.M. Boardman: Infinite loop spaces (to appear in Bull. Amer. Math. Soc.)

8. J.M. Boardman, R.M. Vogt: Homotopy-everything H-spaces (Bull. Amer. Math. Soc. 74 (1968),1117-1122)

9. Th. Bröcker, T. tom Dieck: Kobordismentheorie (Lecture Notes in Math. 178 (1970), Springer Verlag)

10. T. tom Dieck: Glättung äquivarianter Homotopiemengen (Arch. Math. 20 (1969), 288-295)

11. T. tom Dieck: Partitions of unity in homotopy theory (Compositio math. 23 (1971), 159-167)

12. T. tom Dieck, K.H. Kamps, D. Puppe: Homotopietheorie (Lecture Notes in Math. 157 (1970), Springer Verlag)

13. A. Dold: Partitions of unity in the theory of fibrations (Ann. of Math. 78 (1963), 223-255)

14. A. Dold: Halbexakte Homotopiefunktoren (Lecture Notes in Math. 12 (1966), Springer Verlag)

15. A. Dold: Die Homotopieerweiterungseigenschaft ist eine lokale Eigenschaft (Inventiones Math. 6 (1968), 185-189)

16. A. Dold, R.K. Lashof: Principal quasifibrations and fibre homotopy equivalence of bundles (Illinois J. of Math. 3 (1959), 285-305)

17. A. Dold, R. Thom: Quasifaserungen und unendliche symmetrische Produkte (Ann. of Math. 67 (1958), 239-281)

18. J. Dugundji: Topology (Allan and Bacon, Boston 1966)

19. E. Dyer, R.K. Lashof: Homology of iterated loop spaces (Amer. J. Math. 84 (1962), 35-88)

20. S. Eilenberg, G.M. Kelly: Closed categories (Proc. Conf. Categ. Alg. (La Jolla, 1965), 421-562, Springer Verlag 1966)

21. M. Fuchs: Verallgemeinerte Homotopie-Homomorphismen und klassifizierende Räume (Math. Annalen 161 (1965), 197-230)

22. M. Fuchs: A modified Dold-Lashof construction that does classify H-principal fibrations (Math. Annalen 192 (1971), 328-340)

23. P. Gabriel, M. Zisman: Calculus of fractions and homotopy theory (Ergebnisse der Mathematik und ihrer Grenzgebiete 35, Springer Verlag 1967)

24. G.M. Kelly: On MacLane's condition for coherence of natural associativities, commutativities, etc. (J. Algebra 1 (1964), 397-402)

25. F.W. Lawvere: Functional semantics of algebraic theories (Proc. Nat. Acad. Sci. USA 50 (1963), 869-872)

26. F.W. Lawvere: Functional semantics of algebraic theories (Dissertation, Columbia University (1963))

27. J. Lillig: A union theorem of cofibrations (Arch. Math. to appear)

28. S. MacLane: Natural associativity and commutativity (Rice Univ. Studies 49 (1963), 28-46)

29. S. MacLane: Categorial algebra (Bull. Amer. Math. Soc. 71 (1965), 40-106)

30. S. MacLane: The Milgram bar construction as a tensor product of functors (The Steenrod Algebra and Its Applications, Lecture Notes in Math. 168 (1970), 135-152, Springer Verlag)

31. P.J. Malraison, Jr: An equivalence of categories (Preprint)

32. J.P. May: Categories of spectra and infinite loop spaces (Category Theory, Homology Theory, and Their Applications III, Lecture Notes in Math. 99 (1969), 448-479, Springer Verlag)

33. J.P. May: Homology operations on infinite loop spaces (Proceedings of the Summer Institute on Algebraic Topology, University of Wisconsin (1970), 171-185, Amer. Math. Soc.)

34. J.P. May: The geometry of iterated loop spaces (Lecture Notes in Math. 271 (1972), Springer Verlag)

35. M. McCord: Singular homology groups and homology groups of finite topological spaces (Duke Math. J. 33 (1966), 465-474)

36. E. Michael: A note on paracompact spaces (Proc. Amer. Math. Soc. 4 (1953), 831-838)

37. R.J. Milgram: The bar construction and abelian H-spaces (Illinois J. of Math. 11 (1967), 242-250)

38. J. Milnor: Construction of universal bundles I (Ann. of Math. 63 (1956), 272-284)

39. J. Milnor: Construction of universal bundles II (Ann. of Math. 63 (1956), 430-436)

40. J. Milnor: On axiomatic homology theory (Pacific J. of Math. 12 (1962), 337-341)

41. B. Pareigis: Kategorien und Funktoren (Verlag B.G. Teubner, Stuttgart 1969)

42. D. Puppe: Homotopiemengen und ihre induzierten Abbildungen I (Math. Zeitschrift 69 (1958), 299-344)

43. D. Puppe: Some well known weak homotopy equivalences are genuine homotopy equivalences (Symposia Matematica Vol.V, Istituto

Nazionale di Alta Matematica (1970), 363-374)

44. G.B. Segal: Classifying spaces and spectral sequences (Publ. Math. Inst. des Hautes Etudes Scient. (Paris) 34 (1968), 105-112)

45. G.B. Segal: Categories and cohomology theories (To appear)

46. J.D. Stasheff: Homotopy associativity of H-spaces I (Trans. Amer. Math. Soc. 108 (1963),275-292)

47. J.D. Stasheff: Homotopy associativity of H-spaces II (Trans. Amer. Math. Soc. 108 (1963), 293-312)

48. J.D.Stasheff: H-spaces from a homotopy point of view (Lecture Notes in Math. 161, (1970), Springer Verlag)

49. N. Steenrod: A convenient category of topological spaces (Michigan Math. J. 14 (1967), 133-152)

50. N. Steenrod: Milgram's classifying space of a topological group (Topology 7 (1968), 349-368)

51. A. Strøm: Note on cofibrations (Math. Scand. 19 (1966), 11-14)

52. A. Strøm: Note on cofibrations II (Math. Scand. 22 (1968), 130-142)

53. M. Sugawara: On the homotopy-commutativity of groups and loop spaces (Mem. Coll. Sci. Univ. Kyoto Ser. A. 33 (1960), 257-269)

54. R.M. Vogt: Categories of operators and H-spaces (Dissertation, University of Warwick, Coventry (1968))

55. R.M. Vogt: Convenient categories of topological spaces for homotopy theory (Arch. Math. 22 (1971), 545-555)

56. R.M. Vogt: Homotopy limits and colimits (To appear in Math. Zeitschrift)